PHYSICAL PROPERTIES OF III–V SEMICONDUCTOR COMPOUNDS

PHYSICAL PROPERTIES OF III–V SEMICONDUCTOR COMPOUNDS

InP, InAs, GaAs, GaP, InGaAs, and InGaAsP

SADAO ADACHI
Department of Electronic Engineering
Gunma University
Kiryu-shi, Gunma 376, Japan

A Wiley-Interscience Publication
JOHN WILEY & SONS
New York • Chichester • Brisbane • Toronto • Singapore

Library of Congress Cataloging in Publication Data:

Adachi, Sadao, 1950–
Physical properties of III–V semiconductor compounds: InP, InAs, GaAs, GaP, InGaAs, and InGaAsP / Sadao Adachi.
p. cm.
"A Wiley-Interscience publication."
Includes bibliographical references and indexes.
ISBN 0-471-57329-9 (cloth : acid-free)
1. Gallium arsenide semiconductors. 2. Indium alloys. 3. Indium phosphide. I. Title.
QC611.8.G3A33 1992
537.6′226—dc20 92-7286

Printed in the United States of America

10 9 8 7 6 5 4 3 2 1

To
Yūki, Mai, and Kōya

PREFACE

The $In_{1-x}Ga_xAs_yP_{1-y}/InP$ heterostructure system is potentially useful for optoelectronic, high-speed digital, and high-frequency microwave device applications. Even though the basic $In_{1-x}Ga_xAs_yP_{1-y}/InP$ heterostructure concept is understood at this time, some practical device parameters in this system have been hampered by a lack of definite knowledge of many material parameters.

The purpose of this book is twofold: (1) to discuss key properties of InP, $In_{1-x}Ga_xAs_yP_{1-y}$ quaternary, and $In_{0.53}Ga_{0.47}As$ ternary; and (2) to present various material parameters and constants of these semiconductors for a variety of basic research and device applications. A set of the material properties and parameters of these semiconductors are considered in this book. The host of effects associated with the presence of specific impurities and defects is, however, omitted from the coverage. The model used in some cases is based on an interpolation scheme and, therefore, necessitates known values of the material parameters for the related binaries InP, InAs, GaAs, and GaP. Emphasis is, therefore, also placed on material properties not only of InP but also of InAs, GaAs, and GaP binaries.

It is hoped that the book will be useful to both beginning and advanced specialists as well as to workers in related fields, thus contributing to the further development of III–V semiconductor devices.

SADAO ADACHI

Gunma, Japan
May 1992

ACKNOWLEDGMENTS

The author wishes to thank the editors and authors of the following journals and symposium paper for permission to reproduce previously published figures: *Journal of Applied Physics* for Figs. 2.3, 3.1, 4.6, 8.18, 8.19, 8.21, 8.22, 8.25, 8.26, 9.4, 9.5, 9.9, 9.11, 9.12, 9.14, 9.15, 10.13, and 11.4; *Canadian Journal of Physics* for Fig. 3.10; *Zeitschrift für Naturforschung* for Fig. 4.1; *Physical Review* for Figs. 6.1, 6.2, 8.5, 8.8, 8.9, 8.16, and 8.17; *Solid State Communications* for Fig. 6.13; *Journal of Electronic Materials* for Figs. 8.6 and 10.7; *Applied Physics Letters* for Figs. 8.30 and 11.3; *Journal of Crystal Growth* for Figs. 8.31 and 10.3; *IEEE Journal of Quantum Electronics* for Figs. 8.33 and 8.35; and *Institute of Physics Conference Series* (*Gallium Arsenide and Related Compounds*) for Fig. 10.2.

CONTENTS

PRINCIPAL SYMBOLS

A	isotropy factor
a, b, d	hydrostatic (a) and shear deformation potentials (b, d) for holes at the Γ-point valence-band maximum
$a_B^{3D}(a_B^{2D})$	three-dimensional (two-dimensional) exciton Bohr radius
a_0	lattice constant
B	binary material parameter, magnetic field strength
B_o	Born ratio
B_u	bulk modulus
B'_u	pressure derivative of the bulk modulus
C_a	Caucy ratio
C_{ij}	second-order elastic stiffness constant
C_{ijk}	third-order elastic stiffness constant
C_o	elastic compressibility
C_p, C_v	specific heats at constant pressure and constant volume
C_1	deformation potential for electrons at the Γ-point conduction-band minimum
c	bowing parameter, velocity of light
D_{ij}	intervalley deformation potential constant
d	interatomic spacing
d_{ij}	piezoelectric strain constant, nonlinear optical coefficient
d_o	optical-phonon deformation potential
E	energy, electric field strength
E_{ac}	phenomenologic acoustic deformation potential
E_g	band-gap energy

E_g^{ID}, E_g^X, E_g^L	indirect-gap energy, $\Gamma_8^v \rightarrow X_6^c (E_g^X)$, $\Gamma_8^v \rightarrow L_6^c\ (E_g^L)$
$E_{g\lambda}$	lasing emission energy
E_{npo}	phenomenologic optical deformation potential
E_{th}	threshold electric field (Gunn oscillation)
E_0, E_1, . . .	critical point energies at band extrema denoted by 0, 1, . . .
E_1	conduction-band (acoustic) deformation potential
E_γ	cleavage energy
e	electron charge
e_{ij}	elastic strain component, piezoelectric stress constant
e_T^* (e_T)	Born transverse effective charge
f_i	Phillips's ionicity
G^{3D} (G^{2D})	three-dimensional (two-dimensional) exciton Rydberg energy
g	crystal density
$g(\nu)$	phonon density of states
h	Planck's constant
$\hbar$	$h/2\pi$
h_c	critical layer thickness
J_{cv}	joint density-of-states function
K	electromechanical coupling constant
$\tilde{K}_{ij}$	dimensionless phonon deformation potential
k	wave vector
k	extinction coefficient (attenuation index)
k_B	Boltzmann's constant
M	reduced mass of the two atoms ($M^{-1} = M_A^{-1} + M_B^{-1}$)
m_c^α	conductivity effective mass for electrons in the conduction-band minima α ($=\Gamma$, X, or L)
m_c^h, m_d^h	conductivity, density-of-states hole effective masses
m_e^α	density-of-states effective mass for electrons in the conduction-band minima α ($=\Gamma$, X, or L)
m_e^*	cyclotron electron mass
m_{hh}	heavy-hole mass
m_{lh}	light-hole mass
$m_{l\alpha}$, $m_{t\alpha}$	longitudinal, transverse masses for electrons in the conduction-band minima α ($=\Gamma$, X, or L)
m_{so}	spin–orbit-splitoff-band hole mass
m_0	free-electron mass
N	number of unit cells per unit volume, number of atoms per unit volume, number of the equivalent conduction-band minima ($N = 1$ for the Γ minimum, $N = 3$ for the X minima, and $N = 4$ for the L minima)

N, P	number of electrons, holes per unit volume
n	electron concentration, real refractive index
n^*	complex refractive index
P	Poisson's ratio
P, P'	momentum matrix element
p	pressure
p, q, r	phonon deformation potential
p_{ij}	photoelastic constant
Q	quaternary material parameter
q	phonon wave vector
R	normal-incidence reflectivity
R_{ij}	quadratic electrooptic (Kerr) constant
r_{ij}	linear electrooptic (Pockels) constant
S_{ij}	second-order elastic compliance constant
T	temperature, ternary material parameter
t	thickness
V	volume
V_0	volume of the unit cell
v	sound velocity, carrier drift velocity
$\bar{v}$	average sound velocity
v_{h}	average hole velocity
v_{s}	carrier saturation velocity
v_{sw}	velocity of surface acoustic wave
$W(1/W)$	thermal resistivity (conductivity)
X_{ij}	elastic stress component
x, y	mole fractions
x, y, z	coordinates
Y	Young's modulus
α	absorption coefficient
α_{F}	Fröhlich coupling constant
α_{i}	dielectric polarizability
α_{L}, α_{th}	linear expansion coefficient
α_{pe}	photoelastic coefficient
β	elastic nonlinearity parameter
Γ	disorder parameter (phonon scattering), broadening energy (free-electron–hole pair, exciton, plasmon)
γ	average Grüneisen parameter, phonon damping constant
γ_{i}	mode Grüneisen parameter
γ_{i}^{L}	Luttinger's valence-band parameter
ΔE_{c}, ΔE_{v}	conduction-band, valence-band discontinuities
$\Delta E_{\Gamma L}$, $\Delta E_{\Gamma X}$	interconduction-band separation energies [Γ^{c}–L^{c} ($\Delta E_{\Gamma L}$), Γ^{c}–X^{c} ($\Delta E_{\Gamma X}$)]

ΔU_e, ΔU_h	alloy scattering-potentials for electrons, holes
Δ_0, Δ_1, . . .	spin–orbit splitting at band extrema denoted by 0, 1, . . .
ϵ, ϵ_1, ϵ_2	dielectric constant, real part, imaginary part
ϵ_s, ϵ_∞	static, high-frequency dielectric constants
ϵ	one-electron dielectric function
θ_D	Debye temperature
θ_{cal}	"calorimetric" Debye temperature
$\theta_D(0)$	"elastic" Debye temperature
κ_L	compressibility of the linear chain
λ	wavelength of light
μ	exciton reduced mass
μ_H, μ_D	Hall, drift mobilities
μ_0	combined density-of-states mass
$\nu(\omega$ and $\Omega)$	phonon frequency
Ξ_{eff}	effective acoustic-mode deformation potential
σ_{eff}	effective nonlinear cross section
σ_{eh}	nonlinear absorption cross section
τ	thermal time constant
τ_e, τ_h	electron, hole drift times
τ_{er}	energy relaxation time
τ_N, τ_U, τ_D	phonon relaxation times, normal process (τ_N), Umklapp (flipover) process (τ_U), strain and mass point defect (τ_D)
ω	angular frequency
ω_{LO}, ω_{TO}	longitudinal, transverse optican phonon frequencies
ω_p	plasma frequency

1

INTRODUCTION

The progress made in physics and technology of semiconductors depends mainly on two families of materials: the group IV elements and the III–V compounds. The first report of the formation of III–V compounds was published in 1910 by Thiel and Koelsch.[1] They synthesized a compound of indium and phosphorus and reached the conclusion that its formula is very probably InP. The fact that one of the III–V compounds, InSb, is a semiconductor akin to Ge and α-Sn, was reported in 1950 by Blum, et al.[2]

$In_{1-x}Ga_xAs_yP_{1-y}$/InP double-heterostructure lasers emitting in the 1.3–1.7 μm-wavelength region has become a promising candidate for the light source of an optical fiber communication system because of the recent development of low-loss, low-dispersion optical fibers in this spectral region.[3,4] The $In_{1-x}Ga_xAs_yP_{1-y}$/InP system is also necessary for the efficient operation of a number of optoelectronic and transport devices, such as photodetectors, light-emitting diodes, Gunn diodes, and high-speed heterojunction transistors.[5-7] Even though the basic $In_{1-x}Ga_xAs_yP_{1-y}$/InP heterostructure concepts are understood at this time, the determination of some practical device parameters in this system has been hampered by a lack of definite knowledge of many material parameters.

This book provides numeric and graphical information about many of the semiconducting and material properties of InP, $In_{1-x}Ga_xAs_yP_{1-y}$, and $In_{0.53}Ga_{0.47}As$ that are useful for a variety of research works and device applications. The model used in some cases is based on an interpolation scheme and, therefore, necessitates known values of the material parameters for the related binaries, InP, InAs, GaAs, and GaP. Therefore, the emphasis is also placed on material properties of these end-binary compounds.

The material and semiconducting properties reviewed in this book can be classified into 10 groups: (1) structural properties; (2) mechanical, elastic, and lattice vibrational properties; (3) thermal properties; (4) collective effects and some response characteristics; (5) electronic energy-band-structure consequences; (6) electronic deformation potentials; (7) optical properties; (8) elastooptic and electrooptic properties; (9) carrier transport properties; and (10) the strain problem in InGaAs(P)-based heterostructures. The host of effects associated with the presence of specific impurities and defects is, however, omitted from this coverage. The physical quantities and properties considered here are (1) lattice constant; (2) molecular and crystal densities; (3) ordering in InGaAsP alloys; (4) Knoop hardness number; (5) second- and third-order elastic constants and their pressure derivatives; (6) Young's and bulk moduli, Poisson's ratio, and similar; (7) bulk sound velocities and mode Grüneisen parameters; (8) surface acoustic waves; (9) optical phonon frequencies and mode Grüneisen parameters; (10) phonon deformation potentials; (11) specific heat and Debye temperature; (12) thermal expansion coefficient; (13) lattice thermal conductivity; (14) static and high-frequency dielectric constants; (15) piezoelectric stress and strain constants; (16) electromechanical coupling constant; (17) Fröhlich coupling constant; (18) band-gap energies (lowest-direct, lowest-indirect, and higher-lying band gaps); (19) electron and hole effective masses (electron, light-hole, heavy-hole, spin–orbit-splitoff-band, density-of-states, and conductivity effective masses); (20) conduction- and valence-band offsets; (21) temperature and pressure derivatives of the electronic energy-band parameters; (22) electron and hole deformation potentials; (23) optical constants (refractive index, absorption coefficient, etc.); (24) exciton parameters (Rydberg constant and Bohr radius); (25) photoelastic constants; (26) linear and quadratic electrooptic constants; (27) low-field mobilities; (28) alloy scattering potentials; (29) carrier saturation velocities at high electric fields; (30) minority-carrier transport properties; and (31) critical layer thickness for coherency heteroepitaxial growth.

Of particular interest is the deviation of material constants from linearity with respect to the alloy composition y for $In_{1-x}Ga_xAs_yP_{1-y}$. Some material parameters, such as the lattice constant, second-order elastic constant, and dielectric constant, obey Vegard's rule well or exhibit very weak nonlinearity on the alloy composition y. However, some kinds of the material parameters, such as the lattice thermal conductivity, show very strong nonlinearity with respect to y, which arises from the effects of alloy disorder. A chapter on structural properties (Chapter 2) also discusses cleavage properties and ordered alloys (spontaneous long-range ordering). Long-wavelength optical phonon behavior in $In_{1-x}Ga_xAs_yP_{1-y}$ alloys is presented in a chapter on mechanical, elastic, and lattice vibrational properties (Chapter 3). Optical dispersion relations in the reststrahlen and interband transition regions are discussed in detail in a chapter on optical properties (Chapter 8). Free-carrier and intervalence-band optical absorptions are also discussed in that chapter. Majority- and minority-carrier transport properties at low- and high-electric fields in InP, InAs, GaAs, GaP, and $In_{1-x}Ga_xAs_yP_{1-y}$ alloy are presented in a chapter on carrier-transport properties (Chapter 10). Effects of in-plane biaxial strain generated in InGaAs(P)-based, strained-layer heteroepitaxial wafers on their structural and physical properties are discussed in a chapter on the strain problem in InGaAs(P)-based heterostructures (Chapter 11). The material constants and properties derived here are used with wide success to make clear general properties of these semiconductors. Key properties of the material parameters for a variety of $In_{1-x}Ga_xAs_yP_{1-y}$/InP device applications are also discussed in detail in this book.

REFERENCES

1. A. Thiel and H. Koelsch, *Z. Anorg. Chem.* **66,** 288 (1910).
2. A. I. Blum, N. P. Mokrovski, and A. R. Regel', *Seventh All-Union Conference on the Properties of Semiconductors* (Kiev, USSR, 1950); *Izv. Akad. Nauk SSSR Ser. Fiz.* **16,** 139 (1952).
3. T. P. Pearsall, ed., *GaInAsP Alloy Semiconductors*, Wiley, New York, 1982.
4. T.-P. Lee, *Proc. IEEE* **79,** 253 (1991).
5. K. Heime, *InGaAs Field-Effect Transistors*, Research Studies Press, Somerset, England, 1989.
6. S. M. Sze ed., *High-Speed Semiconductor Devices*, Wiley, New York, 1990.
7. H. Morkoç, H. Unlu, and G. Ji, *Principles and Technology of MODFETs*, Wiley, New York, 1991.

2

STRUCTURAL PROPERTIES

2.1 CRYSTAL STRUCTURE AND LATTICE CONSTANT

Some of the III–V compounds, and in particular InP, $In_{1-x}Ga_xAs_yP_{1-y}$, and $In_{1-x}Ga_xAs$, form crystals with the zinc-blende arrangement. The zinc-blende structure is based on the cubic space group $F\bar{4}3m$ in which the lattice atoms are tetrahedrally bound in network arrangements related to those of the group IV (diamond-type) semiconductors. The III–V compounds crystallizing in the cubic zinc-blende structure are the simplest crystals lacking a center of symmetry and, hence, capable of exhibiting piezoelectric and related effects depending on polar symmetry.

The lattice constant a_0 is known to obey Vegard's law well, that is, to vary linearly with alloy composition,[1] and such is the case for InGaAs and InGaAsP alloys.[2,3] The cubic lattice parameter of InP at 300 K is reduced to $a_0 = 5.8688$ Å (5.86875 Å)[4]. Using this law, the lattice-matching relation between the composition fractions x and y for $In_{1-x}Ga_xAs_yP_{1-y}$ to InP can be written as

$$x = \frac{0.1896y}{0.4176 - 0.0125y} \qquad (0 \leq y \leq 1.0) \tag{2.1}$$

In obtaining this expression, we also used the following binary constants:[5] $a_0 = 6.0584$ Å for InAs, $a_0 = 5.4512$ Å for GaP, and $a_0 = 5.6533$ Å for GaAs. Equation (2.1) gives the end-point compositions $x = 0.0$, $y = 0.0$ for InP and $x = 0.468$, $y = 1.0$ for $In_{0.532}Ga_{0.468}As$. The lattice-matching relation is also given in approximated form by

$$x = 0.47y \qquad (0 \leq y \leq 1.0) \tag{2.2}$$

In Fig. 2.1 we show the three-dimensional representation of the lattice constant a_0 for $In_{1-x}Ga_xAs_yP_{1-y}$ over the entire range of compositions ($0 \leq x \leq 1.0$, $0 \leq y \leq 1.0$). The bold line is the locus of a_0 for compositions lattice-matched to InP. Figure 2.2 shows x–y plots for several $In_{1-x}Ga_xAs_yP_{1-y}$ samples nearly lattice-matched to InP as measured by Nahory et al.[3] The solid and dashed lines represent, respectively, the results of Eqs. (2.1) and (2.2).

The lattice constant of a semiconductor can expand or contract when impurity atoms are incorporated. A well-known example is the dilation, or expansion, of the GaAs lattice induced by tellurium.[6,7] Increase in the GaAs lattice constant of ~0.01% has been reported at Te concentrations of $\sim 10^{19}$ cm^{-3}.[7] Kelbert et al.[8] have also measured the change in lattice constant for InP as a function of Te dopant concentration. A dilation of ~0.15% was obtained in the InP lattice constant at a Te concentration of $\sim 10^{20}$ cm^{-3}. These measurements were compared to predictions from Vegard's law. The

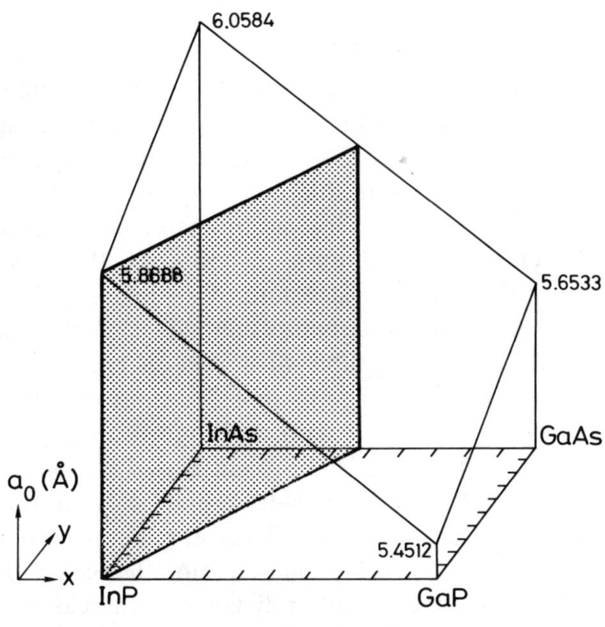

Figure 2.1 Three-dimensional representation of the lattice constant a_0 for $In_{1-x}Ga_xAs_yP_{1-y}$ over the entire range of compositions ($0 \leqq x \leqq 1.0$, $0 \leqq y \leqq 1.0$). The bold line is the locus of a_0 for compositions lattice-matched to InP.

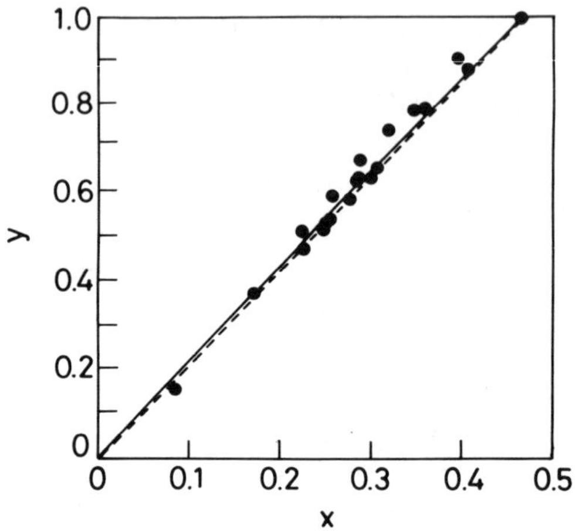

Figure 2.2 Plots of x versus y for $In_{1-x}Ga_xAs_yP_{1-y}$ nearly lattice-matched to InP as measured by Nahory et al.[3] The solid and dashed lines show the results of Eqs. (2.1) and (2.2), respectively.

effect of the impurity doping on the InP lattice constant has also been studied in Refs. 9, and 10.

For InP each In(P) has four nearest neighbors of P(In) at a distance of $\sqrt{3}\,a_0/4$ (= 2.54 Å) at the corners of a regular tetrahedron. The spacing between the near-neighbor In–In (or P–P) atoms is equal to $(1/\sqrt{2})a_0$ (= 4.15 Å). It would be anticipated that for alloys the interatomic bond length of each constituent would remain constant with composition. Unfortunately, however, standard X-ray diffraction techniques give information only on the lattice structure that is averaged over an area larger than the scale of lattice constant. The interatomic structure of these alloys was, therefore, not well understood until recently. In 1982, Mikkelsen and Boyce[11] studied the interatomic structure of an InGaAs alloy by using an extended X-ray-absorption fine-structure method (EXAFS). They found that the Ga–As and In–As near-neighbor distances change by only 0.04 Å as the In composition varies from 1 to 99 mol % despite the fact that this alloy accurately follows Vegard's law with a change in average near-neighbor spacing of 0.17 Å. This result is shown in Fig. 2.3 along with the theoretical estimation based on a valence–force-field calculation.[12] Mikkelsen and Boyce[11] also found that the cation (In,Ga) sublattice approaches a virtual crystal (i.e., an average face-centered cubic lattice) with a broadened single distribution of second-neighbor distances, whereas the anion sublattice exhibits a bimodal anion–anion second-neighbor distribution. Similar EXAFS work on $In_{1-x}Ga_xAs_yP_{1-y}$ quaternary has been done by Oyanagi et al.[13,14] They also found that the cation–anion distances in the quaternary

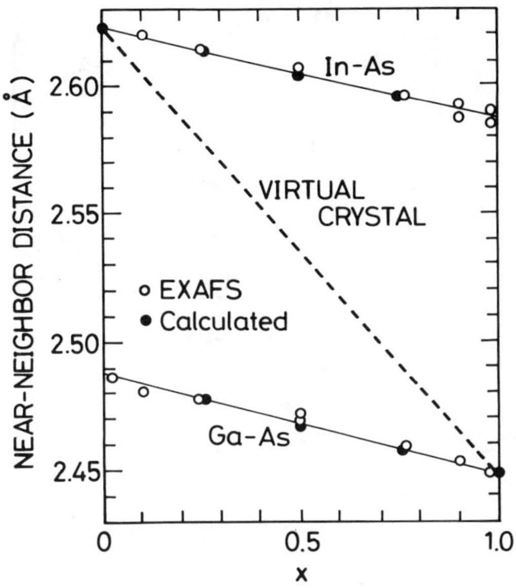

Figure 2.3 Near-neighbor distances d(Ga–As) and d(In–As) versus InAs mole fraction x in the $In_{1-x}Ga_xAs$ ternary alloy measured by EXAFS.[11] The solid circles represent the calculated results based on a valence–force-field approach. The dashed line represents the average cation–anion spacing (Vegard's law). (From Fukui[12]).

alloy deviate from the average interatomic distance, but are rather close to the bond lengths in pure parent crystals. In Fig. 2.4 we plot their cation–anion distance as a function of the y-composition proportion (see also Table 2.1). Detailed theoretical consideration of these results has also been carried out by Ichimura and Sasaki [5] from a thermodynamic analysis.

The lattice constant a_0 is dependent on both the temperature and pressure. The temperature dependence of a_0 is explained by the linear thermal expansion coefficient α_{th} (see Section 4.2). The constant a_0 is related to the pressure p by Murnaghan's equation (see Section 3.2.2). The pressure effect on the structural properties of InP has been studied using resistivity[16] and X-ray measurements.[17,18] These measurements showed that InP undergoes a phase transition at about 11 GPa. The new phase is metallic,[16] and its crystallographic structure has been identified as the rocksalt structure.[17,18] At about 19 GPa another phase is detected that has been identified as body-centered tetragonal.[18] Patel et al.[19] also found changes in the crystal structure and electrical properties for InP and InGaAs after a complete release of pressure from above the semiconductor-to-metal transition. X-ray diffraction data on InP showed differences with respect to the cubic, zinc-blende structure. They considered that these changes might be due to

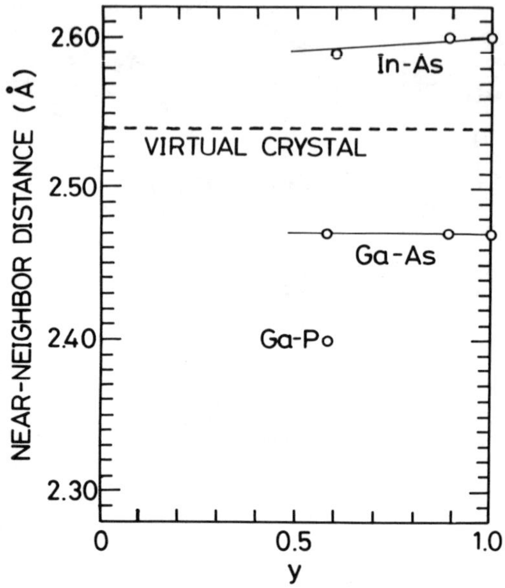

Figure 2.4 Near-neighbor distances d(In–As), d(Ga–As), and d(Ga–P) versus InP mole fraction y in the $In_{1-x}Ga_xAs_yP_{1-y}$ quaternary alloy measured by EXAFS.[13] The dashed line represents the average cation–anion spacing (Vegard's law).

the presence of defects induced after the pressure was released. In contrast, InP quenched from pressures below the semiconductor-to-metal transition did not show any changes in its electronic transport properties and X-ray structure. They also found that the carrier concentration decreased by three orders of magnitude and the electron mobility decreased to a value characteristic of the X-conduction-band minima. Their photoconductivity data are possibly consistent with the X minima being lower than the direct minimum, but the data were not conclusive. Details of the pressure effect on the electronic energy-band structure are discussed in Section 6.4.

2.2 BINDING AND CLEAVAGE PROPERTIES

The cleavage properties of a crystal are related to the atomic arrangement and corresponding electron density map. Group IIIb atoms have three electrons with an s^2p^1 configuration outside a core of closed shells, and group Vb atoms have five electrons in a s^2p^3 configuration. The IIIb and Vb atoms have, therefore, an average of four valence electrons per atom available for binding. We might then expect that the covalent bonds are formed between tetrahedral s^1p^3-hybrid orbitals; for example, for InP

$$\mathrm{In}(5s^25p^1) + \mathrm{P}(3s^23p^3) \rightarrow \mathrm{In}(5s^15p^3)^- + \mathrm{P}(3s^13p^3)^+$$

For such a covalent bonding each V atom donates an electron to a III atom, so that V^+ and III^- ions are formed, each with four valence electrons.

An ionic bond is due to Coulomb attraction between the excess positive and negative charges on ions formed by transfer of electrons from the metallic to the nonmetallic atom:

$$\mathrm{In}(4d^{10}5s^25p^1) + \mathrm{P}(2p^63s^23p^3) \rightarrow \mathrm{In}(4d^{10})^{+3} + \mathrm{P}(2p^63s^23p^6)^{-3}$$

The bonds in most III–V compounds are not adequately described by any of these extreme types, but have characteristics intermediate between those usually associated with the terms *covalent* and *ionic*.

We show in Fig. 2.5 schematic views of the atomic arrangement in the direction along the (111), (110), and (100) planes of InP. This arrangement is the same as for the diamond form except that the two different kinds of atom (III and V) occupy alternate positions in the lattice. For ductile materials, the rate of dislocation formation and propagation away from the

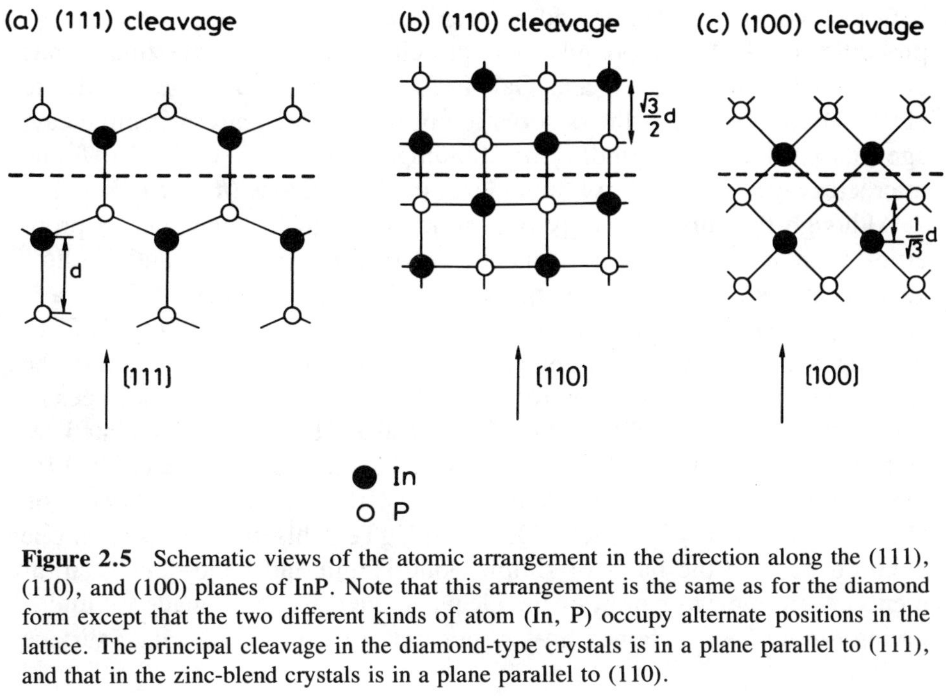

Figure 2.5 Schematic views of the atomic arrangement in the direction along the (111), (110), and (100) planes of InP. Note that this arrangement is the same as for the diamond form except that the two different kinds of atom (In, P) occupy alternate positions in the lattice. The principal cleavage in the diamond-type crystals is in a plane parallel to (111), and that in the zinc-blend crystals is in a plane parallel to (110).

crack tip determines the cleavage and fracture properties. In contrast, for brittle solids, such as tetrahedrally bonded semiconductors, the cleavage energy is dominated by the breaking of bonds across the cleavage surface and the resulting creation of a new surface. In diamond-type crystals, cleavage occurs along (111) surface planes. This is because the (111) surface atoms are only singly bonded to the opposite surface, but the (100) surface atoms are doubly bonded to the opposite surface (Fig. 2.5). The (110) surface atoms are also singly bonded to the opposite surface, but the plane spacing is shorter than that of the (111) planes. It is more difficult to separate the planes of shorter spacing. In fact, it was found experimentally that for Si the cleavage energy (E_γ) for the (110) surface is considerably larger than that for the (111) surface [$E_\gamma = 1.9\ \mathrm{J/m^2}$ for the (110) surface,[20] $E_\gamma = 1.24 \sim 1.14\ \mathrm{J/m^2}$ for the (111) surface].[20–22] Note that the cleavage energy E_γ is the energy per unit surface area necessary to separate a crystal along a given plane.

In the case of zinc-blende crystals, we must take into account the effects of surface polarity and electron density distribution.[23] The III–V compounds have two types of (111) surface polarities—namely, (111)*A* and (111)*B*—and hence there will be an electrostatic attraction between these different planes. Such an attractive force will make it difficult to separate along the (111) planes. However, the (110) surfaces are composed of equal numbers of III and V atoms, so there will be no overall electrostatic force between the planes. Wolff and Broder[24] have investigated the cleavage properties of III–V compounds. The principal cleavage in the zinc-blende crystals InP, InAs, GaAs, and GaP is found to be in a plane parallel to (110). Microcleavage is the observation of light figure patterns from cleavage pits produced by grinding or abrading the surface. They also found microcleavages in InP, InAs, and GaAs in (*hhk*) planes ($h \geq k$).[24]

Although the surface energy of a solid is one of its fundamental properties, there is a dearth of experimental values for it. Messmer and Bilello[20] have measured the surface energy E_γ for the {110} planes of GaAs and GaP. The measured values are $0.86\ \mathrm{J/m^2}$ for GaAs and $1.9\ \mathrm{J/m^2}$ for GaP. Berding et al.[25] have recently presented a method for the calculation of the surface and cleavage energies for the (100), (110), and (111) surfaces of some semiconductors (Si, GaAs, CdTe, and HgTe). Their calculated results give an ordering of $E_\gamma(\mathrm{Si}) > E_\gamma(\mathrm{GaAs}) > E_\gamma(\mathrm{CdTe}) > E_\gamma(\mathrm{HgTe})$ for the experimentally observed cleavage faces; that is, for the cleavage on (111) in Si and (110) in GaAs, CdTe, and HgTe. This ordering is what one would expect on the basis of simple bond-length and bond-density arguments. A simple estimation of the cleavage energy can be made by multiplying the number of bonds broken per unit area on the cleavage surface times the energy per bond. Thus, although—for a given bond length—the bond density on the (110) surface (on which GaAs cleaves) is higher than on the (111) surface (on which Si cleaves), the shorter bond length and the larger bond strength of Si compared to GaAs combine to produce a larger cleavage energy in Si. In turn, both the shorter bond length and the greater bond strength of GaAs compared to HgTe and CdTe result in a larger cleavage energy in GaAs. Finally, the weaker HgTe bond, compared to CdTe, and their nearly equal bond lengths predict a smaller cleavage energy in HgTe.

2.3 MOLECULAR AND CRYSTAL DENSITIES

The crystal density g is one of the simplest and most important material parameters. There are four molecules in a unit cell of the zinc-blende lattice. If an accurate lattice constant is available, the calculation of g gives in principle a good, reliable value. The lattice parameters mentioned in Section 2.1 carry connotations for molecular and crystal densities as presented in Table 2.1. The calculated crystal density for stoichiometric InP is 4.791 g/cm^2. The values for InAs, GaAs, and GaP are, respectively, 5.668, 5.317, and 4.129 g/cm^3. The quaternary parameter (Q) can, thus, be derived from these binary parameters (B values) by using the interpo-

TABLE 2.1 Crystal Structure, Lattice Constant, Near-Neighbor Distance, and Crystal Density for Stoichiometric InP, $In_{1-x}Ga_xAs_yP_{1-y}$, and $In_{0.53}Ga_{0.47}As$ (300 K)

Parameter	InP	$In_{1-x}Ga_xAs_yP_{1-y}$	$In_{0.53}Ga_{0.47}As$
Space (point) group	$F\bar{4}3m(T_d^2)$	$F\bar{4}3m(T_d^2)$	$F\bar{4}3m(T_d^2)$
Lattice constant a_0 (Å)	5.8688	5.8688	5.8688
d(cation–anion) (Å)	2.54	d(In–As) = 2.59 ($y = 0.58$) d(In–As) = 2.60 ($y = 0.89, 1.0$) d(Ga–As) = 2.47 ($y = 0.58, 0.89, 1.0$) d(Ga–P) = 2.40 ($y = 0.58$)	d(In–As) = 2.61 d(Ga–As) = 2.47
d(cation–cation) (Å)	4.15		
Molecular density (cm^{-3})	1.979×10^{22}	1.979×10^{22}	1.979×10^{22}
Crystal density g (g/cm^3)	4.791	$4.791 + 0.575y + 0.138y^2$	5.504

lation scheme

$$Q(x, y) = (1 - x)yB_{\text{InAs}} + (1 - x)(1 - y)B_{\text{InP}} + xyB_{\text{GaAs}} + x(1 - y)B_{\text{GaP}} \tag{2.3}$$

Introducing Eq. (2.1) into Eq. (2.3), we obtain g as a function of y for $In_{1-x}Ga_xAs_yP_{1-y}$. As shown in Fig. 2.6, the calculated g values (solid line) vary almost quadratically with composition y. We find that the following expression gives a good representation for $g(y)$ of this alloy system:

$$g(y) = 4.791 + 0.575y + 0.138y^2 \quad \text{g/cm}^3 \tag{2.4}$$

Nahory et al.[3] also measured the lattice parameters a_0 values for four binary end members of InGaAsP. Their obtained values are 5.8696 Å (InP), 6.0590 Å (InAs), 5.6536 Å (GaAs), and 5.4512 Å (GaP) with an accuracy of ±0.0001 Å. These lattice data provide $g(\text{InP}) = 4.789$ g/cm^3

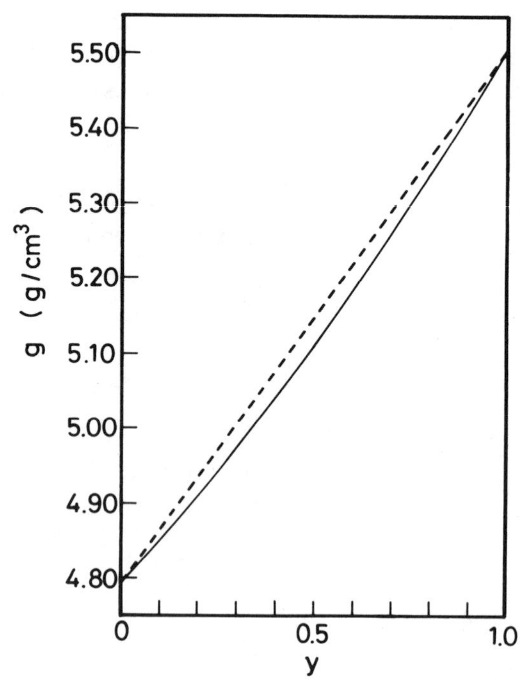

Figure 2.6 Crystal density g as a function of y for $In_{1-x}Ga_xAs_yP_{1-y}$ lattice-matched to InP (solid line). The dashed line is linearly interpolated between the InP and $In_{0.53}Ga_{0.47}As$ end-point values.

$g(\mathrm{InAs}) = 5.666\ \mathrm{g/cm^3}$, $g(\mathrm{GaAs}) = 5.317\ \mathrm{g/cm^3}$, and $g(\mathrm{GaP}) = 4.129\ \mathrm{g/cm^3}$. The difference from our results is negligibly small.

2.4 ORDERED ALLOYS

In recent years, spontaneous long-range ordering into the CuAuI-, CuPt-, chalcopyrite-, famatinite-, and luzonite-type structures have been observed in several normally disordered isovalent III–V alloy semiconductors[26–33] (InGaAs,[29–31] InGaAsP).[31–33] These ordered compounds have the stoichiometric compositions expressed by $A_nB_{4-n}C_4$ (or $A_4B_nC_{4-n}$) ($n = 1$, 2, or 3). At intermediate compositions, one can expect to find ordered- and disordered-phase coexistence regions; the variation probably depends on the temperature and growth condition. In Fig. 2.7 we show the unit cell of (*a*) disordered $\mathrm{In_{0.5}Ga_{0.5}As}$ (zinc-blende lattice), (*b*) perfectly ordered $\mathrm{InGa_3As_4}$ (famatinite), and (*c*) layered tetragonal $\mathrm{In_2Ga_2As_4}$ alloys. A thermodynamic calculation based on a first-principle theory predicts that an ordered phase with a large lattice–constant mismatch is more stable than the corresponding random phase.[34,35] Some experimental data, on the other hand, show that the kinetics of crystal growth plays an important role in the formation of an ordered structure.[27] Kondow et al.[27] have recently shown that this breaking of face-centered cubic symmetry observed in

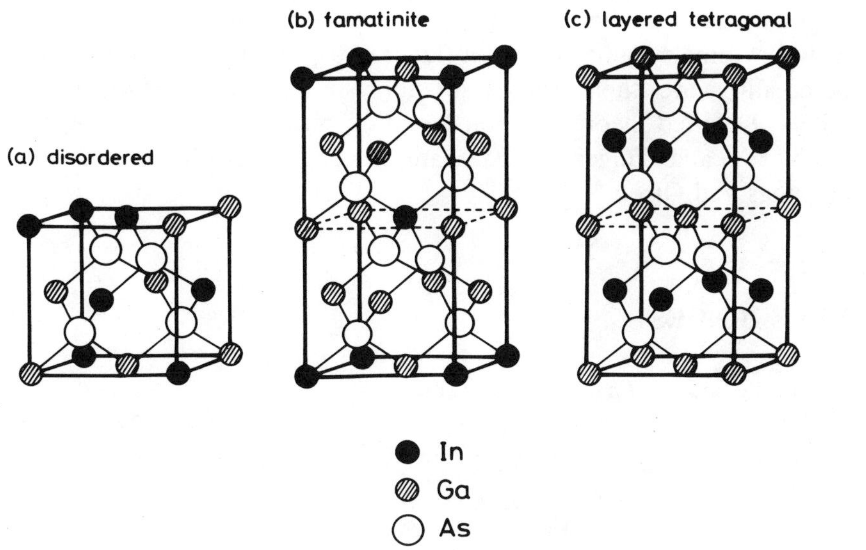

Figure 2.7 Crystal structures of (*a*) disordered $\mathrm{In_{0.5}Ga_{0.5}As}$ (zinc-blend), (*b*) ordered $\mathrm{InGa_3As_4}$ (famatinite), and (*c*) layered tetragonal $\mathrm{In_2Ga_2As_4}$ alloys.

$Ga_2In_2P_4$ is not a consequence of thermodynamic stability. Bernard et al.[36] have also discussed the physical origins of stable and metastable ordering in bulk and epitaxial semiconductor alloys and superlattices based on a thermodynamic first-principle theory of stability.

If the ordered phase is the true equilibrium phase, the physics of semiconducting alloys will be completely changed. This is because it has been assumed that the physics is thoroughly based on a random atomic arrangement of the crystal lattice. One of the most commonly used models for alloys is the virtual-crystal approximation (VCA).[37] In this model not only are all the atoms located randomly on ideal lattice sites of the average unit cell but also the substituted sublattice assumes an average value of bond length, bond ionicity, atomic potential, and so on. The VCA has been widely used up to the present and has given considerable success, such as in the electronic-energy-band structures of semiconductor alloys.[38] However, it is natural to think that many material properties of an ordered ternary compound alloy differ from those of the disordered alloy of the same composition. In fact, preliminary experimental results indicate noticeable changes in the fundamental band gaps for the material of interest here, for InGaAs (layered tetragonal $InGaAs_2$),[39] and for other III–V ternary alloys.[26] Either perfect or partial ordering lowers the structural symmetry from the face-centered cubic symmetry space group $F\bar{4}3m$ to that of its body-centered tetragonal symmetry subgroup $I\bar{4}2m$ (famatinite) or $I\bar{4}2d$ (chalcopyrite), simple tetragonal symmetry subgroup $P\bar{4}m2$ (layered tetragonal), simple cubic symmetry subgroup $P\bar{4}3m$ (luzonite), or rhombohedral symmetry subgroup $R3m$ (layered trigonal). The energy gaps of ordered alloys are, then, usually smaller than the band gaps of fully random alloys. Detailed theoretical consideration on this subject has been carried out by Wei and Zunger,[26,40] Mascarenhas et al.,[28] Koiller et al.,[41] and Mascarenhas and Olson.[42]

REFERENCES

1. L. Vegard, *Z. Phys.* **5,** 17 (1921).
2. J. C. Woolley and B. C. Smith, *Proc. Phys. Soc. London* **72,** 214 (1958).
3. R. E. Nahory, M. A. Pollack, W. D. Johnston, Jr., and R. L. Barns, *Appl. Phys. Lett.* **33,** 659 (1978).
4. G. Giesecke and H. Pfister, *Acta Cryst.* **11,** 369 (1958).
5. G. Giesecke, in *Semiconductors and Semimetals*, Vol. 2, R. K. Willardson and A. C. Beer, eds., Academic, New York, 1966, p. 63; H. Kressel and J. K. Butler, *Semiconductor Lasers and Heterojunction LED's*, Academic, New York, 1977.

6. J. B. Mullin, B. W. Straughan, C. M. H. Driscoll, and A. F. W. Willoughby, *J. Appl. Phys.* **47,** 2584 (1976).
7. P. S. Dobson, P. F. Fewster, D. T. J. Hurle, P. W. Hutchinson, J. B. Mullin, B. W. Straughan, and A. F. W. Willoughby, in *Gallium Arsenide and Related Compounds*, Institute of Physics, Bristol, UK, 1979, p. 163.
8. F. G. Kellert, K. T. Chan, J. E. Turner, and V. M. Robbins, *J. Electron. Mater.* **19,** 1425 (1990).
9. J. P. Farges, C. Schiller, and W. J. Bartels, *J. Cryst. Growth* **83,** 159 (1987).
10. A. Knauer, J. Kräusslich, R. Kittner, R. Staske, and A. Bärwolff, *Cryst. Res. Technol.* **25,** 441 (1990).
11. J. C. Mikkelsen, Jr. and J. B. Boyce, *Phys. Rev. Lett.* **49,** 1412 (1982); *Phys. Rev. B* **28,** 7130 (1983).
12. T. Fukui, *Jpn. J. Appl. Phys.* **23,** L208 (1984); *J. Appl. Phys.* **57,** 5188 (1985).
13. H. Oyanagi, Y. Takeda, T. Matsushita, T. Ishiguro, and A. Sasaki, in *Gallium Arsenide and Related Compounds*, Institute of Physics, Bristol, UK, 1986, p. 295.
14. Y. Takeda, H. Oyanagi, and A. Sasaki, *J. Appl. Phys.* **68,** 4513 (1990).
15. M. Ichimura and A. Sasaki, *J. Cryst. Growth* **98,** 18 (1989).
16. S. Minomura and H. G. Drickamer, *J. Phys. Chem. Solids* **23,** 451 (1962).
17. J. C. Jamieson, *Science* **139,** 845 (1963).
18. C. S. Menoni and I. L. Spain, *Phys. Rev. B* **35,** 7520 (1987).
19. D. Patel, C. S. Menoni, and I. L. Spain, *J. Appl. Phys.* **66,** 1658 (1989).
20. C. Messmer and J. C. Bilello, *J. Appl. Phys.* **52,** 4623 (1981).
21. J. J. Gilman, *J. Appl. Phys.* **31,** 2208 (1960).
22. R. J. Jaccodine, *J. Electrochem. Soc.* **110,** 524 (1963).
23. C. Hilsum and A. C. Rose-Innes, *Semiconducting III–V Compounds*, Pergamon, New York, 1961.
24. G. A. Wolff and J. D. Broder, *Acta Cryst.* **12,** 313 (1959).
25. M. A. Berding, S. Krishnamurthy, A. Sher, and A.-B. Chen, *J. Appl. Phys.* **67,** 6175 (1990).
26. S.-H. Wei and A. Zunger, *Phys. Rev. B* **39,** 3279 (1989) and references cited therein.
27. M. Kondow, H. Kakibayashi, T. Tanaka, and S. Minagawa, *Phys. Rev. Lett.* **63,** 884 (1989).
28. A. Mascarenhas, S. Kurtz, A. Kibbler, and J. M. Olson, *Phys. Rev. Lett.* **63,** 2108 (1989).
29. T. S. Kuan, W. I. Wang, and E. L. Wilkie, *Appl. Phys. Lett.* **51,** 51 (1987).
30. O. Ueda, Y. Nakata, and T. Fujii, *Appl. Phys. Lett.* **58,** 705 (1991).
31. M. A. Shahid, S. Mahajan, D. E. Laughlin, and H. M. Cox, *Phys. Rev. Lett.* **58,** 2567 (1987).
32. M. A. Shahid and S. Mahajan, *Phys. Rev. B* **38,** 1344 (1988).

33. W. E. Plano, D. W. Nam, J. S. Major, Jr., K. C. Hsieh, and N. Holonyak, Jr., *Appl. Phys. Lett.* **53,** 2537 (1988).

34. A. A. Mbaye, L. G. Ferreira, and A. Zunger, *Phys. Rev. Lett.* **58,** 49 (1987).

35. G. P. Srivastava, J. L. Martins, and A. Zunger, *Phys. Rev. B* **31,** 2561 (1985); erratum, ibid. **38,** 12694 (1988).

36. J. E. Bernard, R. G. Dandrea, L. G. Ferreira, S. Froyen, S.-H. Wei, and A. Zunger, *Appl. Phys. Lett.* **56,** 731 (1990).

37. L. Nordheim, *Ann. Phys.* (Leipzig) **9,** 607, 641 (1931).

38. J. A. Van Vechten and T. K. Bergstresser, *Phys. Rev. B* **1,** 3351 (1970).

39. T. Fukui and H. Saito, *Jpn. J. Appl. Phys.* **23,** L521 (1984).

40. S.-H. Wei and A. Zunger, *Appl. Phys. Lett.* **56,** 662 (1990).

41. B. Koiller, M. A. M. Davidovich, and L. M. Falicov, *Phys. Rev. B* **41,** 3670 (1990).

42. A. Mascarenhas and J. M. Olson, *Phys. Rev. B* **41,** 9947 (1990).

3

MECHANICAL, ELASTIC, AND LATTICE VIBRATIONAL PROPERTIES

3.1 KNOOP HARDNESS

The hardness test has been used for a long time as a simple means of characterizing the mechanical behavior of solids.[1] The interest in the hardness of III–V compounds stems principally from their use in optoelectronic and electron-device applications. There are many indications that dislocations induced during the fabrication of these devices subsequently have a very damaging effect on their operational performance and degradation behavior.[2]

The Knoop hardness anisotropy is a well-known phenomenon[3] and has been measured for InP[4–7] $In_{1-x}Ga_xAs_yP_{1-y}$, and $In_{0.53}Ga_{0.47}As$ lattice-matched to InP.[7] The Knoop hardness number (KHN) not only varies with both the plane and direction of indentation but also depends on the type of dopant. For InP,[6,7] S-doped crystal is the hardest among S-, Se-, and Zn-doped crystals. These crystals are all harder than undoped InP. On the other hand, the addition of Sn apparently softens the lattice.

In Fig. 3.1 the Knoop hardness anisotropy of undoped InP, and nominally undoped $In_{0.53}Ga_{0.47}As$ and $In_{1-x}Ga_xAs_yP_{1-y}$ ($y = 0.745$) alloys lattice-matched to InP[7] are plotted. The ternary and quaternary hardness anisotropy curves share a general shape, fundamentally different from that of InP, having two maxima and one minimum instead of one maximum and

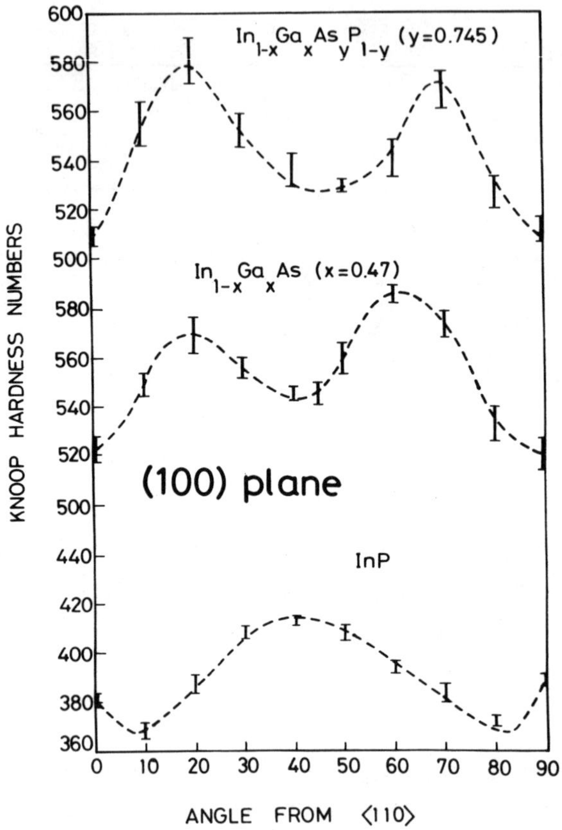

Figure 3.1 Knoop hardness anisotropy on the (100) plane of undoped InP, and nominally undoped $In_{0.53}Ga_{0.47}As$ and $In_{1-x}Ga_xAs_yP_{1-y}$ ($y = 0.745$) alloys. (From Watts and Willoughby.[7])

two minima. The major slip system in the diamond and zinc-blende structures is known to be $\frac{1}{2}\langle 110 \rangle \{111\}$.[8] Dislocations generated by plastic deformation, therefore, lie on $\{111\}$ planes with $\frac{1}{2}\langle 110 \rangle$ Burgers vectors. The radically different shape of the anisotropy curves observed may suggest significant differences between the deformation behavior of the alloys and InP, which might be associated with different dislocation dynamics and even, perhaps, a primary slip system different from the $\langle 110 \rangle$, $\{111\}$ system.

In Fig. 3.2 we plot the KHNs measured on the (100) plane along the $\langle 110 \rangle$ direction as a function of y for the undoped or the nominally undoped $In_{1-x}Ga_xAs_yP_{1-x}$ quaternary ($y = 0.0$, 0.745, and 1.0).[7] The mi-

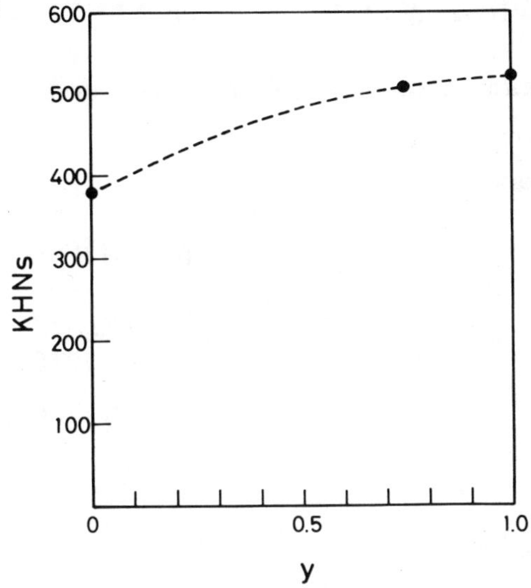

Figure 3.2 Knoop hardness numbers (KHNs) measured on the (100) plane along the ⟨110⟩ direction as a function of y for the undoped or the nominally undoped $In_{1-x}Ga_xAs_yP_{1-y}$ quaternary ($y = 0.0, 0.745, 1.0$).[7]

crohardness as a function of composition for the $InAs_yP_{1-y}$ alloy has also been reported by Vigdorovich and Nashel'skii.[9] They found that a microhardness–composition curve does not show a maximum, as is usually found in ordinary metallic alloy systems. However, after taking account of cracking near the indentations, they proposed that the curve would contain a maximum near the harder material, specifically, InP ($y \sim 0.1$). The data of Watts and Willoughby[7] indicate that the KHNs for the alloys ($y = 0.745$, 1.0) are about 40% higher than the value for InP. It is, however, difficult to recognize a peak in the curve of hardness versus composition (see Fig. 3.2), since their data did not cover the whole quaternary range of composition.

Microhardness investigations on both doped and undoped InP wafers have been carried out by Arivuoli et al.[10] using a Vickers pyramidal indentor. Their results suggested that the addition of dopants (Zn, S, Sn, or Cd/S) has two main effects on the crystal lattice: (1) pure solution hardening and (2) a reduction of the dislocation density. It was noted, however, that the two effects do not seem to be related to each other, as proved by the fact that low dislocation densities are associated with softer crystals and vice versa.

3.2 ELASTIC PROPERTIES AND SOUND VELOCITIES

3.2.1 Elastic Stiffness and Compliance Constants

The quaternary InGaAsP and the related binaries crystallize into the zinc-blende structure, and so the elastic stiffness tensor [C] takes in the form[11]

$$[C] = \begin{bmatrix} C_{11} & C_{12} & C_{12} & 0 & 0 & 0 \\ C_{12} & C_{11} & C_{12} & 0 & 0 & 0 \\ C_{12} & C_{12} & C_{11} & 0 & 0 & 0 \\ 0 & 0 & 0 & C_{44} & 0 & 0 \\ 0 & 0 & 0 & 0 & C_{44} & 0 \\ 0 & 0 & 0 & 0 & 0 & C_{44} \end{bmatrix} \tag{3.1}$$

The macroscopic theory of the elastic properties of solids is described in detail in tensor notation by Nye.[11] The elastic compliance tensor [S], which has the same form as Eq. (3.1), is connected reciprocally with the tensor [C] through Hooke's relation. Explicit equations for the component S_{ij} in terms of C_{ij} can be given by

$$S_{11} = \frac{C_{11} + C_{12}}{(C_{11} - C_{12})(C_{11} + 2C_{12})} \tag{3.2a}$$

$$S_{12} = \frac{-C_{12}}{(C_{11} - C_{12})(C_{11} + 2C_{12})} \tag{3.2b}$$

$$S_{44} = \frac{1}{C_{44}} \tag{3.2c}$$

Theoretical work on elastic moduli for semiconductors has been reported by many authors. Keyes[12] showed that the dimensionless reduced bulk modulus and shear moduli (i.e., the elastic constants multiplied by the fourth power of the nearest-neighbor distance) exhibited smaller variations for different compounds of the same structure (diamond, zinc-blende, or wurtzite) than did the original constants. Adachi[13] used a linear relation between the elastic moduli and lattice constants of the III–V compounds to predict values for AlAs, which have no experimental data. Other calculations of the elastic constants for AlAs have also been reported by Kagaya and Soma[14] and Chetty et al.,[15] who used an empirical pseudopotential method of the Ashcroft and variations of the local Heine-Abarenkov form[14] and an ab initio pseudopotential method.[15] Schilfgaarde and Sher[16] have estimated a complete set of the elastic constants of tetrahedral semiconductors as a vehicle to test several aspects of Harrison's tight-binding theory. Theoretical prediction of the elastic moduli of some mixed material systems (KCl–KBr, CaF_2–SrF_2, and GaP–GaAs) have also been carried out by Giri and Mitra[17] on the basis of a classical valence–force approach.

Although the elastic stiffness constants C_{ij} of the III–V binaries have been extensively measured, little is known about semiconductor alloys. A set of the C_{ij} values for InP have been determined by Hickernell and Gayton[18] and Nichols et al.[19] by measuring ultrasonic wave velocities. The values they obtained are (at room temperature; in 10^{11} dyn/cm^2) $C_{11} = 10.22$, $C_{12} = 5.76$, and $C_{44} = 4.60$[18] and $C_{11} = 10.11$, $C_{12} = 5.61$, and $C_{44} = 4.56$.[19] To our knowledge, however, only one has been reported to date for the semiconductor alloy ($Al_xGa_{1-x}As$).[13] The elastic constant can be attributed to the long-range Coulomb forces in solids. Thus, it seems that the C_{ij} value is a linear variation of alloy composition. In fact, the data for $Al_xGa_{1-x}As$,[13] which were estimated indirectly from experimental information on the surface acoustic wave properties, are successfully interpreted by a linear interpolation scheme. A theoretical estimate for the $GaAs_yP_{1-y}$ alloy[17] also showed a linear variation of C_{ij} with respect to the mole fraction y.

The elastic stiffness and compliance constants as a function of y for $In_{1-x}Ga_xAs_yP_{1-y}$ lattice-matched to InP are shown in Figs. 3.3 and 3.4,

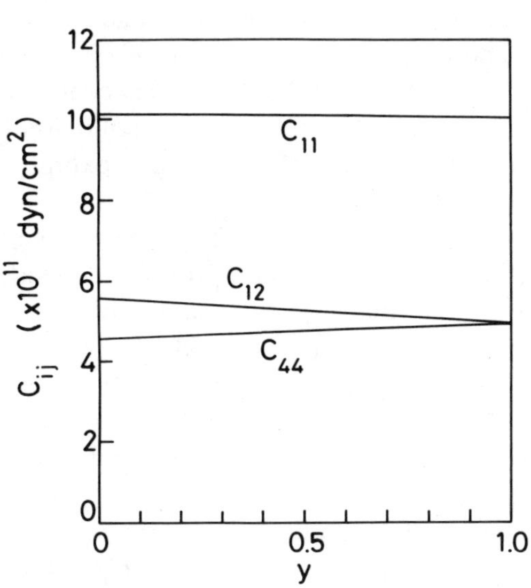

Figure 3.3 Elastic stiffness constants, C_{11}, C_{12}, and C_{44}, as a function of y for $In_{1-x}Ga_xAs_yP_{1-y}$ lattice-matched to InP.

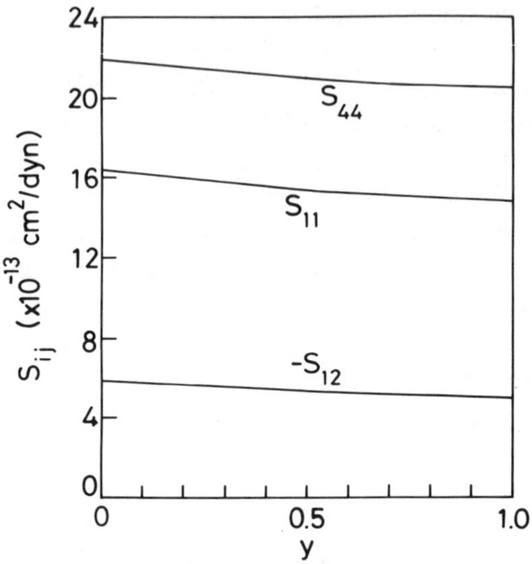

Figure 3.4 Elastic compliance constants, S_{11}, S_{12}, and S_{44}, as a function of y for $In_{1-x}Ga_xAs_yP_{1-y}$ lattice-matched to InP.

respectively. By using the interpolation scheme, these plots were obtained from the following binary data: for InP, from Ref. 19; for InAs, from Ref. 20; for GaAs, from Ref. 21; and for GaP, from Ref. 22. All the curves exhibit a very weak nonlinear dependence on the y-composition fraction and can, therefore, be successfully approximated by a linear relationship (see Table 3.1). The effect of the variation of composition on the elastic properties of an alkali–halide solid solution has been studied by Slagle and McKinstry,[23] who investigated the KCl–KBr solid solution over the complete range of composition. The stiffness moduli exhibited a weak quadratic variation, as in the present case.

The elastic moduli are dependent on both the temperature and pressure. Nichols et al.[19] studied the hydrostatic-pressure dependence of C_{ij} for InP up to 4 kbar. They found that the stiffness moduli vary linearly with pressure up to the highest value reached ($\sim$4 kbar). Their results are also summarized in Table 3.1 along with the linearly interpolated quaternary ($In_{1-x}Ga_xAs_yP_{1-y}$) and ternary ($In_{0.53}Ga_{0.47}As$) values.

Third-order elastic constants play an important role in solid-state physics. For the cubic point groups, O, O_h, and T_d, the third-order elastic-constant tensor has six independent components: C_{111} ($= C_{222} = C_{333}$), C_{112} ($= C_{223} = C_{133} = C_{113} = C_{122} = C_{233}$), C_{123}, C_{144} ($= C_{255} = C_{366}$), C_{166} ($= C_{155} = C_{244} = C_{344} = C_{266} = C_{355}$), and C_{456}.[24] Elastic constant–hydrostatic-pressure data allow the determination of the third-order elastic stiffness constants. Nichols et al.[19] have determined these values for InP

TABLE 3.1 Elastic Parameters for InP, $In_{1-x}Ga_xAs_yP_{1-y}$, and $In_{0.53}Ga_{0.47}As$ (300 K)

Parameter	InP	$In_{1-x}Ga_xAs_yP_{1-y}$	$In_{0.53}Ga_{0.47}As$
Elastic stiffness constant C_{ij} ($\times 10^{11}$ dyn/cm^2)			
C_{11}	10.11	$10.11 - 0.12y$	9.99
C_{12}	5.61	$5.61 - 0.68y$	4.93
C_{44}	4.56	$4.56 + 0.33y$	4.89
Elastic compliance constant S_{ij} ($\times 10^{-13}$ cm^2/dyn)			
S_{11}	16.38	$16.38 - 1.54y$	14.84
S_{12}	−5.84	$-5.84 + 0.94y$	−4.90
S_{44}	21.93	$21.93 - 1.43y$	20.50
Pressure derivation dC_{ij}/dp			
dC_{11}/dp	4.17	$4.17 + 0.28y$	4.45
dC_{12}/dp	4.80	$4.80 - 0.12y$	4.68
dC_{44}/dp	0.36	$0.36 + 0.38y$	0.74
Third-order elastic constant C_{ijk} ($\times 10^{12}$ dyn/cm^2)			
C_{111}	−8.6		
C_{112}	−1.85		
C_{123}	−5.1		
C_{144}	−6.5		
C_{166}	+1.6		
C_{456}	−0.042		
Young's modulus Y ($\times 10^{11}$ dyn/cm^2)			
{100} plane			
⟨001⟩ direction	6.106	$6.106 + 0.976y - 0.344y^2$	6.738
⟨011⟩ direction	9.303	$9.303 + 1.037y - 0.426y^2$	9.914
{110} plane			
⟨001⟩ direction	6.106	$6.106 + 0.976y - 0.344y^2$	6.738
⟨111⟩ direction	11.27	$11.27 + 0.95y - 0.46y^2$	11.76
{111} plane	9.303	$9.303 + 1.037y - 0.426y^2$	9.914
Poisson's ratio P			
{100} plane			
$m = \langle 010 \rangle$, $n = \langle 001 \rangle$	0.357	$0.357 - 0.033y + 0.006y^2$	0.330
$m = \langle 011 \rangle$, $n = \langle 0\bar{1}1 \rangle$	0.020	$0.020 - 0.006y$	0.014
{110} plane			
$m = \langle 001 \rangle$, $n = \langle 1\bar{1}0 \rangle$	0.357	$0.357 - 0.033y + 0.006y^2$	0.330
$m = \langle 1\bar{1}1 \rangle$, $n = \langle 1\bar{1}\bar{2} \rangle$	0.236	$0.236 - 0.044y + 0.012y^2$	0.204
{111} plane	0.369	$0.369 - 0.052y + 0.012y^2$	0.329
Bulk modulus B_u ($\times 10^{11}$ dyn/cm^2)	7.110	$7.110 - 0.516y + 0.020y^2$	6.614
Isotropy factor A	0.493	$0.493 + 0.035y - 0.010y^2$	0.518
Compressibility C_o ($\times 10^{-12}$ cm^2/dyn)	1.406	$1.406 + 0.102y + 0.004y^2$	1.512
Born ratio B_o	1.101	$1.101 - 0.014y + 0.004y^2$	1.091
Cauchy ratio C_a	1.230	$1.230 - 0.306y + 0.084y^2$	1.008

from their measured pressure dependence of the elastic constants (see Table 3.1).

3.2.2 Young's Modulus, Poisson's Ratio, and Similar

There is considerable interest in the effect of mechanical stresses resulting from crystal growth and device processing on the behavior and reliability of semiconductor devices. Precise computation of such stresses requires knowledge of Young's modulus and Poisson's ratio, in particular for specific orientations within the crystallographic plane defining the surface of the semiconductor. Young's modulus Y is not isotropic in cubic zinc-blende-type crystals.[25] The modulus Y for an arbitrary crystallographic direction m can now be given by

$$\frac{1}{Y} = S_{11} - 2(S_{11} - S_{12} - \tfrac{1}{2}S_{44})(m_1^2 m_2^2 + m_2^2 m_3^2 + m_1^2 m_3^2) \quad (3.3)$$

where the S_{ij} values are the elastic compliance constants and the m_i values are the direction cosines for m. Poisson's ratio P also varies with orientation. If a longitudinal stress in the direction m and the transverse strain along the orthogonal direction n are under consideration, then the ratio P can be given by

$$P = -\frac{S_{12} + (S_{11} - S_{12} - \tfrac{1}{2}S_{44})(m_1^2 n_1^2 + m_2^2 n_2^2 + m_3^2 n_3^2)}{S_{11} - 2(S_{11} - S_{12} - \tfrac{1}{2}S_{44})(m_1^2 m_2^2 + m_2^2 m_3^2 + m_1^2 m_3^2)} \quad (3.4)$$

The modulus Y for the direction of the cube axes ($\langle 100 \rangle$) is readily given by $Y = 1/S_{11}$. The ratio P, in this case, is written as $P = -S_{12}/S_{11}$. We plot in Fig. 3.5 the $\langle 100 \rangle$-direction Y and P values as a function of y for $In_{1-x}Ga_xAs_yP_{1-y}$ lattice-matched to InP. The variations of Y and P for directions within the important crystallographic planes $\{100\}$, $\{110\}$, and $\{111\}$ are also listed in Table 3.1. It should be noted that the Y and P are invariant within the $\{111\}$ plane.[26]

The bulk modulus B_u and isotropy factor A for zinc-blende-type crystals are given by

$$B_u = \frac{C_{11} + 2C_{12}}{3} \quad (3.5)$$

$$A = \frac{C_{11} - C_{12}}{2C_{44}} \quad (3.6)$$

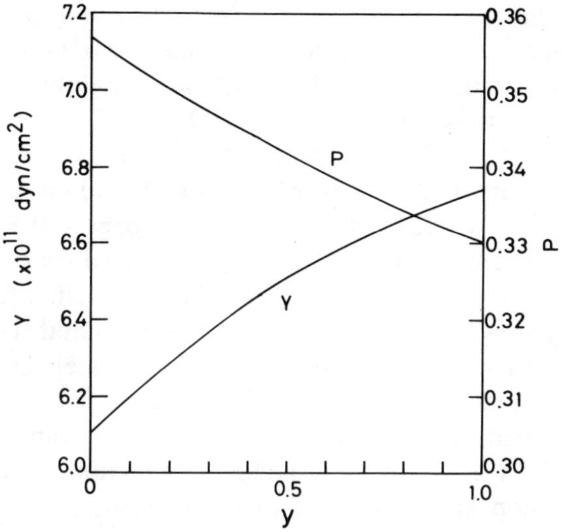

Figure 3.5 The ⟨100⟩-axis Young's modulus Y and Poisson's ratio P as a function of y for $In_{1-x}Ga_xAs_yP_{1-y}$ lattice-matched to InP.

The compressibility C_o is also given by

$$C_o = \left(\frac{C_{11} + 2C_{12}}{3}\right)^{-1} \tag{3.7}$$

The Cauchy (C_a) and Born ratios (B_o) are written as

$$C_a = \frac{C_{12}}{C_{44}} \tag{3.8}$$

$$B_o = \frac{(C_{11} + C_{12})^2}{4C_{11}(C_{11} - C_{44})} \tag{3.9}$$

Numerical values of these parameters are also listed in Table 3.1.

Bulk moduli of semiconductors are of physical interest and also of practical importance in the interpretation of high-pressure experimental results. In semiconductor heterostructures, different bulk moduli in the different layers give rise in hydrostatic pressure experiments to large axial strains that can have a considerable effect on the result of experiments. Conversely, observation of such effects can provide an accurate method of measuring the relative bulk moduli of the layers. Menoni and Spain[27] have carried out X-ray diffraction experiments on InP at room temperature up to hydrostatic pressures of 190 kbar. From these measurements they determined the bulk modulus of InP to be $B_u = (7.6 \pm 0.4) \times 10^{11}$ dyn/cm^2. Prins and Dunstan[28,29] have also determined the bulk moduli of an $In_{1-x}Ga_xAs_yP_{1-y}$ lattice-matched to InP from high-hydrostatic-pressure photoluminescence measurements up to 100 kbar. $In_{1-x}Ga_xAs_yP_{1-y}$ quaternary was found to be the same as InP to within 1%, while the value for $In_{0.53}Ga_{0.47}As$ ternary was 5% lower [$B_u = (7.25 \pm 0.08) \times 10^{11}$ dyn/cm^2 for $y = 0.51$; $(7.25 \pm 0.35) \times 10^{11}$ dyn/cm^2 for $y = 0.89$; and $(6.90 \pm 0.08) \times 10^{11}$ dyn/cm^2 for $In_{0.53}Ga_{0.47}As$]. These values are slightly larger than those estimated from the linear interpolation scheme. Kushwaha[30] calculated the compressibility values B_o for some III–V and II–VI compounds based on an eight-parameter bond-bending force model. Our result for InP is comparable to that of Kushwara ($B_o = 1.46 \times 10^{-12}$ cm^2/dyn). He also calculated the mean-square displacements for such compounds and used them to check the validity of the Lindemann's melting criteria.

The compression is now related to the pressure p with Murnaghan's equation[31]

$$p = \frac{B_u}{B_u'}\left\{\left[\frac{a_0(0)}{a_0(p)}\right]^{3B_u'} - 1\right\} \tag{3.10}$$

where B_u' is the pressure derivative of the bulk modulus [$B_u' = \frac{1}{3}(dC_{11}/d_p) + \frac{2}{3}(dC_{12}/dp) = 4.59$], and $a_0(0)$ and $a_0(p)$ are the lattice constants at pressures $p = 0$ and $p \neq 0$, respectively. We show in Fig. 3.6 the pressure dependence of the lattice constant for InP as calculated from Eq. (3.10). This calculation indicates that the pressure-induced change in a_0 at 50 kbar is about −2.0%.

3.2.3 Sound Velocities and Mode Grüneisen Parameters

A perfectly isotropic material is one for which A (isotropy factor) = 1.0. As seen in Table 3.1, InP and InGaAsP quaternary are not perfectly isotropic in their elastic properties. Sound velocities in these materials are, thus, strongly dependent on the propagation directions. If the crystal density g and the stiffness constant C_{ij} of a solid are known, one can calculate the bulk sound velocity v (long-wavelength, nondispersive) from the following general relation:

$$v = \left(\frac{C_{ij}}{g}\right)^{1/2} \tag{3.11}$$

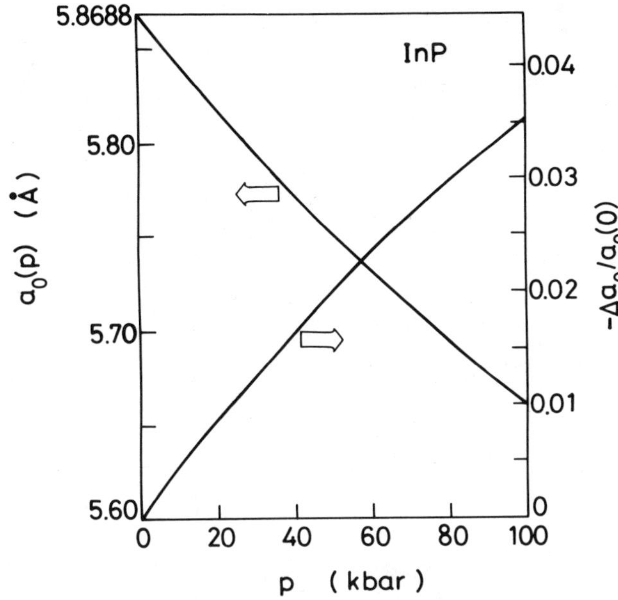

Figure 3.6 Pressure dependence of the lattice constant a_0 for InP; $a_0(0)$ and $a_0(p)$ are, respectively, the lattice constants at pressures $p = 0$ and $p \neq 0$, and $\Delta a_0 = a_0(p) - a_0(0)$.

If we neglect nonlinear terms in the equation of motion, pure longitudinal sound waves may propagate in the [100], [110], and [111] directions. Pure transverse waves may propagate not only in these three directions, but also in the crystallographic directions [*hkl*] ($h = k$; l arbitrary) and [*hk*0] (h and k arbitrary). We can see in Table III of Ref. 32 definition of sound velocity expressed in terms of these constants along the high-symmetry directions [100], [110], and [111] in cubic, zinc-blende crystals (GaAs, InP, etc.). In Fig. 3.7 we show the calculated sound velocities along the [110] direction in $In_{1-x}Ga_yAs_yP_{1-y}$ lattice-matched to InP; v_{LA} corresponds to the longitudinal mode, and v_{TA1} and v_{TA2} correspond to the transverse modes (polarizations along the [001] and [$1\bar{1}0$] directions, respectively). The numeric values for g used in the calculations are taken from Table 2.1; and for the C_{ij} values, from Table 3.1. Table 3.2 lists expressions in terms of the y-composition fraction for longitudinal and transverse wave velocities along the [100], [110], and [111] directions in InP, $In_{1-x}Ga_xAs_yP_{1-y}$ and $In_{0.53}Ga_{0.47}As$ lattice-matched to InP.

If we include the nonlinear term, we find that pure transverse modes do not exist. The transverse waves are always accompanied by a longitudinal wave. On the other hand, pure longitudinal modes continue to exist for all three principal directions [100], [110], and [111]. For these directions, the

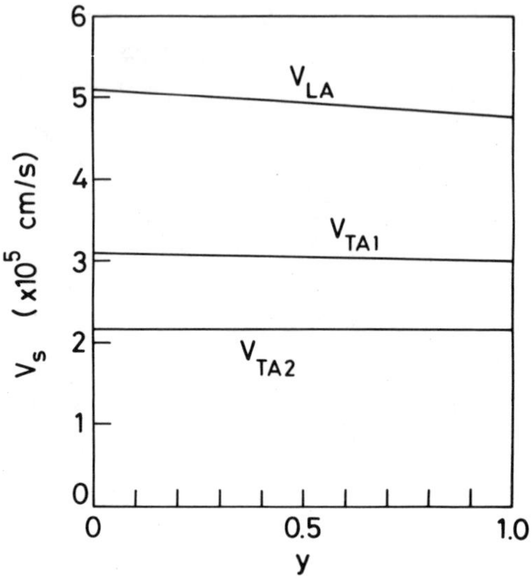

Figure 3.7 Variation of the sound velocities propagating in the [110] direction in $In_{1-x}Ga_xAs_yP_{1-y}$ lattice-matched to InP.

nonlinear wave equation can be reduced to the form[33]

$$\frac{\partial^2 u}{\partial t^2} = \frac{1}{g}\frac{\partial^2 u}{\partial x^2}\left(M_2 + M_3\frac{\partial u}{\partial x} + \text{higher-order terms}\right) \tag{3.12}$$

where u is the displacement velocity. The constants M_2 and M_3 can be written in terms of the second- and third-order elastic coefficients K_2 and

TABLE 3.2 Sound Velocities for Major Directions in the Cubic, Zinc-Blend Lattice of InP, $In_{1-x}Ga_xAs_yP_{1-y}$, and $In_{0.53}Ga_{0.47}As$, as Controlled by the Second-Order Elastic Constants C_{ij} (300 K)

Propagation Direction	Direction or Plane of Polarization	Sound Velocity ($\times 10^5$ cm/s)		
		InP	$In_{1-x}Ga_xAs_yP_{1-y}$	$In_{0.53}Ga_{0.47}As$
[100]	[100][a]	4.594	$4.594 - 0.334y$	4.260
	(100)[b]	3.085	$3.085 - 0.104y$	2.981
[110]	[110][a]	5.092	$5.092 - 0.355y$	4.737
	[001][b]	3.085	$3.085 - 0.104y$	2.981
	$[1\bar{1}0]$[b]	2.167	$2.167 - 0.023y$	2.144
[111]	[111][a]	5.247	$5.247 - 0.362y$	4.885
	(111)[b]	2.511	$2.511 - 0.056y$	2.455

[a]Longitudinal acoustic waves.
[b]Transverse acoustic waves.

K_3:

$$M_2 = K_2, \qquad M_3 = K_3 + 2K_2 \tag{3.13}$$

The coefficients K_2 and K_3 are, in turn, related to the more familiar C_{ij} and C_{ijk} and are given by

$$K_2 = C_{11}, \qquad K_3 = C_{111} \tag{3.14}$$

for the [100] direction,

$$K_2 = \tfrac{1}{2}(C_{11} + C_{12} + 2C_{44})$$

$$K_3 = \tfrac{1}{4}(C_{111} + 3C_{112} + 12C_{166}) \tag{3.15}$$

for the [110] direction, and

$$K_2 = \tfrac{1}{3}(C_{11} + 2C_{12} + 4C_{44})$$

$$K_3 = \tfrac{1}{9}(C_{111} + 6C_{112} + 12C_{144} + 24C_{166} + 2C_{123} + 16C_{456}) \tag{3.16}$$

for the [111] direction. The nonlinearity parameter β is also defined by[33]

$$\beta = -\frac{3K_2 + K_3}{K_2} \tag{3.17}$$

The amplitude of the generated second harmonic is proportional to this parameter.

The mode Grüneisen parameter represents the strain derivative of the lattice vibrational frequencies and is an important quantity in describing the anharmonic properties of solids.[34] We present in Table 3.3 the mode Grüneisen parameters for long-wavelength ($\boldsymbol{q} \to 0$) acoustic phonons in the [100], [110], and [111] directions of InP. Theoretical values of these parameters as calculated by Soma and Kagaya[35] are also listed in the table in parentheses.

3.2.4 Surface Acoustic Waves

A surface acoustic wave (SAW) is a mode of propagation of elastic energy along the free boundary of an infinite half space. The amplitude of the

TABLE 3.3 Mode Grüneisen Parameters for Long-Wavelength ($q \rightarrow 0$) Acoustic Phonons in Three Major Directions [100], [110], and [111] of InP (300 K)[a]

Propagation Direction	Direction or Plane of Polarization	Phonon Mode	Grüneisen Parameter
[100]	[100]	LA	1.30 (1.07–1.33)
[100]	(100)	TA	0.11 (0.08–0.28)
[110]	[110]	LA	1.22 (1.08–1.27)
[110]	[001]	TA	0.11 (0.08–0.28)
[110]	$[1\bar{1}0]$	TA	−0.67 (−0.99 to −0.64)
[111]	[111]	LA	1.20 (1.08–1.27)
[111]	(111)	TA	−0.27 (−0.45 to −0.18)

[a]Theoretical values as calculated by Soma and Kagaya[35] are also presented in parentheses.

displacement undergoes an exponential decay with depth below the free surface and vanishes within a distance of the order of two wavelengths.[36] The velocities of the surface wave values v_{sw} in a zinc-blende-type crystal are given by[13]

$$C_{11}\left(v_{sw}^2 - \frac{C_{44}}{g}\right)\left(v_{sw} - \frac{C_{11}}{g} + \frac{C_{12}^2}{C_{11}g}\right)^2 = C_{44}v_{sw}^4\left(v_{sw}^2 - \frac{C_{11}}{g}\right) \tag{3.18a}$$

$$C_{11}\left(v_{sw}^2 - \frac{C_{44}}{g}\right)\left(v_{sw}^2 - \frac{2C_{44} + C_{12} + C_{11}}{2g} + \frac{C_{12}^2}{C_{11}g}\right)^2 = C_{44}v_{sw}^4\left(v_{sw}^2 - \frac{2C_{44} + C_{12} + C_{11}}{2g}\right) \tag{3.18b}$$

Equations (3.18a) and (3.18b), which are of third degree in v_{sw}^2, refer to the [100] and [110] propagation directions, respectively. In Fig. 3.8 we plot the calculated v_{sw} as a function of y for propagation on a (100) plane of $In_{1-x}Ga_xAs_yP_{1-y}$ quaternary. One easily finds that the curve exhibits a weak nonlinear dependence on the y-composition fraction for both the [100] and [110] propagation directions. The v_{sw} variations versus y plotted in Fig. 3.8 can now be expressed as (in units of 10^5 cm/s)

$$v_{sw} = 2.465 - 0.040y - 0.028y^2 \tag{3.19}$$

for the [100] direction and

$$v_{sw} = 2.628 - 0.060y - 0.028y^2 \tag{3.20}$$

for the [110] direction.

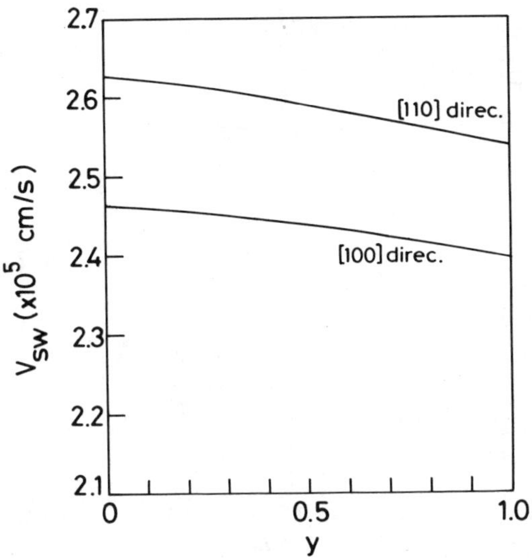

Figure 3.8 Surface-acoustic-wave velocity versus composition y propagating in the [100] and [110] directions in $In_{1-x}Ga_xAs_yP_{1-y}$ lattice-matched to InP.

To our knowledge, no experimental work has been published to date on the SAW properties of the $In_{1-x}Ga_xAs_yP_{1-y}$ alloy. We can see experimental results relating to these properties only on the $Al_xGa_{1-x}As$ ternary.[37,38] Sapriel et al.[37] have measured the velocity of SAWs for $Al_xGa_{1-x}As$ as a function of x using a Brillouin scattering technique. They found a lack of agreement with the theoretical prediction at higher Al concentrations and attributed this to "softening" of the elastic constants due to substitution of Ga by Al. Steel et al.[38] have more recently studied SAW properties (such as SAW velocities, SAW attenuation, and piezoelectric coupling coefficients) on $Al_xGa_{1-x}As$ with $x = 0.2$ and 0.4. The measured SAW velocities are found to be substantially higher than those for GaAs, in agreement with the theoretical predictions of Adachi.[13]

3.3 LATTICE VIBRATIONAL PROPERTIES

3.3.1 Dispersion Curves and Phonon Density of States

Research on the crystal dynamics of perfect lattices has aroused considerable interest in compounds having a zinc-blende structure. This development is the outcome of many experimental results, particularly inelastic neutron-scattering data for phonon dispersion curves. Borcherds et al.[39] made thermal neutron scattering measurements on InP. The application of a force model was used to describe this experimental result correctly.[40–42]

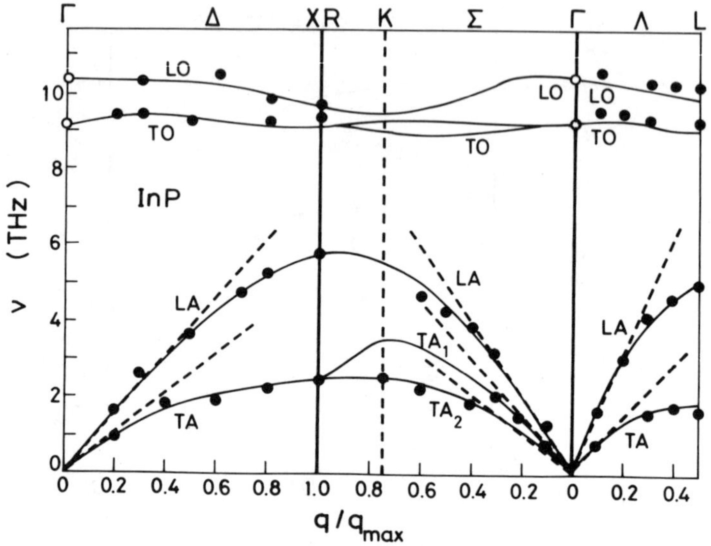

Figure 3.9 Phonon dispersion curve of InP. The solid circles show the neutron scattering data taken from Ref. 39, and the solid lines indicate calculated results of Ref. 42. The open circles show the values of the zone-center LO and TO phonon frequencies obtained from Raman scattering data.[43] The dashed lines near $\boldsymbol{q} \to 0$, $\nu \to 0$ represent the initial slopes of the acoustic modes as calculated from the sound velocities listed in Table 3.2.

We reproduce in Fig. 3.9 the phonon frequency (ν) versus reduced wave vector (q) results for $\boldsymbol{q}$ along the principal symmetry directions [100], [110], and [111] of InP at room temperature, as obtained from these studies. The solid circles are the neutron scattering data,[39] and the solid lines are calculated results of Ref. 42. Also shown in Fig. 3.9 (open cirlces) are the values of the zone-center LO and TO phonon frequencies obtained from Raman scattering data.[43] The dashed lines near $\boldsymbol{q} \to 0$, $\nu \to 0$ represent the initial slopes of the acoustic modes as calculated from the sound velcoities listed in Table 3.2. It is recognized that the dispersion curves of InP exhibit major features similar to those of other zinc-blende crystals, such as GaAs,[44] GaP,[45] InSb,[46] GaSb,[47] CdTe,[48] ZnTe,[49] and ZnSe.[50]

The properties of critical points and their consequences for the phonon dispersion curves in crystals have first been discussed by Van Hove[51] by making use of topologic considerations. The phonon density of states $g(\nu)$ increases greatly when $\nabla_q(\nu) \to 0$ (critical points) for one of the various branches, and this happens predominantly when $\boldsymbol{q}$ for that branch reaches a zone boundary.[32] A calculation of the phonon density of states $g(\nu)$ (one phonon) for InP has been made by Kushwaha and Kushwaha[40] using an eight-parameter bond-bending force model. Their result is reproduced in

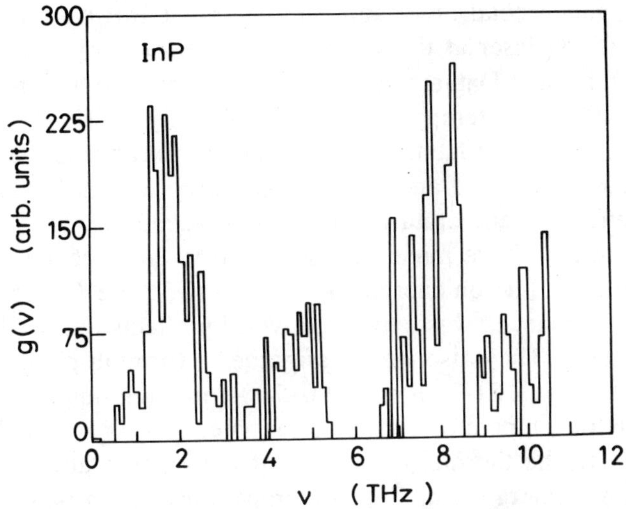

Figure 3.10 Phonon density of states $g(\nu)$ for InP. (From Kushwaha and Kushwaha.[40])

Fig. 3.10. The frequency interval they used was $\Delta\nu = \nu_{LO}(\Gamma)/100$. The $g(\nu)$ curve shows a first maximum near 1.5 THz associated with TA modes. Shorter-wavelength LA modes give rise to a relatively smaller peak near 5 THz, and the massive concentration of optical phonons near 8 (and 10) THz also produces notable peak(s).

There has been no detailed experimental study on the phonon dispersion relations in the $In_{1-x}Ga_xAs_yP_{1-y}$ alloy except the long-wavelength optical phonon behavior, as will be discussed in the next subsection. Theoretical work on the lattice dynamics of mixed III–V crystals has been done by several authors.[52–54] O'Hara et al.[52] have evaluated the spectral density of phonon states for one-dimensional mass-disordered quaternary alloys using a negative eigenvalue theorem. Similar work, but on ternary alloys, has been performed by Myles[53] using a generalized embedded-cluster method and by Sinai et al.[54] using a modified local Green's function method.

3.3.2 Phonon Spectra and Mode Grüneisen Parameters

First-order Raman scattering (RS) measurements give information only about phonon states with $\boldsymbol{q} = 0$. This severe limitation can be overcome by using second-order Raman spectra; two phonons with $\boldsymbol{q}_1 + \boldsymbol{q}_2 \simeq 0$ are then produced, and the complete phonon spectrum may become accessible. Infrared optical measurements can also give information about phonon states both with $\boldsymbol{q} = 0$ and $\boldsymbol{q} \neq 0$.

The first- and second-order Raman spectra of InP was first measured with an Nd:YAG laser at the wavelength 1.064 μm by Alfrey and Borcherds.[43] Koteles and Datars[55] have measured the two-phonon infrared absorption in InP at low temperature (approximately 20 K). We tabulate in Table 3.4 these results together with the neutron scattering data.[39] Agreement between the infrared and neutron scattering data is found to be very good if the effect of temperature is taken into account.

Sinyukov et al.[56] have made of resonant first- and second-order RS by phonons in InP for photon energies from 1.75 to 3.05 eV. The behavior of all observed resonances was explained well by effects due to the $E_0 + \Delta_0$ and E_1 band gaps. They also made assignments for multiphonon structures that appeared in the 210–270- and 600–700-cm^{-1} regions.

The effects of temperature on the phonon energies measured by RS are due primarily to the thermal expansion of the crystal lattice. The dependence of phonon energy $\omega(\boldsymbol{q}, T)$ with temperature T can be written as

$$\omega(\boldsymbol{q}, T) \simeq \omega_0(\boldsymbol{q}) \left(1 - \frac{\alpha_L^2 a_0}{\kappa_L k_B} T\right) \tag{3.21}$$

where $\omega_0(\boldsymbol{q})$ is the phonon dispersion of the harmonic linear chain, a_0 is the lattice constant, α_L is the coefficient of linear expansion (see Section 4.2), κ_L is the compressibility of the linear chain, k_B is the Boltzmann's constant, and $\boldsymbol{q}$ is the phonon wave vector. This expression predicts a linear decrease in phonon energy with an increase of T. Shealy and Wicks[57] found that this is nearly the case for the LO phonons in GaAs, $Al_xGa_{1-x}As$, and $In_{1-x}Ga_xAs$ alloys. However, to our knowledge, detailed experimental data have not yet been reported for InP and $In_{1-x}Ga_xAs_yP_{1-y}$ quaternary.

The mode Grüneisen parameter γ_i is defined as

$$\gamma_i = \frac{\partial \ln \omega_i}{\partial \ln V} = \frac{B_u}{\omega_i} \frac{d\omega_i}{dp} \tag{3.22}$$

where ω_i is the phonon energy of mode i, V is the volume, B_u is the bulk modulus, and p is the pressure. In a zinc-blende semiconductor the Born's transverse effective charge is related to the splitting of the long-wavelength LO and TO phonon modes by

$$e_T^{*2} = \frac{\epsilon_\infty a_0^3 M}{16\pi} (\omega_{LO}^2 - \omega_{TO}^2) \tag{3.23}$$

where ϵ_∞ is the high-frequency dielectric constant and M is the reduced mass of the two atoms ($M^{-1} = M_{In}^{-1} + M_P^{-1}$).

TABLE 3.4 One- and Two-Phonon Raman and Infrared Peak Frequencies Together with Neutron Scattering Data for InP[a]

Critical Point	Phonon Frequency (cm^{-1}) Raman[b]	Infrared[c]	Neutron[d]	Assignment (Attribution)
Γ	346.4 ± 0.3	351		LO
	304.5 ± 0.3		307 ± 7	TO
	620			2TO
	653			LO + TO
X	315 ± 2	328.5	332 ± 3	LO
	282 ± 2	326.5	324 ± 7	TO
	125 ± 1	190.5	193 ± 10	LA
	67 ± 1	67.5	68 ± 3	TA
	135			2TA
	163			TO − LA
	189			LO − LA
				LA + TA
	214			TO − TA
	251			LO − TA
				2LA
	382			LO + TA
	438			LO + LA
	596			LO + TO
	630			2LO
L	331 ± 2	340.5	340 ± 10	LO
	246 ± 2	315.5	317 ± 5	TO
	82 ± 1	167.5	167 ± 3	LA
	54 ± 1	53.5	55.0 ± 0.7	TA
	85			LO − TO
	135			LA + TA
	163			2LA
				TO − LA
	189			TO − TA
	251			LO − LA
	281			LO − TA
	382			LO + TA
	412			LO + LA
	490			2TO
	573			LO + TO

[a]Assignment in terms of critical point phonon frequencies is also given.
[b]G. F. Alfrey and P. H. Borcherds, *J. Phys. C* **5,** L275 (1972) (300 K).
[c]E. S. Koteles and W. R. Datars, *Solid State Commun.* **19,** 221 (1976) (20 K).
[d]P. H. Borcherds et al., *J. Phys. C* **8,** 2022 (1975) (300 K).

The dependence of first- and second-order Raman spectra on pressure in InP has been studied by Trommer et al.[58] in a diamond anvil cell at pressures up to 100 kbar. The phonon frequencies decreased almost linearly with an increase of relative lattice compression $-\Delta a_0/a_0$. They determined the Grüneisen parameters of the LO and TO phonons at Γ, and also those of several optical and acoustic phonons corresponding to critical points at the edge of the zone. The effects of uniaxial stress on the long-wavelength optical phonons (LO and TO) in InP have been studied by Anastassakis et al.[59] using Raman and far-infrared spectroscopic techniques. They also determined the mode Grüneisen parameters of these phonon modes. We summarize these results in Table 3.5.

Trommer et al.[58] have found a nonlinear decrease of e_T^* with decreasing lattice constant. This decrease in e_T^* is characteristic of most zinc-blende-type semiconductors measured for GaP, GaAs, ZnSe, and similar.[58] The dynamic charge obtained from this splitting also decreased with decreasing lattice constant, a fact that reflects an increase in covalency on compression.

TABLE 3.5 Mode Grüneisen Parameters γ for a Number of Critical Points and Born's Transverse Effective Charge e_T^* of InP (300 K)

Mode Grüneisen Parameter γ	
LO (Γ)	1.24 ± 0.02[a]
	1.19 ± 0.04[b]
	0.82 ± 0.05[c]
TO (Γ)	1.44 ± 0.02[a]
	1.48 ± 0.04[b]
	0.96 ± 0.05[c]
LO − TO (Γ)	0.20 ± 0.05[a]
TO (*L*)	1.42 ± 0.1[a]
TO (*X*)	1.42 ± 0.1[a]
TA (*X*)	−2.08 ± 0.1[a]
TA (*L*)	−2.00 ± 0.1[a]
e_T^* (at 1 Pa)	2.54[a]
de_T^*/dp ($\times 10^{-2}$ e/GPa)	−2.4[a,b]
	−2.1[b]

[a]R. Trommer et al., *Phys. Rev. B* **21,** 4869 (1980).
[b]Obtained from infrared-laser Raman scattering [E. Anastassakis et al., *Phys. Rev. B* **38,** 7702 (1988)].
[c]Obtained from far-infrared reflectivity [E. Anastassakis et al., *Phys. Rev. B* **38,** 7702 (1988)].

3.3.3 Long-Wavelength Optical Phonons in InGaAsP

There are many theoretical approaches to understand the long-wavelength optical phonon behaviors in multinary semiconductor alloys.[60–66] Long-wavelength optical phonons in III–V ternary compounds exhibit either a "one-mode" or a "two-mode" behavior (or more rigorously, three different types of mode behavior, viz., "one-," "two-," and "one–two-mode" behaviors).[67–69] In a one-mode system, such as most I–VII alloys, a single set of long-wavelength optical phonons appears. The phonon spectrum is approximately constant in strength but varies in energy as the alloy composition is varied from one end member to the other. The two-mode systems, common for III–V alloys, exhibit two distinct sets of optical modes with frequencies characteristic of each end member and strengths roughly proportional to the respective concentration. The one–two-mode system exhibits a single mode over only a part of the composition range, with two modes observed over the remaining range of composition.

InGaP ternary is the only alloy for which one-mode behavior is currently assumed.[70–72] However, Jusserand and Slempkes[73] have concluded from RS study that the alloy shows a somewhat modified two-mode behavior. The results of the studies reported in Refs. 70–72 were obtained on polycrystalline samples while the authors of Ref. 73 used monocrystalline layers grown on InP. Abdelouhab et al.[74] have recently studied optical phonon modes in monocrystalline InGaP grown on GaAs substrates. They concluded that the presence of internal stress in epitaxial layers could induce a switchover from one- to two-mode behavior. Galtier et al.[75] have also studied the hydrostatic-pressure dependence of phonon-mode behavior in not only InGaP but also GaAsP alloys (polycrystalline and epitaxially grown samples) by means of Raman spectroscopy. This comparative study showed that the concept of a one-mode system is not appropriate for describing the physics of this kind of alloy.

Long-wavelength optical phonon behaviors in $In_{1-x}Ga_xAs_yP_{1-y}$ quaternary lattice-matched to InP have been studied by many workers using Raman or far-infrared spectroscopy.[68,73,76–80] Pinczuk et al.[76] have reported Raman spectra and identified two sets of modes, one InP-like that extrapolated to a local mode for $y = 1.0$ and the other InGaAs-like. Similar results from RS measurements were found by Portal et al.,[77] but these authors identified independent GaAs- and InAs-like modes in the $y = 1.0$ case, and InP- and InAs-like modes in the $y = 0.23$ composition. The four-mode behavior of $In_{1-x}Ga_xAs_yP_{1-y}$ has also been confirmed by Jusserand and Slempkes,[73] Pickering,[79] and Soni et al.[80] In Figs. 3.11 and 3.12, respectively, we plot the LO and TO phonon frequencies as a function of y

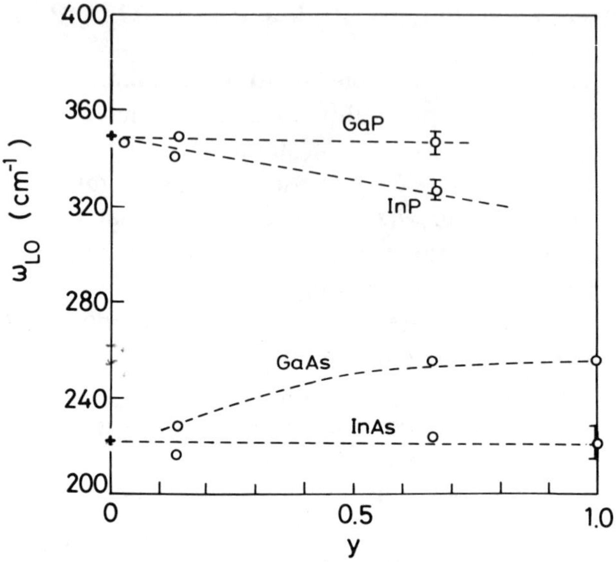

Figure 3.11 Longitudinal optical (LO) phonon frequency ω_{LO} versus composition y for $In_{1-x}Ga_xAs_yP_{1-y}$ lattice-matched to InP. The experimental data are taken from Ref. 80. The impurity modes of Ga and As in InP (crosses) are also taken from Refs. 70 and 81, respectively. The dashed lines mark a possible explanation of the experimental findings.

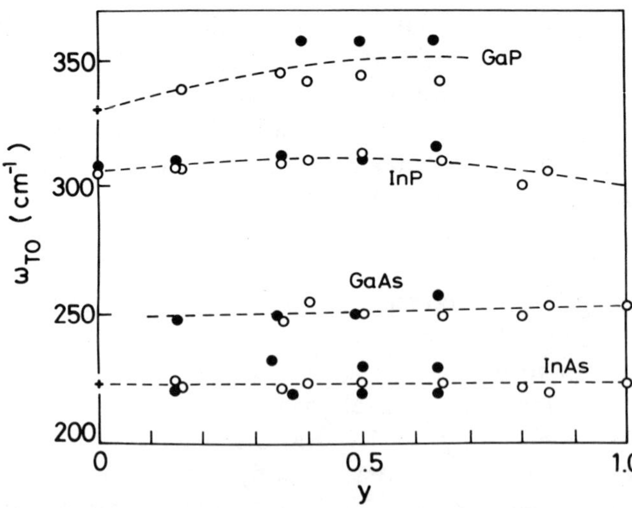

Figure 3.12 Transverse optical (TO) phonon frequency ω_{TO} versus composition y for $In_{1-x}Ga_xAs_yP_{1-y}$ lattice-matched to InP. The experimental data are taken from Refs. 76 (solid circles) and 79 (open circles). The impurity modes of Ga and As in InP (crosses) are also taken from Refs. 70 and 81, respectively. The dashed lines mark a possible explanation of the experimental findings.

for $In_{1-x}Ga_xAs_yP_{1-y}$. The data are taken for the LO mode from Ref. 80 and for the TO mode from Refs. 76 and 79. The impurity modes of Ga and As in InP are also taken from Refs. 70 and 81, respectively. It can be seen that the data fall into four separate bands, that is, a four-mode category, except for the GaP mode at $y > 0.7$ and for the GaAs mode at $y < 0.1$.

The long-wavelength lattice dynamics has also been studied for other III–V quaternary systems, such as InGaAsP lattice-matched to GaAs,[82] InGaAsSb lattice-matched to InAs and GaSb,[83] and InAlGaAs lattice-matched to InP.[84] The InGaAsSb system showed three-phonon modes with GaAs, InSb, and mixed InAs/GaSb characteristics. Pickering[83] considered that the behavior of this system may be termed "three-mode" because of the degeneracy of the GaSb- and InAs-like modes. A similar type of behavior has also been found in AlGaAsP system due to the degeneracy of the AlP and GaP modes.[85] The InAlGaAs system showed the three-mode behavior as expected (InAs-, AlAs-, and GaAs-like modes).[84]

3.3.4 Phonon Deformation Potentials

As we can easily recognize from Ref. 58, phonon frequency is a strong function of hydrostatic pressure. This quantity is also strongly influenced by uniaxial stress. Such stress-induced frequency shifts of phonons can be essentially explained by the phonon deformation potentials (PDPs). The PDPs are important anharmonicity parameters useful in a number of physical and technological applications. Detailed expressions of PDPs and their consequences for the first-order Raman shifts in semiconductors were first presented by Cerdeira et al.[86]

The PDPs are components of a fourth-rank tensor K_{ij} in suppressed notation [see Eq. (3.1) as the same tensor form]. For a zinc-blende crystal there are only three independent components of the PDP tensor:

$$K_{11} = K_{22} = K_{33} = Mp \tag{3.24a}$$

$$K_{12} = K_{23} = K_{13} = Mq \tag{3.24b}$$

$$K_{44} = K_{55} = K_{66} = Mr \tag{3.24c}$$

where M is the reduced mass of the two atoms. According to Cerdeira et al.,[86] the long-wavelength optical phonon frequencies in the presence of strain (e) obey the following secular equation:

$$\begin{vmatrix} pe_{xx} + q(e_{yy} + e_{zz}) - \lambda & 2re_{xy} & 2re_{xz} \\ 2re_{xy} & pe_{yy} + q(e_{xx} + e_{zz}) - \lambda & 2re_{yz} \\ 2re_{xz} & 2re_{yz} & pe_{zz} + q(e_{xx} + e_{yy}) - \lambda \end{vmatrix} = 0 \tag{3.25}$$

where $\lambda = \Omega^2 - \omega_i^2$ and $\Omega \simeq \omega_i + \lambda/2\omega_i$ is the strain-dependent phonon frequency (ω_i = phonon frequency without strain).

The strain (e) and stress (X) relation is now given by

$$e_{ij} = \sum_{kl} S_{ijkl} X_{kl} \tag{3.26}$$

where S_{ijkl} is the elastic compliance component. This relation yields the nonzero strain components

$$e_{xx} = e_{yy} = e_{zz} = (S_{11} + 2S_{12}) \frac{X}{3} \tag{3.27a}$$

for the hydrostatic pressure,

$$e_{xx} = S_{11}X, \qquad e_{yy} = e_{zz} = S_{12}X \tag{3.27b}$$

for the [100] uniaxial stress, and

$$e_{xx} = e_{yy} = e_{zz} = (S_{11} + 2S_{12}) \frac{X}{3}, \qquad e_{xy} = e_{yz} = e_{zx} = S_{44} \frac{X}{6} \tag{3.27c}$$

for the [111] uniaxial stress.

Introducing Eq. (3.27) into Eq. (3.25), we obtain a shift of the phonon frequency.[86] For the hydrostatic stress, the shift of the phonon frequency is given by

$$\Delta\Omega_{\mathrm{H}} = \frac{X}{6\omega_i} (p + 2q)(S_{11} + 2S_{12}) \tag{3.28}$$

For the [100] and [111] uniaxial stresses, the threefold degeneracy of the $\boldsymbol{q} = 0$ optical phonons is split into a singlet (Ω_{s}) with an eigenvector parallel to the stress and a doublet (Ω_d) with eigenvectors perpendicular to the stress, where Ω_s and Ω_d are given by

$$\Omega_s = \omega_i + \Delta\Omega_H + \frac{2}{3}\Delta\Omega \qquad (3.29a)$$

$$\Omega_d = \omega_i + \Delta\Omega_H - \frac{1}{3}\Delta\Omega \qquad (3.29b)$$

The splitting energy ($\Delta\Omega = \Omega_s - \Omega_d$) is written as

$$\Delta\Omega = \Omega_s - \Omega_d = \begin{cases} \dfrac{X}{2\omega_i}(p - q)(S_{11} - S_{12}) & \text{for [100] stress} \quad (3.30a) \\ \dfrac{X}{2\omega_i} rS_{44} & \text{for [111] stress} \quad (3.30b) \end{cases}$$

The PDPs could then be determined experimentally by measuring either a shift of the phonon frequency for each mode ($\Delta\Omega_s = \omega_i - \Omega_s$ or $\Delta\Omega_d = \omega_i - \Omega_d$) or the splitting energy $\Delta\Omega$. It is noted that the hydrostatic-pressure component $-(p + 2q)/6\omega_i (= \gamma_i)$ corresponds to the mode Grüneisen parameter (Section 3.3.2).

Under (100) coplanar stress X (i.e., biaxial stress along the [010] and [001] directions is assumed), the nonzero strain components in the layer are

$$e_{xx} = 2S_{12}X, \qquad e_{yy} = e_{zz} = (S_{11} + S_{12})X \qquad (3.31)$$

The energy shift $\Delta\omega_{LO}$ for the allowed LO phonon mode is then written as

$$\Delta\omega_{LO} = \omega_{LO}(X = 0) - \omega_{LO}(X \neq 0) = 2\Delta\Omega_H - \frac{2}{3}\Delta\Omega \qquad (3.32)$$

where $\Delta\Omega_H$ and $\Delta\Omega$ are given by Eqs. (3.28) and (3.30a). Typically, for light scattering from a (100) surface (zinc-blend-type crystals), only the LO mode is allowed and the TO mode is forbidden in the backscattering configuration. This coplanar strain configuration is just the case where a large residual strain is usually induced in lattice-mismatched (strained) superlattices and heterostructures grown on (100) substrates (see Chapter 11).

There have been many reports on the effects of uniaxial stress on the $\boldsymbol{q} = 0$ optical phonon frequencies and their degeneracies in semiconductors.[59,86–94] The information obtained from such studies concerns primarily the PDPs. Anastassakis et al.[59] have studied the effects of uniaxial stress on the $\boldsymbol{q} = 0$ optical phonons of InP using Raman and far-infrared spectroscopic techniques. They determined the dimensionless PDPs, $\tilde{K}_{11} = p/\omega_i^2$, $\tilde{K}_{12} = q/\omega_i^2$, and $\tilde{K}_{44} = r/\omega_i^2$, for both the LO and TO phonons. If the stress dependence of the LO–TO splitting is significantly large, as in the case of InP (see Ref. 58), we must consider two different sets of values for p, q, and r ($\tilde{K}_{11}$, $\tilde{K}_{12}$, and $\tilde{K}_{44}$), one for the LO phonons and the other for the TO phonons.[86] In Table 3.6 we list the determined PDPs for InP together with those for InAs, GaAs, and GaP reported by many workers. It is noted that the data of GaAs cover a large range of PDP values by different authors. Anastassakis and Cardona[94] have also investigated the effect of a uniaxial stress on the LO and TO phonon frequencies of In-hardened GaAs. They found that for a stress along [111] the singlet–doublet phonon frequency splittings are smaller than those in pure GaAs. They considered that a possible explanation of this effect may lie in a hardening of the internal stress parameter due, at least in part, to the increase in the lattice constant.

The PDP parameters have not yet been determined for InGaAsP quaternary. Experimentally, the relatively well-known quantity is the mode Grüneisen parameter $\gamma_i = -(p + 2q)/(6\omega_i^2)[= -(\tilde{K}_{11} + 2\tilde{K}_{12})/6]$. Since all known γ_i values of III–V and II–VI compounds are close to 1.0 (and especially for InP; see Table 3.5), we can take the same value for InGaAsP quaternary. The shear deformation parameter $(p\text{–}q)/(2\omega_i^2)$ has not yet been measured for the InGaAsP quaternary. A general theoretical argument[86] indicates that this parameter should be smaller than 1.0, and increase as a function of ionicity. Phillips's ionicity f_i,[95] however, does not vary much for III–V materials. Hence we can use InP values for InGaAsP quaternary. Of course, we can also use some sort of an interpolation scheme based on the experimentally determined PDP values for the InGaAsP-related binary end members listed in Table 3.6.

Once the PDP values are known, a straightforward determination of strain in semiconductors is possible by measuring the strain-induced shifts in the $\boldsymbol{q} = 0$ optical phonon frequencies. Indeed, this type of study has been extensively done on bulk or epitaxial films of Si, II–VI, and III–V materials in conjunction with surface treatment,[96,97] laser annealing,[98] internal bond stress (in alloys),[99] and misfit strain (in strained-layer materials).[74,100–108] Shen and Pollak[96] and Hang et al.[97] studied the polish-induced strain in InP and GaAs using RS from the LO phonon modes. They obtained the magnitude and sign of the surface strain as well as the strain skin depth and strain inhomogeneity for various polishing conditions.

TABLE 3.6 Phonon Deformation Potentials (PDPs) for InP, InAs, GaAs, and GaP

PDP	InP	InAs	GaAs	GaP
$\tilde{K}_{11}$ (LO)	-1.6 ± 0.1[a]		-2.0 ± 0.1[d]	
	-1.0 ± 0.1[b]		-1.0 ± 0.2[e]	
			-1.1 ± 0.2[f]	
$\tilde{K}_{11}$ (TO)	-2.5 ± 0.1[a]		-2.4 ± 0.2[d]	
	-1.6 ± 0.1[b]		-1.4 ± 0.2[e]	
			-1.8 ± 0.1[f]	
$\tilde{K}_{12}$ (LO)	-2.8 ± 0.1[a]		-2.7 ± 0.1[d]	
	-1.9 ± 0.1[b]		-1.5 ± 0.2[e]	
			-1.6 ± 0.2[f]	
$\tilde{K}_{12}$ (TO)	-3.2 ± 0.1[a]		-3.0 ± 0.1[d]	
	-2.1 ± 0.1[b]		-1.6 ± 0.2[e]	
			-1.9 ± 0.1[f]	
$\tilde{K}_{44}$ (LO)	-0.18 ± 0.02[a]		-0.60 ± 0.16[g]	-0.50 ± 0.10[i]
	-0.18 ± 0.04[b]		-0.60 ± 0.20[d]	
			-0.54 ± 0.12[e]	
			-0.50 ± 0.10[f]	
			-0.53 ± 0.03[h]	
$\tilde{K}_{44}$ (TO)	-0.47 ± 0.02[a]	-0.76 ± 0.15[c]	-0.20 ± 0.20[c]	-0.58 ± 0.10[i]
	-0.38 ± 0.04[b]		-0.80 ± 0.40[g]	
			-0.80 ± 0.40[d]	
			-0.80 ± 0.12[e]	
			-0.70 ± 0.10[f]	
			-0.88 ± 0.03[h]	
$\tilde{K}_{11}$–$\tilde{K}_{12}$ (LO)	1.20 ± 0.02[a]		0.70 ± 0.16[g]	1.06 ± 0.10[i]
	0.90 ± 0.04[b]		0.45 ± 0.12[e]	
			0.50 ± 0.10[f]	
			0.70 ± 0.03[h]	
$\tilde{K}_{11}$–$\tilde{K}_{12}$ (TO)	0.69 ± 0.02[a]	1.14 ± 0.24[c]	0.20 ± 0.20[c]	0.60 ± 0.10[i]
	0.42 ± 0.04[b]		0.60 ± 0.30[g]	
			0.24 ± 0.14[e]	
			0.16 ± 0.06[f]	
			0.30 ± 0.03[h]	

[a] Obtained from infrared-laser Raman scattering [E. Anastassakis et al., *Phys. Rev. B* **38,** 7702 (1988)].
[b] Obtained from far-infrared reflectivity [E. Anastassakis et al., *Phys. Rev. B* **38,** 7702 (1988)].
[c] F. Cerdeira et al., *Phys. Rev. B* **5,** 580 (1972).
[d] E. Anastassakis, *J. Phys. C* **16,** 3329 (1983).
[e] A. K. Sood et al., *Phys. Status Solidi B* **129,** 505 (1985).
[f] M. Hünermann et al., *Phys. Rev. B* **34,** 5381 (1986).
[g] B. A. Weinstein and M. Cardona, *Phys. Rev. B* **5,** 3120 (1972).
[h] P. Wickboldt et al., *Phys. Rev. B* **35,** 1362 (1987).
[i] I. Balslev, *Phys. Status Solidi B* **61,** 207 (1974).

REFERENCES

1. N. A. Goryunova, A. S. Borschevskii, and D. N. Tretiakov, in *Semiconductors and Semimetals*, Vol. 4, R. K. Willardson and A. C. Beer, Academic, New York, 1968, p. 3.
2. See, for instance, S. N. G. Chu and S. Nakahara, *Appl. Phys. Lett.* **56,** 434 (1990).
3. C. A. Brookes, J. B. O'Neill, and B. A. W. Redfern, *Proc. Roy. Soc. London A* **322,** 73 (1971).
4. D. Brasen, *J. Mater. Sci.* **11,** 791 (1976).
5. D. Brasen, *J. Mater. Sci.* **13,** 1776 (1978).
6. S. Mahajan and A. K. Chin, *J. Cryst. Growth* **54,** 138 (1981).
7. D. Y. Watts and A. F. W. Willoughby, *J. Appl. Phys.* **56,** 1869 (1984).
8. F. R. N. Nabarro, *Theory of Crystal Dislocations*, Dover, New York, 1987.
9. V. N. Vigdorovich and A. Y. Nashel'skii, *Sov. Powder Metall. Metal Ceram.* **2,** 123 (1963).
10. D. Arivuoli, R. Fornari, and J. Kumar, *J. Mater. Sci. Lett.* **10,** 559 (1991).
11. J. F. Nye, *Physical Properties of Crystals*, Clarendon, Oxford, 1972.
12. R. W. Keyes, *J. Appl. Phys.* **33,** 3371 (1962).
13. S. Adachi, *J. Appl. Phys.* **58,** R1 (1985).
14. H.-M. Kagaya and T. Soma, *Phys. Status Solidi B* **127,** 89 (1985).
15. N. Chetty, A. Muñoz, and R. M. Martin, *Phys. Rev. B* **40,** 11934 (1989).
16. M. V. Schilfgaarde and A. Sher, *Phys. Rev. B* **36,** 4375 (1987).
17. A. K. Giri and G. B. Mitra, *Phys. Status Solidi B* **134,** K11 (1986).
18. F. S. Hickernell and W. R. Gayton, *J. Appl. Phys.* **37,** 462 (1966).
19. D. N. Nichols, D. S. Rimai, and R. J. Sladek, *Solid State Commun.* **36,** 667 (1980).
20. D. Gerlich, *J. Appl. Phys.* **34,** 2915 (1963).
21. T. B. Bateman, H. J. McSkimin, and J. M. Whelan, *J. Appl. Phys.* **30,** 544 (1959).
22. R. Weil and W. Groves, *J. Appl. Phys.* **39,** 4049 (1968).
23. O. D. Slagle and H. A. McKinstry, *J. Appl. Phys.* **38,** 446 (1967).
24. R. N. Thurston and K. Brugger, *Phys. Rev.* **133,** A1604 (1964); erratum, ibid. **135,** AB3 (1964).
25. W. A. Brantley, *J. Appl. Phys.* **44,** 534 (1973).
26. T. D. Riney, *J. Appl. Phys.* **32,** 454 (1961).
27. C. S. Menoni and I. L. Spain, *Phys. Rev. B* **35,** 7520 (1987).
28. A. D. Prins and D. J. Dunstan, *Semicond. Sci. Technol.* **4,** 239 (1989).
29. A. D. Prins and D. J. Dunstan, *Phil. Mag. Lett.* **58,** 37 (1988).
30. M. S. Kushwaha, *Phys. Rev. B* **24,** 2115 (1981).

31. F. D. Murnaghan, *Proc. Natl. Acad. Sci. USA* **30,** 244 (1944).
32. J. S. Blakemore, *J. Appl. Phys.* **53,** R123 (1982).
33. D. N. Joharapurkar and M. A. Breazeale, *J. Appl. Phys.* **67,** 76 (1990).
34. J. Philip and M. A. Breazeale, *J. Appl. Phys.* **54,** 752 (1983).
35. T. Soma and H.-M. Kagaya, *Phys. Status Solidi B* **135,** K103 (1986).
36. See, for example, G. W. Farnell, in *Physical Acoustics*, Vol. 6, W. P. Mason and R. N. Thurston, eds., Academic, New York, 1970, p. 109.
37. J. Sapriel, J. C. Michel, J. C. Tolédano, R. Vacher, J. Kervarec, and A. Regreny, *Phys. Rev. B* **28,** 2007 (1983).
38. V. E. Steel, W. D. Hunt, M. A. Emanuel, J. J. Coleman, and B. J. Hunsinger, *J. Appl. Phys.* **66,** 90 (1989).
39. P. H. Borcherds, G. F. Alfrey, D. H. Saunderson, and A. D. B. Woods, *J. Phys. C* **8,** 2022 (1975).
40. M. S. Kushwaha and S. S. Kushwaha, *Can. J. Phys.* **58,** 351 (1980).
41. S. Yip and Y.-C. Chang, *Phys. Rev. B* **30,** 7037 (1984).
42. R. K. Ram, S. S. Kushwaha, and A. Shukla, *Phys. Status Solidi B* **154,** 553 (1989).
43. G. F. Alfrey and P. H. Borcherds, *J. Phys. C* **5,** L275 (1972).
44. J. L. T. Waugh and G. Dolling, *Phys. Rev.* **132,** 2410 (1963).
45. J. L. Yarnell, J. L. Warren, R. G. Wenzer, and P. J. Dean, in *Neutron Inelastic Scattering*, Vol. 1, International Atomic Energy Agency, Vienna, 1968, p. 301.
46. D. L. Price, J. M. Rowe, and R. M. Nicklow, *Phys. Rev. B* **3,** 1268 (1971).
47. M. K. Farr, J. G. Traylor, and S. K. Sinha, *Phys. Rev. B* **11,** 1587 (1975).
48. J. M. Rowe, R. M. Nicklow, D. L. Price, and K. Zanio, *Phys. Rev. B* **10,** 671 (1974).
49. N. Vagelatos, D. Wehe, and J. S. King, *J. Chem. Phys.* **60,** 3613 (1974).
50. J. C. Irwin and J. LaCombe, *Can. J. Phys.* **50,** 2596 (1972).
51. L. Van Hove, *Phys. Rev.* **89,** 1189 (1953).
52. M. J. O'Hara, C. W. Myles, J. D. Dow, and R. D. Painter, *J. Phys. Chem. Solids* **42,** 1043 (1981).
53. C. W. Myles, *Phys. Rev. B* **28,** 4519 (1983).
54. J. J. Sinai, S. Y. Wu, and Z. Zheng, *Phys. Rev. B* **31,** 3721 (1985).
55. E. S. Koteles and W. R. Datars, *Solid State Commun.* **19,** 221 (1976).
56. M. Sinyukov, R. Trommer, and M. Cardona, *Phys. Status Solidi B* **86,** 563 (1978).
57. J. R. Shealy and G. W. Wicks, *Appl. Phys. Lett.* **50,** 1173 (1987).
58. R. Trommer, H. Müller, M. Cardona, and P. Vogl, *Phys. Rev. B* **21,** 4869 (1980).
59. E. Anastassakis, Y. S. Raptis, M. Hünermann, W. Richter, and M. Cardona, *Phys. Rev. B* **38,** 7702 (1988).

60. A. P. G. Kutty, *Solid State Commun.* **14,** 213 (1974).
61. P. N. Sen and G. Lucovsky, *Phys. Rev. B* **12,** 2998 (1975).
62. R. Bonneville, *Phys. Rev. B* **24,** 1987 (1981).
63. H. C. Gupta, G. Ahuja, S. Prakash, L. M. Tiwari, and B. B. Tripathi, *Phys. Rev. B* **28,** 7191 (1983).
64. S. Prakash, H. C. Gupta, and B. B. Tripathi, *Phys. Rev. B* **28,** 7390 (1983).
65. R. Bonneville, *Phys. Rev. B* **29,** 907 (1984).
66. Y. Brada and L. Samuel, *Phys. Rev. B* **35,** 8260 (1987).
67. S. Yamazaki, A. Ushirokawa, and T. Katoda, *J. Appl. Phys.* **51,** 3722 (1980).
68. G. M. Zinger, I. P. Ipatova, and A. I. Ryskin, *Sov. Phys. Semicond.* **18,** 13 (1984).
69. Y. T. Cherng, D. H. Jaw, M. J. Jou, and G. B. Stringfellow, *J. Appl. Phys.* **65,** 3285 (1989).
70. G. Lucovsky, M. H. Brodsky, M. F. Chen, R. J. Chicotka, and A. T. Ward, *Phys. Rev. B* **4,** 1945 (1971).
71. R. Beserman, C. Hirliman, M. Balkanski, and J. Chevallier, *Solid State Commun.* **20,** 485 (1976).
72. E. Jahne, W. Pilz, M. Giehler, and L. Hildisch, *Phys. Status Solidi B* **91,** 155 (1979).
73. B. Jusserand and S. Slempkes, *Solid State Commun.* **49,** 95 (1984).
74. R. M. Abdelouhab, R. Braunstein, K. Bärner, M. A. Rao, and H. Kroemer, *J. Appl. Phys.* **66,** 787 (1989).
75. P. Galtier, J. Chevallier, M. Zigone, and G. Martinez, *Phys. Rev. B* **30,** 726 (1984).
76. A. Pinczuk, J. M. Worlock, R. E. Nahory, and M. A. Pollack, *Appl. Phys. Lett.* **33,** 461 (1978).
77. J. C. Portal, P. Perrier, M. A. Renucci, S. Askenazy, R. J. Nicholas, and T. P. Pearsall, in *Proc. 14th Int. Conf. Phys. Semicond.*, Institute of Physics, Bristol, UK, 1979, p. 829.
78. P. M. Amirtharaj, G. D. Holah, and S. Perkowitz, *Phys. Rev. B* **21,** 5656 (1980).
79. C. Pickering, *J. Electron. Mater.* **10,** 901 (1981).
80. R. K. Soni, S. C. Abbi, K. P. Jain, M. Balkanski, S. Slempkes, and J. L. Benchimol, *J. Appl. Phys.* **59,** 2184 (1986).
81. N. P. Kekelidze, G. P. Kekelidze, and Z. D. Makharadze, *J. Phys. Chem. Solids* **34,** 2117 (1973).
82. T. Inoshita and A. Usui, *Jpn. J. Appl. Phys.* **23,** L135 (1984).
83. C. Pickering, *J. Electron. Mater.* **15,** 51 (1986).
84. R. Borroff, R. Merlin, A. Chin, and P. K. Bhattacharya, *Appl. Phys. Lett.* **53,** 1652 (1988).

85. G. Lucovsky, R. D. Burnham, A. S. Alimanda, and H. A. Six, in *Proc. 12th Int. Conf. Phys. Semicond.*, Teuber, Stuttgart, 1974, p. 326.
86. F. Cerdeira, C. J. Buchenauer, F. H. Pollak, and M. Cardona, *Phys. Rev. B* **5,** 580 (1972).
87. B. A. Weinstein and M. Cardona, *Phys. Rev. B* **5,** 3120 (1972).
88. I. Balslev, *Phys. Status Solidi B* **61,** 207 (1974).
89. M. Chandrasekhar, J. B. Renucci, and M. Cardona, *Phys. Rev. B* **17,** 1623 (1978).
90. A. K. Sood, E. Anastassakis, and M. Cardona, *Phys. Status Solidi B* **129,** 505 (1985).
91. M. Hünermann, W. Richter, J. Saalmüller, and E. Anastassakis, *Phys. Rev. B* **34,** 5381 (1986).
92. P. Wickboldt, E. Anastassakis, R. Sauer, and M. Cardona, *Phys. Rev. B* **35,** 1362 (1987).
93. E. Anastassakis and M. Cardona, *Solid State Commun.* **63,** 893 (1987).
94. E. Anastassakis and M. Cardona, *Solid State Commun.* **64,** 543 (1987).
95. J. C. Phillips, *Bonds and Bands in Semiconductors*, Academic, New York, 1973.
96. H. Shen and F. H. Pollak, *Appl. Phys. Lett.* **45,** 692 (1984).
97. Z. Hang, H. Shen, and F. H. Pollak, *J. Appl. Phys.* **64,** 3233 (1988).
98. H. S. Tan, M. H. Kuok, S. C. Ng, C. K. Ong, and S. H. Tang, *J. Appl. Phys.* **55,** 1116 (1984).
99. K. Kakimoto and T. Katoda, *Jpn. J. Appl. Phys.* **24,** 1022 (1985).
100. B. Jusserand, P. Voisin, M. Voos, L. L. Chang, E. E. Mendez, and L. Esaki, *Appl. Phys. Lett.* **46,** 678 (1985).
101. M. Nakayama, K. Kubota, T. Kanata, H. Kato, S. Chika, and N. Sano, *J. Appl. Phys.* **58,** 4342 (1985).
102. J. Menéndez, A. Pinczuk, J. P. Valladares, R. D. Feldman, and R. F. Austin, *Appl. Phys. Lett.* **50,** 1101 (1987).
103. T. Kato, T. Matsumoto, M. Hosoki, and T. Ishida, *Jpn. J. Appl. Phys.* **26,** L1597 (1987).
104. G. Landa, R. Carles, C. Fontaine, E. Bedel, and A. Muñoz-Yagüe, *J. Appl. Phys.* **66,** 196 (1989).
105. M. K. Jackson, R. H. Miles, T. C. McGill, and J. P. Faurie, *Appl. Phys. Lett.* **55,** 786 (1989).
106. T. P. Humphreys, J. B. Posthill, K. Das, C. A. Sukow, R. J. Nemanich, N. R. Parikh, and A. Majeed, *Jpn. J. Appl. Phys.* **28,** L1595 (1989).
107. A. Freundlich, J. C. Grenet, G. Neu, G. Landa, and R. Carles, *Appl. Phys. Lett.* **55,** 1558 (1989).
108. B. Roughani, M. Kallergi, J. L. Aubel, and S. Sundaram, *J. Appl. Phys.* **66,** 4946 (1989).

4

THERMAL PROPERTIES

4.1 SPECIFIC HEAT AND DEBYE TEMPERATURE

Investigation of the thermal properties in solids is an old topic that arises in connection with the fundamental physical properties of the solids. A major step forward in our knowledge concerns the thermal energy content of a solid. This leads us to one of the most essential thermal parameters, the specific heat or heat capacity of a solid.[1–3] In Fig. 4.1 the specific heat at constant pressure C_p versus temperature for InP is shown. The data are taken from a calorimetric investigation by Piesbergen.[4] He reported both C_p and C_v (at constant volume) for several III–V semiconductors and Ge in the temperature range 12–273.2 K. The measured C_p for InP is nearly the same as C_v at temperatures below 70 K, but exceeds C_v above 70 K as a result of thermal expansion of the crystal lattice. Unfortunately, there has been no detailed report on the specific heat of alloys. Since alloying produced no significant effect on its lattice dynamic properties (see Chapter 3), it seems that for semiconductor alloys the linear interpolation method would give generally acceptable specific-heat values. Table 4.1 presents the interpolated C_p value at 300 K for $In_{1-x}Ga_xAs_yP_{1-y}$ quaternary together with those of the end binary members (InP,[4] InAs,[5] GaAs,[6] and GaP[1]).

The Debye temperature θ_D is a useful parameter in solid-state problems because of its inherent relationship to lattice vibration. The parameter θ_D can be used in characterizing the excitation of phonons and to describe various lattice thermal phenomena. The Debye temperatures of many ma-

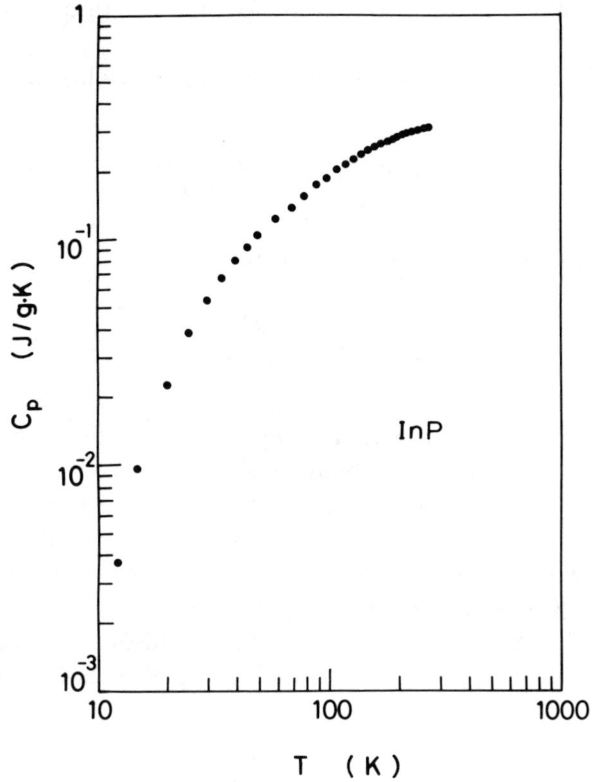

Figure 4.1 Specific heat at constant pressure C_p versus temperature for InP taken from a calorimetric investigation. (From Piesbergenl.[4])

TABLE 4.1 Specific Heat C_p (at 300 K), Debye Temperature θ_D (at 300 K), and $\theta_D(0)$ (at 0 K) for InP, InAs, GaAs, GaP, $In_{1-x}Ga_xAs_yP_{1-y}$, and $In_{0.53}Ga_{0.47}As$

Material	C_p [J/(g · K)]	θ_D (K)	$\theta_D(0)$ (K)
InP	0.322	422	320.5
InAs	0.352	280^b	262
GaAs	0.327	370	344
GaP	0.313	495	446
$In_{1-x}Ga_xAs_yP_{1-y}$[a]	$0.322 + 0.026y - 0.008y^2$	$422 - 100y$	$320.5 - 2.0y - 18.0y^2$
$In_{0.53}Ga_{0.47}As$[a]	0.340	322	300.5

[a]Linearly interpolated value using Eq. (2.3).
[b]At 273 K.

terials are known very precisely from low-temperature measurements of specific heat.

The Debye model for lattice vibrational energy yields in the relation[6]

$$C_p \simeq C_v = C_{cl} F\left(\frac{\theta_D}{T}\right) \tag{4.1}$$

where F is the Debye function defined by

$$F(y) = \int_0^y \frac{3x^4 e^x}{y^3(e^x - 1)^2}\, dx \tag{4.2}$$

Figure 4.2 plots the temperature dependence of the effective "calorimetric" Debye variable θ_{cal}, which results from inverting Eq. (4.1) and using the values of Piesbergen (Fig. 4.1);[4] θ_{cal} has a constant value ($= \theta_D$) from higher T to $T \simeq 180$ K and decreases gradually, showing a minimum at $T \simeq 20$ K, and then increases with further decrease of T.

Marcus and Kennedy[7] have investigated the relation between the elastic constant and the Debye temperature at 0 K ["elastic" Debye temperature $\theta_D(0)$] in the Debye approximation. Using the Marcus-Kennedy formula,

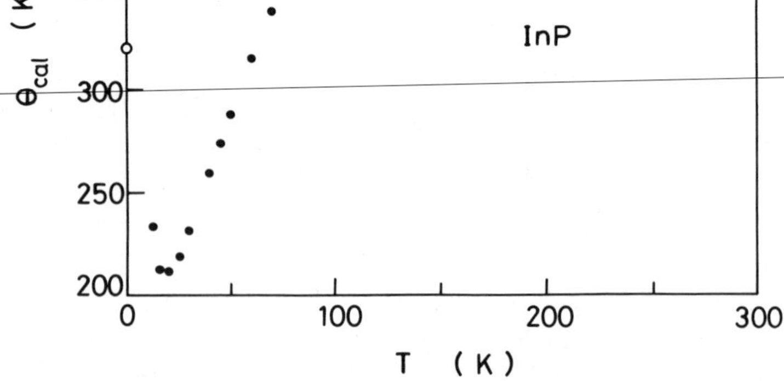

Figure 4.2 Temperature dependence of the effective "calorimetric" Debye variable θ_{cal}, which results from inverting Eq. (4.1).[4] The open circle is the zero-temperature asymptotic limit value as deduced by Steigmeier.[8]

Steigmeier[8] has shown that it would be possible to estimate unknown Debye temperatures using simple parameters such as the atomic mass or the lattice constant. This consideration is based on the observation of Keyes,[9] who estimated the unknown elastic constant (see Section 3.2.1). Steigmeier determined the Debye temperatures $\theta_D(0)$ of the compounds InAs, InP, AlAs, AlP, and GaP from their lattice constants. The open circle in Fig. 4.2 is the value obtained by him [$\theta_D(0) = 320.5$ K] (see also Table 4.1).

4.2 THERMAL EXPANSION COEFFICIENT

If the temperature of a crystal is changed, the resulting deformation may be specified by the strain tensor [e]. The thermal expansion coefficient is a second-rank symmetric tensor relating the temperature T (scalar quantity) and the second-rank strain tensor [e] by[10]

$$[e] = [\alpha]T \tag{4.3}$$

The thermal expansion tensor [α] in zinc-blende-type crystal has the form

$$[\alpha] = \begin{bmatrix} \alpha_{th} & 0 & 0 \\ 0 & \alpha_{th} & 0 \\ 0 & 0 & \alpha_{th} \end{bmatrix} \tag{4.4}$$

The linear expansion coefficient α_{th} [$= (1/a_0)(\partial a_0/\partial T)_p$] is known to be proportional to the specific heat C_v (Grüneisen's rule):

$$\alpha_{th} = \frac{\gamma C_v C_0}{3V} \tag{4.5}$$

where γ is the average Grüneisen parameter, C_0 is the compressibility (see Chapter 3), and V is the volume of the crystal. The expansion coefficients depend markedly on the temperature and are positive for most crystals.

The coefficient α_{th} is usually determined by measuring the temperature dependence of the lattice constant. Experimental evaluations of α_{th} for InP have been done below room temperature by Sirota et al.[11,12] and Deus et al.[13] and above room temperature by Bernstein and Beals,[14] Kudman and Paff,[15] Bisaro et al.,[16] and Pietsch et al.[17] We plot in Fig. 4.3 the linear expansion coefficient α_{th} as a function of temperature for InP. The data are derived from Refs. 11 and 12 (solid circles), 13 (open circles), 15 (open square), and 16 (open triangle). It can be seen in the figure that α_{th} decreases from its positive value on cooling, passing through zero for $T \simeq$

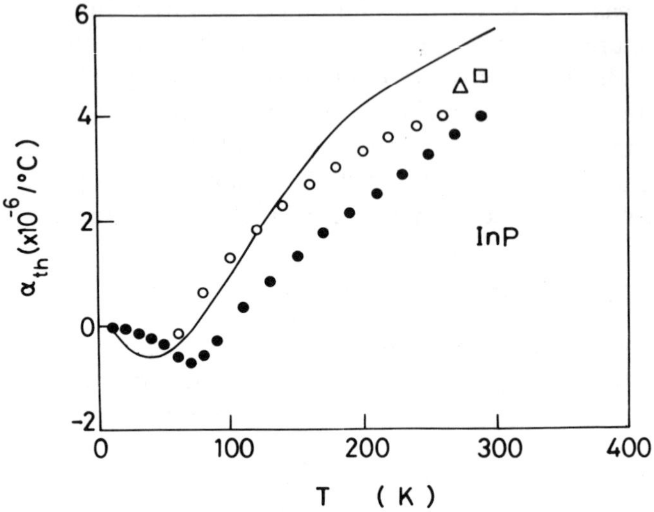

Figure 4.3 Linear thermal expansion coefficient α_{th} as a function of temperature for InP. The experimental data are taken from Refs. 11 and 12 (solid circles), 13 (open circles), 15 (open square), and 16 (open triangle). The solid line is the calculated result of Kagaya and Soma[21].

100 K (solid circles). Deus et al.[13] also found the negative character of α_{th} at temperatures below 64 K (open circles). They, however, reported no detailed information below 40 K because of the lower limit of the temperature interval in this region.

It is well known that not only zinc-blende materials but also diamond-type semiconductors show an unusual negative thermal expansion below ~100 K.[18–20] For InP Kagaya and Soma[21] have predicted theoretically a negative thermal expansion below 75 K using lattice dynamic and homogeneous deformation methods. The expansion parameter α_{th} was given by means of Eq. (4.5). They calculated the temperature-dependent average Grüneisen parameter γ by summing the phonon modes over the irreducible $\frac{1}{48}$th part of the Brillouin zone. The obtained α_{th} showed negative values at low temperatures. Their result is also reproduced in Fig. 4.3 by the solid line. Similar theoretical consideration, but based on density- functional-theory calculations of thermodynamical potentials (total energies and entropies) have been performed by Biernacki and Scheffler.[22] Their result showed excellent agreement with published experimental data of Si, and they concluded that the origin of the negative expansion effect is traced back to the entropy contribution of the Gibbs free energy.

The $In_{1-x}Ga_xAs_yP_{1-y}$/InP quaternary system is a promising material system for high-speed transistors and optoelectronic device applications. The efficiency and performance of such devices depend on the defect level present in the material and how likely those defects are to be mobile under device operating conditions. A difference in the thermal expansion coefficient usually exists between the two different heterojunction materials, and thus the lattice can be strained and dislocations may be produced. The difference in the thermal expansion can also produce a driving force that moves the defects in heterojunctions.

The composition dependence of the expansion coefficient α_{th} has been measured for many semiconducting alloys, including InGaP,[23] GaAsP[24,25] AlGaP,[26] and InGaAsP alloys.[16,17,27,28] In Table 4.2 and Fig. 4.4 we present the expansion coefficient difference $\Delta\alpha_{th}$ between $In_{1-x}Ga_xAs_yP_{1-y}$ and InP versus composition parameter y. Bert et al.[27] have measured the difference in α_{th} in the $In_{1-x}Ga_xAs_yP_{1-y}$/InP heterostructure system at temperatures from 23 to 600°C. These results are shown in Fig. 4.4 by the open circles. Biksaro et al.[16] have also determined α_{th} values of $In_{1-x}Ga_xAs_yP_{1-y}$ lattice-matched to InP ($y = 0.0$, 0.6, and 1.0) in the temperature range 25–400°C. The values they obtained are 4.56 ± 0.10 for InP ($y = 0.0$), 5.42 ± 0.10 for $y = 0.60$, and 5.66 ± 0.10 for $In_{0.53}Ga_{0.47}As$ ($y = 1.0$) (all in 10^{-6}/°C). Their results are also indicated in the figure by the solid circles. It is clear from the figure that $\Delta\alpha_{th}$ increases with increasing y. The solid line is obtained from the linear interpolation scheme of Eq. (2.3). The binary data used are as follows: 4.56×10^{-6}/°C for InP, 5.16×10^{-6}/°C for InAs, 6.40×10^{-6}/°C for GaAs, and 5.91×10^{-6}/°C for GaP. As clearly seen in Fig. 4.4, the values estimated from the interpolation scheme are in very good agreement with Bert's experimental data. The calculated $\Delta\alpha_{th}$, as a function of y, is almost

TABLE 4.2 Thermal Expansion Difference $\Delta\alpha_{th} = \alpha_{th}(In_{1-x}Ga_xAs_yP_{1-y}) - \alpha_{th}(InP)$ as a Function of Composition Parameter y (~300 K) (α_{th} for InP Assumed to be 4.56×10^{-6}/°C)

y	$\Delta\alpha_{th}$ ($\times 10^{-6}$/°C)
0	0.0
0.39	0.46[a]
0.50	0.63[a]
0.60	0.73[a]
0.60	0.86[b]
0.74	0.85[a]
0.84	0.95[a]
1.00	1.08[a]
1.00	1.10[b]

[a]N. A. Bert et al., *Sov. Phys. Tech. Phys.* **26,** 610 (1981).
[b]R. Bisaro et al., *Appl. Phys. Lett.* **34,** 100 (1979).

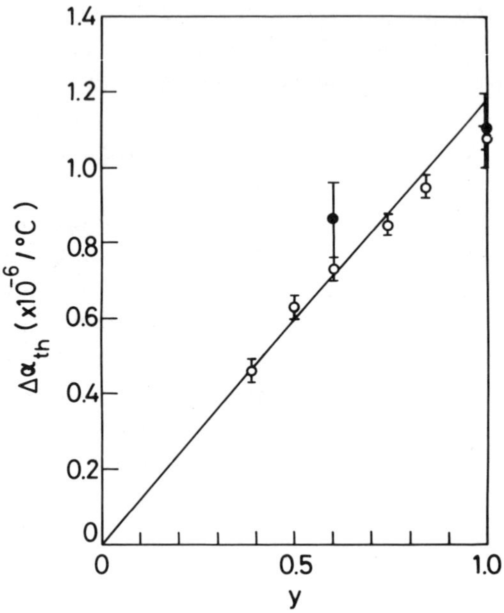

Figure 4.4 Thermal expansion difference $\Delta\alpha_{th} = \alpha_{th}\,(In_{1-x}Ga_xAs_yP_{1-y}) - \alpha_{th}(InP)$ versus composition y ($\sim$300 K). The solid and open circles are taken from Refs. 16 and 27, respectively. The solid line is obtained from the linear interpolation scheme of Eq. (2.3).

a linear relationship given by (in $10^{-6}/°C$)

$$\Delta\alpha_{th}(y) = 1.18y \tag{4.6}$$

Pietsch[29] has studied theoretically the thermal expansion coefficients of alloys by means of a perturbation treatment with the use of a Keating-like approach for the strain energy of binary members. He found that the calculated bowing parameters for α_{th} coincide with the published experimental findings. The α_{th} in a strained material system, GaAs films on Si(001) substrates, has been studied by Lucas et al.[30] between 20 and 450°C. They observed that the measured α_{th} of GaAs in the direction parallel to the film plane follows the thermal expansion of the Si substrate and is therefore smaller than in bulk GaAs. The α_{th} values for GaAs films on Si(001) substrates in the direction parallel and perpendicular to the film plane were 3.46×10^{-6} and $8.91 \times 10^{-6}/°C$, respectively. Pietsch and Marlow[28] have also discussed the effects of a thermally created elastic strain on α_{th} in tetragonally strained epitaxial layers of $In_{1-x}Ga_xAs_yP_{1-y}$ on InP(001) substrates. They demonstrated that the thermally created strain is higher than expected for a cubic "relaxed" material. The anisotropy of α_{th} in

$In_{0.53}Ga_{0.47}As/InP$ thin multilayers has also been determined by measuring the temperature dependence of the lattice mismatch with respect to the substrate.[31] This study suggested that the influence of the elastic interaction through the interfaces on α_{th} is not noticeable.

4.3 LATTICE THERMAL CONDUCTIVITY

Lattice thermal conductivity ($1/W$), or thermal resistivity (W), results essentially from interactions between phonons and from the scattering of phonons by crystalline imperfections. Knowledge of the thermal conductivity of semiconductors plays an important role in the design of power-dissipating devices (e.g., Peltier devices). A useful description of the theoretical and the practical aspects of thermal conductivity in semiconductors is given by Steigmeier and Kudman,[32] Holland,[33] Maycock,[34] and Bhandari and Rowe,[35] who review some of the work done on III–V compounds, including some of the ternary and quaternary alloys.

An exact calculation of the lattice thermal conductivity for the III–V binaries is possible, in principle, but lack of knowledge of various parameters (e.g., anharmonic forces and lattice vibration spectra) and the difficulty of obtaining exact solution of phonon–phonon interactions are formidable barriers. In Fig. 4.5 we show the temperature dependence of the thermal conductivity $1/W$ for InP, InAs, GaAs, and GaP. These data are taken for InP from Ref. 36, for InAs from Refs. 37 and 38, for GaAs from Refs. 39 and 40, and for GaP from Ref. 32. In most III–V semiconductors[33] we observe experimentally that the thermal conductivity of a pure single crystal is zero at 0 K and rises approximately exponentially to a maximum near 10 K, falls somewhat faster than $1/T$, and then varies approximately as $1/T$ to the melting point.

In the case of alloy compounds, an additional contribution, which is the result of a random distribution of constituent atoms in sublattice sites, should be taken into consideration. A phenomenologic model of the thermal conductivity for ternary alloys was first proposed by Abeles.[41] He has done calculations using an analysis of the lattice thermal conductivity that was reasonably successful for semiconducting alloys. His theory is based on the model developed by Klemens[42] and Callaway,[43] and starts from three kinds of relaxation times: $\tau_N^{-1} = B_1\omega^2$ (three-phonon normal process), $\tau_U^{-1} = B_2\omega^2$ (three-phonon Umklapp process), and $\tau_D^{-1} = A\Gamma\omega^2$ (strain and mass point defect), where ω is the phonon frequency, B_1, B_2, and A are constants independent of ω, and Γ is the disorder parameter and is a function of the masses and radii of the constituent atoms. The thermal resistivity

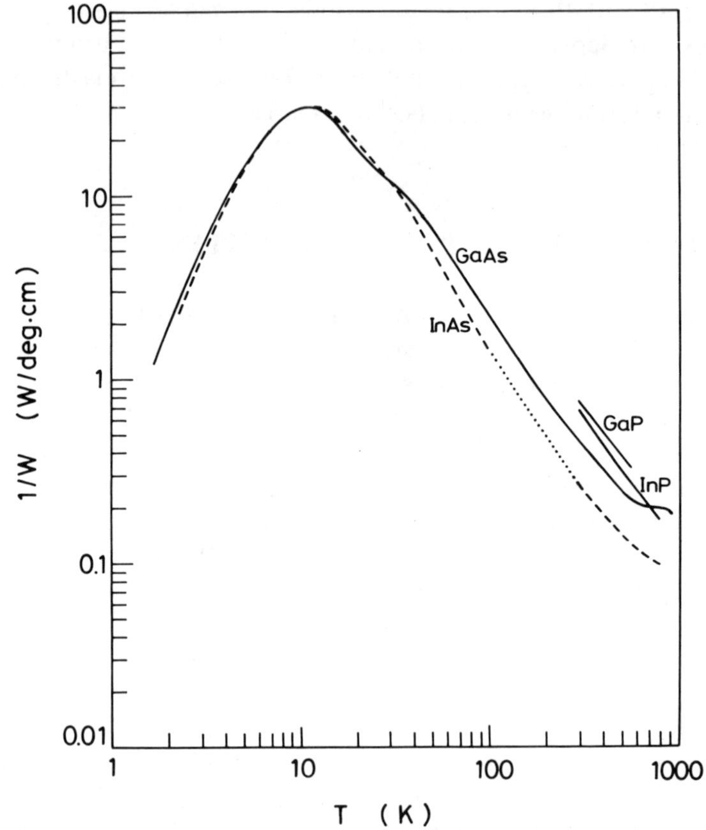

Figure 4.5 Temperature dependence of the thermal conductivity $1/W$ for InP, InAs, GaAs, and GaP. The experimental data are taken for InP from Ref. 36, for InAs from Refs. 37 and 38, for GaAs from Refs. 39 and 40, and for GaP from Ref. 32.

$W(x)$ obtained by Abeles is written as

$$\frac{W(x)}{W_p(x)} = \left(1 + \frac{5}{9}\alpha\right)\left\{\frac{\tan^{-1} U}{U} + \frac{[1 - (\tan^{-1} U/U)]^2}{[(1+\alpha)/5\alpha]U^4 - (1/3)U^2 - (\tan^{-1} U/U) + 1}\right\}^{-1} \tag{4.7}$$

with

$$U(x)^2 = U_0(x)^2 \left(1 + \tfrac{5}{9}\alpha\right)^{-1} \tag{4.8a}$$

$$U_0(x)^2 = A_1 \Gamma(x) W_p(x)^{-1} \tag{4.8b}$$

In Eqs. (4.7) and (4.8), $W_p(x)$ is the thermal resistivity of the crystal in which the disorder lattice is replaced by an ordered virtual crystal, α is a normal-process : Umklapp-process scattering ratio (i.e., $\alpha = B_1/B_2$), and A_1 is nearly constant within a family of the III–V compounds. Then $\Gamma(x)$ can be expressed by

$$\Gamma(x) = x(1 - x)\left\{\left[\frac{\Delta M}{M(x)}\right]^2 + \epsilon\left[\frac{\Delta\delta}{\delta(x)^2}\right]\right\} \tag{4.9}$$

where $\Delta M = M_A - M_B$, $\Delta\delta = \delta_A - \delta_B$, $M(x) = xM_A + (1 - x)M_B$, and $\delta(x) = x\delta_A + (1 - x)\delta_B$ [M_i and δ_i (i = A, B) are the masses and radii of the constituent atoms and ϵ is regarded as a phenomenologic, adjustable parameter]. The first and second terms in the curly bracket of Eq. (4.9) correspond to the contributions from the mass-defect and strain scatterings, respectively.

Abeles's model has been successfully used for many III–V ternaries by Adachi,[44] Nakwaski,[45] and Szmulowicz et al.[46] Adachi[44] has shown that the simple expression

$$W(x) = xW_{AC} + (1 - x)W_{BC} + C_{A-B}x(1 - x) \tag{4.10}$$

is essentially the same as that of Abeles's result [i.e., Eq. (4.7)], where W_{AC} and W_{BC} are the values for the binary members and C_{A-B} is a contribution arising from the lattice disorder generated in the ternary $A_xB_{1-x}C$ system by the random distribution of A and B atoms in one of the two sublattice sites. In an $A_{1-x}B_xC_yD_{1-y}$ quaternary system, not only should the A–B disorder due to the random distribution of the A and B atoms in the cationic sublattice and the C–D disorder due to the random distribution of the C and D atoms in the anionic sublattice be taken into consideration, but also the fact that both the cationic and anionic sublattices are disordered as well.[45]

Various experimental data on the III–V ternaries have led to an increased understanding of the thermal conductivity phenomena in these materials. It is important to point out that when large numbers of foreign atoms are added to a host lattice as in alloying, the thermal conductivity may decrease significantly. In fact, the published experimental data on the III–V ternaries, AlGaAs,[47] InGaAs,[34,48–50] InGaP,[51] InAsP,[34,52] and GaAsP,[34,53] exhibited strong nonlinearity with respect to alloy composition. Such compositional variations can be successfully explained by the quadratic function of Eq. (4.10).

Figure 4.6 shows a schematic drawing of the room-temperature thermal resistivity of the four ternary alloy boundaries.[44] The bowing parameters

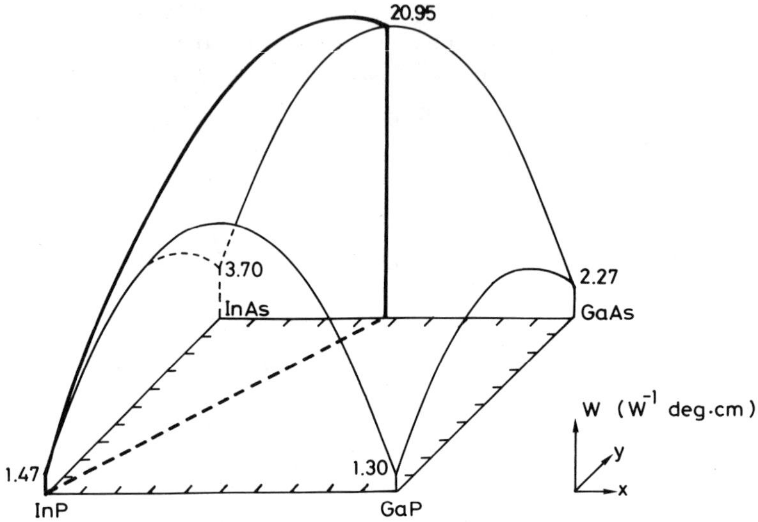

Figure 4.6 Three-dimensional representation of the room-temperature thermal resistivity W for $In_{1-x}Ga_xAs_yP_{1-y}$ quaternary alloy over the entire range of compositions. (From Adachi.[44])

used are as follows: $C_{In\text{-}Ga} = 72$ (deg · cm)/W for $In_{1-x}Ga_xAs$ and $In_{1-x}Ga_xP$ and $C_{As\text{-}P} = 25$ (deg · cm)/W for $InAs_yP_{1-y}$ and $GaAs_yP_{1-y}$. The bold line in the figure is the locus of $W(x, y)$ for compositions of $In_{1-x}Ga_xAs_yP_{1-y}$ lattice-matched to InP.

Yet despite the fundamental importance of the $In_{1-x}Ga_xAs_yP_{1-y}$ quaternary system to semiconductor laser applications, little has been reported concerning thermal-conductivity properties of the system.[44–46, 51, 54] Both et al.[54] have reported the thermal resistivity for $In_{1-x}Ga_xAs_yP_{1-y}$ alloys of target compositions $y = 0.0$–1.0 in steps of 0.2 measured by a thermal-transient method. The samples were grown by liquid phase epitaxy (2–3.5-μm thick). Their measured InP value is somewhat larger than that reported by Maykock,[34] but the $In_{0.53}Ga_{0.47}As$ value agrees well with that of Abrahams et al.[48] (see Table 4.3).

We plot in Fig. 4.7 the room-temperature thermal resistivity W as a function of y for $In_{1-x}Ga_xAs_yP_{1-y}$ lattice-matched to InP. The solid circles are the experimental data taken from Both et al.[54] The solid and dashed lines are, respectively, the calculated results of Adachi's[44] and Nakwaski's[45] models. It is clear that both the models explain the experimental findings well. The theoretical curve (solid line) shows that W increases markedly with alloying and exhibits a maximum of about 24 (deg · cm)/W at an alloying composition of $y \simeq 0.75$. Both the theory and the experiment indicate that the thermal resistivity of the $In_{1-x}Ga_xAs_yP_{1-y}$ alloy in the

TABLE 4.3 Room-Temperature Thermal Resistivity W of $In_{1-x}Ga_xAs_yP_{1-y}$ Lattice-Matched to InP

	W [(deg · cm)/W]	
y	Measured	Calculated[d]
0.0	1.14,[a] 1.47[b]	1.47
0.1		7.05
0.2	12.3–16.6[a]	11.84
0.3		15.83
0.4	18.0–30.3[a]	19.02
0.5		21.40
0.6	20.4–26.5[a]	22.96
0.7		23.71
0.8		23.63
0.82	20.5–25.0[a]	
0.9		22.71
1.0	16.4–23.4,[a] 20.4[c]	20.95

[a]W. Both et al., *Cryst. Res. Technol.* **21,** K85 (1986).
[b]P. D. Maycock, *Solid-State Electron.* **10,** 161 (1967).
[c]M. S. Abrahams et al., *J. Phys. Chem. Solids* **10,** 204 (1959).
[d]S. Adachi, *J. Appl. Phys.* **54,** 1844 (1983).

composition range of fiber optical-communication ($y \simeq 0.6$ for 1.3-μm wavelength and $y \simeq 0.9$ for 1.55-μm wavelength) is about 20 times higher than that of InP.

We limited our discussion to room temperature. The available data suggested that for any sample at higher temperatures, a law $1/W = AT^{-n}$ is followed. The factors n obtained at 300 K from Fig. 4.5 are 1.55 for InP, 1.20 for InAs, 1.25 for GaAs, and 1.30 for GaP (i.e., $1.20 \leqq n \leqq 1.55$). These are stronger temperature dependences than is predicted for the three-phonon processes, and probably indicate the presence of higher-order processes. If we take $n = 1.375$ for $In_{1-x}Ga_xAs_yP_{1-y}$ quaternary, the maximum error introduced in the temperature range 150–600 K is 13% which is less than the other uncertainties. The value of A is known to be dependent on the doping level. One should note that the binary data used in the calculations[44,45] were for undoped materials. Very heavily doped samples could have significantly lower thermal conductivities (i.e., smaller A values).[39,40]

Regarding the thermal time constant of the laser–diode chip itself, Fujita[55] described a method of investigation. It relies on current-modulated injection as the source of modulated heating. He made use of the fact that the output light power depends on the injected current and junction temperature. The thermal time constant τ can be related to the thermal resis-

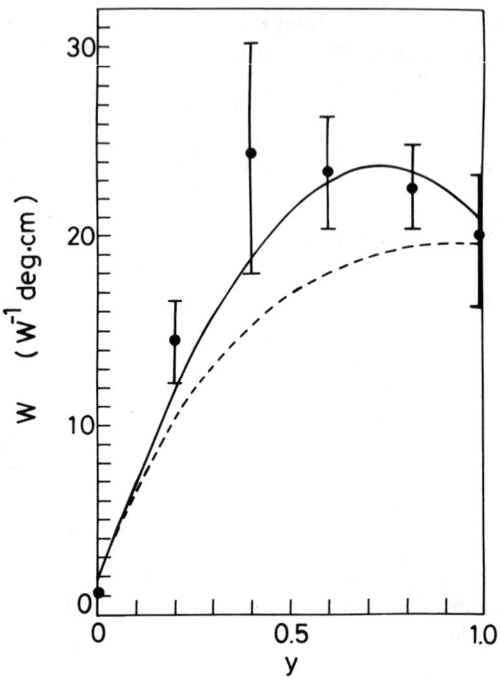

Figure 4.7 Room-temperature thermal resistivity W versus composition y for $In_{1-x}Ga_xAs_yP_{1-y}$ lattice-matched to InP. The solid circles are the experimental data taken from Ref. 54. The solid and dashed lines are the calculated results of Adachi's[44] and Nakwaski's[45] models, respectively.

tivity W by L^2C_vW, where L is a characteristic length (from source-to-sensor) and C_v is the specific heat. He reported a value of $\tau = 11.7 \times 10^{-6}$ s for an $In_{1-x}Ga_xAs_yP_{1-y}$ laser of 1.3-μm lasing wavelength. The thermal conductivity of heterostructure laser devices can be calculated using a model of Joyce and Dixon[56] in which two-dimensional heat flow is assumed to proceed from a uniformly excited stripe to a constant-temperature heat sink on one face.

REFERENCES

1. U. Piesbergen, in *Semiconductors and Semimetals*, Vol. 2, R. K. Willardson and A. C. Beer, eds., Academic, New York, 1966, p. 49.
2. V. V. Tarassov and A. F. Demidenko, *Phys. Status Solidi* **30,** 147 (1968).
3. J. C. Irwin and J. LaCombe, *J. Appl. Phys.* **45,** 567 (1974).
4. U. Piesbergen, *Z. Naturforsch.* **18a,** 141 (1963).

5. B. D. Lichter and P. Sommelet, *Trans. Metal. Soc. AIME* **245,** 1021 (1969).
6. J. S. Blakemore, *J. Appl. Phys.* **53,** R123 (1982).
7. P. M. Marcus and A. J. Kennedy, *Phys. Rev.* **114,** 459 (1959).
8. E. F. Steigmeier, *Appl. Phys. Lett.* **3,** 6 (1963).
9. R. W. Keyes, *J. Appl. Phys.* **33,** 3371 (1962).
10. J. F. Nye, *Physical Properties of Crystals*, Clarendon, Oxford, 1972.
11. N. N. Sirota, A. M. Antyukhov, and A. A. Sidorov, *Dokl. Akad. Nauk SSSR* **277,** 1379 (1984).
12. N. N. Sirota and A. A. Sidorov, *Dokl. Akad. Nauk SSSR* **284,** 1111 (1985).
13. P. Deus, H. A. Schneider, U. Voland, and K. Stiehler, *Phys. Status Solidi A* **103,** 443 (1987).
14. L. Bernstein and R. J. Beals, *J. Appl. Phys.* **32,** 122 (1961).
15. I. Kudman and R. J. Paff, *J. Appl. Phys.* **43,** 3760 (1972).
16. R. Bisaro, P. Merenda, and T. P. Pearsall, *Appl. Phys. Lett.* **34,** 100 (1979).
17. U. Pietsch, J. Bak-Misiuk, and V. Gottschalch, *Phys. Status Solidi A* **82,** K137 (1984).
18. S. I. Novikova, *Sov. Phys. Solid State* **3,** 129 (1961).
19. H. Ibach, *Phys. Status Solidi* **31,** 625 (1969).
20. K. Haruna, H. Maeta, K. Ohashi, and T. Koike, *J. Phys. C* **19,** 5149 (1986).
21. H.-M. Kagaya and T. Soma, *Solid State Commun.* **58,** 479 (1986).
22. S. Biernacki and M. Scheffler, *Phys. Rev. Lett.* **63,** 290 (1989).
23. I. N. Arsent'ev, D. Akhmedov, S. G. Konnikov, V. A. Mishurnyi, and V. E. Umanskii, *Sov. Phys. Semicond.* **14,** 1389 (1980).
24. E. D. Pierron, D. L. Parker, and J. B. McNeely, *J. Appl. Phys.* **38,** 4669 (1967).
25. M. E. Straumanis, J.-P. Krumme, and M. Rubenstein, *J. Electrochem. Soc.* **114,** 640 (1967).
26. V. N. Bessolov, S. G. Konnikov, V. E. Umanskii, and Y. P. Yakovlev, *Sov. Phys. Tech. Phys.* **28,** 257 (1983).
27. N. A. Bert, A. T. Gorelenok, S. G. Konnikov, V. E. Umanskii, and A. S. Usikov, *Sov. Phys. Tech. Phys.* **26,** 610 (1981).
28. U. Pietsch and D. Marlow, *Phys. Status Solidi A* **93,** 143 (1986).
29. U. Pietsch, *Phys. Status Solidi B* **133,** 483 (1986).
30. N. Lucas, H. Zabel, H. Morkoç, and H. Unlu, *Appl. Phys. Lett.* **52,** 2117 (1988).
31. J. Bak-Misiuk, J. Wolfe, and U. Pietsch, *Phys. Status Solidi A* **118,** 209 (1990).
32. E. F. Steigmeier and I. Kudman, *Phys. Rev.* **141,** 767 (1966).
33. M. G. Holland, in *Semiconductors and Semimetals*, Vol. 2, R. K. Willardson and A. C. Beer, eds., Academic, New York, 1966, p. 3.

34. P. D. Maycock, *Solid-State Electron.* **10,** 161 (1967).
35. C. M. Bhandari and D. M. Rowe, *Thermal Conduction in Semiconductors*, Wiley, New York, 1988.
36. I. Kudman and E. F. Steigmeier, *Phys. Rev.* **133,** A1665 (1964).
37. S. S. Shalyt, *Sov. Phys. Solid State* **4,** 1403 (1963).
38. E. F. Steigmeier and I. Kudman, *Phys. Rev.* **132,** 508 (1963).
39. M. G. Holland, *Phys. Rev.* **134,** A471 (1964).
40. A. Amith, I. Kudman, and E. F. Steigmeier, *Phys. Rev.* **138,** A1270 (1965).
41. B. Abeles, *Phys. Rev.* **131,** 1906 (1963).
42. P. G. Klemens, *Phys. Rev.* **119,** 507 (1960).
43. J. Callaway, *Phys. Rev.* **113,** 1046 (1959).
44. S. Adachi, *J. Appl. Phys.* **54,** 1844 (1983).
45. W. Nakwaski, *J. Appl. Phys.* **64,** 159 (1988).
46. F. Szmulowicz, F. L. Madarasz, P. G. Klemens, and J. Diller, *J. Appl. Phys.* **66,** 252 (1989).
47. M. A. Afromowitz, *J. Appl. Phys.* **44,** 1292 (1973).
48. M. S. Abrahams, R. Braunstein, and F. D. Rosi, *J. Phys. Chem. Solids* **10,** 204 (1959).
49. M. C. Ohmer, W. C. Mitchel, G. A. Graves, D. E. Holmes, H. Kuwamoto, and P. W. Yu, *J. Appl. Phys.* **64,** 2775 (1988).
50. D. G. Arasly, R. N. Ragimov, and M. I. Aliev, *Sov. Phys. Semicond.* **24,** 225 (1990).
51. W. Both and F. P. Herrmann, *Cryst. Res. Technol.* **17,** K117 (1982).
52. R. Bowers, J. E. Bauerle, and A. J. Cornish, *J. Appl. Phys.* **30,** 1050 (1959).
53. R. O. Carlson, G. A. Slack, and S. J. Silverman, *J. Appl. Phys.* **36,** 505 (1965).
54. W. Both, V. Gottschalch, and G. Wagner, *Cryst. Res. Technol.* **21,** K85 (1986).
55. O. Fujita, *J. Appl. Phys.* **57,** 978 (1985).
56. W. B. Joyce and R. W. Dixon, *J. Appl. Phys.* **46,** 855 (1975).

5

COLLECTIVE EFFECTS AND SOME RESPONSE CHARACTERISTICS

5.1 STATIC AND HIGH-FREQUENCY DIELECTRIC CONSTANTS

The concept of the dielectric behavior of solid is an old one that is important for several electron-device properties.[1] Below the reststrahlen range in optical spectra (see Section 8.2), the real part of the dielectric constant asymptotically approaches the static or low-frequency dielectric constant ϵ_s. The optical constant connecting the reststrahlen–near-infrared range is called the high-frequency (or optical) dielectric constant ϵ_∞. The dielectric constants are second-rank tensors having the following form in the zinc-blende crystal:[2]

$$[\epsilon] = \begin{bmatrix} \epsilon & 0 & 0 \\ 0 & \epsilon & 0 \\ 0 & 0 & \epsilon \end{bmatrix} \tag{5.1}$$

The dielectric constant ϵ_s can be written as

$$\epsilon_s = \epsilon_\infty + \frac{4\pi N e_T^2}{\omega_{TO}^2 M} \tag{5.2}$$

where M is the reduced mass of the crystal, N is the number of unit cells per unit volume, and e_T is the effective charge. The dielectric constants ϵ_s

and ϵ_∞ are also related to the long-wavelength optical phonon frequencies by the Lyddane–Sachs–Teller relationship:

$$\frac{\epsilon_s}{\epsilon_\infty} = \left(\frac{\omega_{LO}}{\omega_{TO}}\right)^2 \tag{5.3}$$

The static dielectric constant ϵ_s of InP has been reported by several authors. Neidert et al.[3] measured microwave resonant frequencies in semiinsulating InP at 5–18 GHz and obtained an average value of $\epsilon_s = 12.55$ at 300 K. Meiners[4] also found a value of $\epsilon_s = 12.56 \pm 0.2$ (at 300 K) from low-frequency capacitance measurements. Seeger[5] has recently applied a microwave transmission interference method developed by Howell[6] for the determination of ϵ_s of semiinsulating InP at temperatures of 77 and 300 K. He obtained $\epsilon_s = 12.2$ at 77 K and 12.5 at 300 K.

The term ϵ_∞ is the dielectric constant for frequencies well above ω_{LO} (Γ) and below the electronic absorption edge. Pikhtin and Yas'kov[7] reported ϵ_∞ of InP to be 9.3 at 77 K and 9.5 at 298 K. Newman[8] also determined the values $\epsilon_s = 15$ and $\epsilon_\infty = 10.6$ for InP from the reflectance spectrum in the reststrahlen range. These values are, however, considerably larger than those widely used (e.g., $\epsilon_\infty = 9.55$,[9] 9.61,[10] and 9.79[11]). The calculated values from the Lyddane–Sachs–Teller relationship of Eq. (5.3) with $\omega_{LO} = 346.4$ cm^{-1}, $\omega_{TO} = 304.5$ cm^{-1} (see Table 3.4), and $\epsilon_s = 12.5$ is $\epsilon_\infty = 9.66$.

Some data on the temperature dependence of the dielectric constants at atmospheric pressure have been reported. These include results on InP in the temperature range 100–300 K,[4] on GaAs in the ranges 100–300 K[12,13] and 4.2–350 K,[14] and values of ϵ_s at a few temperatures for GaP.[15,16] These data showed that the values of ϵ_s decrease with decreasing temperature. For InP[4] a value of 12.56 ± 0.2 was obtained at 300 K and an extrapolated value of 11.93 ± 0.2 was obtained at 77 K. Below about ~220 K it can be expressed by the relationship $\epsilon_s = 11.76 + 2.66 \times 10^{-3}\,T$.

Samara[14] has studied the hydrostatic-pressure effects on ϵ_s for GaAs and GaP for pressures up to 0.7 GPa. He reported the values of ϵ_s and its logarithmic pressure derivatives for these materials. Anastassakis et al.[10] also reported the volume coefficient $\partial \ln \epsilon_\infty / \partial \ln V$ ($= 0.9 \pm 0.3$) of the high-frequency dielectric constant for InP. This value for GaAs is reported to be 0.7 ($d\epsilon_\infty/dp = -0.09$ GPa^{-1}).[17] It is known that the dominant contribution to the decrease of ϵ_∞ in covalent semiconductors under pressure arises from the change in the microscopic polarizability, which is a decreasing function of volume.

Adachi has studied in Ref. 9 the general properties of ϵ_s and ϵ_∞ for a specific family of compounds, namely, III–V and II–VI compounds, and concluded that the constants could be deduced by the use of the linear interpolation method. The simplest linear interpolation method is the use of Eq. (2.3). The relative permittivity ϵ of a medium containing a mixture of dipoles of ith type having polarizability α_i and occurring with a density N_i per unit volume, is given by the Clausius–Mosotti relation

$$\frac{\epsilon - 1}{\epsilon + 2} = \frac{1}{3} \sum_i N_i \alpha_i \tag{5.4}$$

From this relation[18,19] we can obtain $\epsilon(x, y)$ for $In_{1-x}Ga_xAs_yP_{1-y}$ quaternary

$$\frac{\epsilon(x, y) - 1}{\epsilon(x, y) + 2} = (1 - x)y \frac{\epsilon(\mathrm{InAs}) - 1}{\epsilon(\mathrm{InAs}) + 2} + (1 - x)(1 - y) \frac{\epsilon(\mathrm{InP}) - 1}{\epsilon(\mathrm{InP}) + 2} + xy \frac{\epsilon(\mathrm{GaAs}) - 1}{\epsilon(\mathrm{GaAs}) + 2} + x(1 - y) \frac{\epsilon(\mathrm{GaP}) - 1}{\epsilon(\mathrm{GaP}) + 2} \tag{5.5}$$

In Fig. 5.1 we show the interpolated ϵ_s and ϵ_∞ as a function of y for $In_{1-x}Ga_xAs_yP_{1-y}$ quaternary. The solid and dashed lines are, respectively, calculated from Eqs. (2.3) and (5.5). It is clear from the figure that these methods give nearly the same interpolated values. In Table 5.1 we sum-

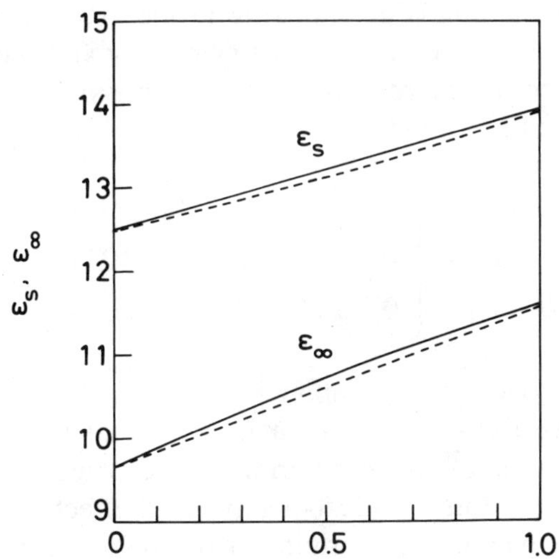

Figure 5.1 Static and high-frequency dielectric constants, ϵ_s and ϵ_∞, as a function of y for $In_{1-x}Ga_xAs_yP_{1-y}$ lattice-matched to InP. The solid and dashed lines are, respectively, calculated from Eqs. (2.3) and (5.5).

TABLE 5.1 Dielectric Constants ϵ_s and ϵ_∞ at 300 K (Atmospheric Pressure) for InP, InAs, GaAs, GaP, $In_{1-x}Ga_xAs_yP_{1-y}$, and $In_{0.53}Ga_{0.47}As$

Material	ϵ_s	ϵ_∞
InP	12.50	9.66
InAs	14.60	12.25
GaAs	13.18	10.89
GaP	11.11	9.11
$In_{1-x}Ga_xAs_yP_{1-y}$[a]	$12.50 + 1.44y$	$9.66 + 1.95y$
$In_{0.53}Ga_{0.47}As$[a]	13.94	11.61

[a]Linearly interpolated value using Eq. (2.3).

marize the values of ϵ_s and ϵ_∞ for InP, InAs, GaAs, GaP, $In_{1-x}Ga_xAs_yP_{1-y}$, and $In_{0.53}Ga_{0.47}As$.

5.2 PIEZOELECTRICITY

5.2.1 Piezoelectric Stress and Strain Constants

Piezoelectricity is the generation of electronic polarization by application of stress to a crystal lacking a center of symmetry. An important aspect of the zinc-blende structure is the absence of a center of symmetry or inversion. The III–V compounds crystallizing in the zinc-blende structure are the simplest crystals lacking a center of symmetry and, hence, capable of exhibiting piezoelectric and related effects. The piezoelectric tensor in zinc-blende crystals has the form

$$[e] = \begin{bmatrix} 0 & 0 & 0 & e_{14} & 0 & 0 \\ 0 & 0 & 0 & 0 & e_{14} & 0 \\ 0 & 0 & 0 & 0 & 0 & e_{14} \end{bmatrix} \tag{5.6}$$

that is, it has only one tensor component e_{14}.

In Table 5.2 we list the experimental and calculated values of e_{14} for InP. Arlt and Quadflieg[20] have determined piezoelectric constants for a variety of semiconductors by piezoelectric Hall-effect measurements and have proposed the macroscopic origins of piezoelectricity as being due to such as ionic polarization, strain-dependent ionicity, and electronic polarization. They did not, however, carry out measurements on InP. Boyle and Sladek[21] have measured the attenuation of ultrasonic longitudinal waves (30, 90, and 150 MHz) traveling in the [110] direction in Cr-doped, *n*-type

TABLE 5.2 Piezoelectric Constant e_{14} (C/m^2) of InP

Measured	Calculated
$\lvert 0.040 \pm 0.002 \rvert$[a]	$\lvert 0.29 \rvert$[b]
	-0.035[c]
	-0.115[d]
	-0.093[e]
	$+0.056$[f]

[a]W. F. Boyle and R. J. Sladek, *Solid State Commun.* **16,** 323 (1975).
[b]F. S. Hickernell, *IEEE Trans. Sonics Ultrasonics* **SU-13,** 73 (1966).
[c]J. C. Phillips and J. A. Van Vechten, *Phys. Rev. Lett.* **23,** 1115 (1969).
[d]K. Hübner, *Phys. Status Solidi B* **57,** 627 (1973).
[e]M. Miura et al., *J. Phys. Chem. Solids* **42,** 931 (1981).
[f]S. D. Gironcoli et al., *Phys. Rev. Lett.* **62,** 2853 (1989).

InP, and determined the piezoelectric constant $|e_{14}|$ to be 0.040 ± 0.002 C/m^2 at atmospheric pressure. Nichols and Sladek[22] also subjected InP samples up to 2 kbar of hydrostatic pressure and found that the pressure did not change the piezoelectric constant significantly ($|e_{14}| = 0.042$ C/m^2 at 2 kbar).

We shall discuss the piezoelectric constants of specific families of binary compounds—namely, the III–V and II–VI compounds—from a simplified point of view. The plots of e_{14} versus Phillips's ionicity f_i for some of the III–V and II–VI compounds are shown in Fig. 5.2. Many theoretical works[20,23–30] have been performed to understand piezoelectric properties of crystals. It is clear from the figure that e_{14} passes through zero at $f_i \simeq 0.5$ while undergoing a reversal in sign (i.e., smaller f_i values of the III–V compounds possess negative e_{14} values and larger f_i values of the II–VI compounds possess positive ones). Theoretical calculations of the piezoelectric constant e_{14} of InP have yielded widely different values ranging from $|0.29|$ to $+0.056$ C/m^2 (see Table 5.2). Phillips and Van Vechten[23] showed the reversal in sign of e_{14} on going from $N = 1, 2$ to $N = 3$ in tetrahedrally-coordinated $A^N B^{8-N}$ crystals based on the lattice-dynamic treatment with the addition of the concept of their ionicity scale. Hübner[25] also predicted e_{14} values for a number of the III–V, II–VI, and I–VII compounds and found them to be in good numeric agreement with experimental data including sign. He gave the value of $e_{14} = -0.115$ C/m^2 for InP. Similar consideration has been made by Miura et al.,[26] who obtained $e_{14} = -0.093$ C/m^2 for this material.

More recently, Gironcoli et al.[30] have presented an *ab initio* approach to piezoelectricity in semiconductors. The piezoelectric tensor was given

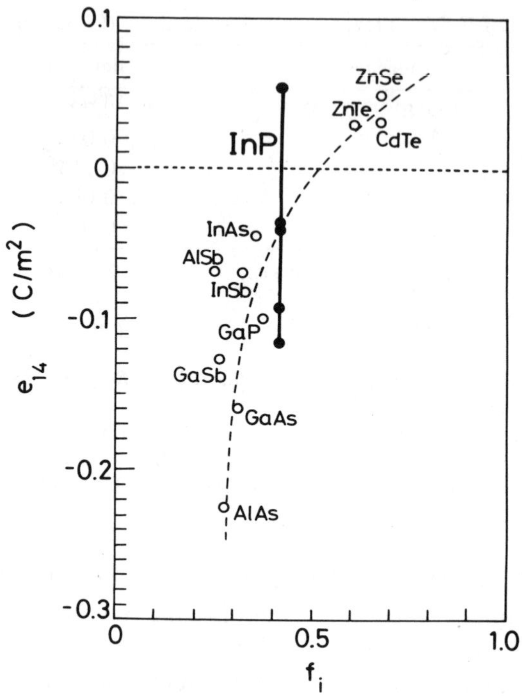

Figure 5.2 Plots of e_{14} versus Phillips's ionicity f_i for some of the III–V and II–VI binaries. The values for InP (solid circles) are taken from Table 5.2.

in their model by the stress induced by a homogeneous electric field, and the perturbation was treated self-consistently by linear response, thus avoiding both supercells and numeric differentiation. They calculated e_{14} of nine III–V binary semiconductors (Al, Ga, In)(P, As, Sb) and found that, contrary to common belief, two III–V semiconductors AlP and InP have positive e_{14} values (the value of e_{14} for InP they obtained is $+0.056$ C/m^2). This result is, thus, in direct contrast to those obtained by Hübner[25] and Miura et al.[26] Unfortunately, however, no experimental determination of the e_{14} sign of InP has been reported to date.

The piezoelectric effects are significant especially for the study of transport properties in semiconductors. The piezoelectric scattering of electrons in semiconductors is usually not of major practical importance, except in high-quality crystals, because of impurity scattering. At low temperatures ($\leqq 50$ K), however, lattice scattering of electrons is known to be dominated by the piezoelectric interaction, which causes elastic scattering due to relatively low-energy acoustic phonons. Rode[31] has used the value of $|e_{14}| = 0.035$ C/m^2 to analyze electron-transport properties of InP. This value corresponds to that predicted by Phillips and Van Vechten.[23] The piezoelectric constant e_{14} of $In_{1-x}Ga_xAs_yP_{1-y}$ quaternary is thought to be successfully estimated by the linear interpolation scheme, because the Phillips's ionicity does not vary much for the end binaries.[32] In Fig. 5.3 we show the interpolated $|e_{14}|$ as a function of y for $In_{1-x}Ga_xAs_yP_{1-y}$ lattice-matched to InP. The experimental data of the end binaries used in the calculation are 0.040 (InP[21]), 0.045 (InAs[20]), 0.16 (GaAs[20]), and 0.10 (GaP[33]) (all in C/m^2). The interpolated values show a weak nonlinearity with respect to the composition y (see Table 5.3).

The piezoelectric strain constant $d_{k,ij}$, which has the same tensor form as Eq. (5.6), is connected reciprocally with the piezoelectric stress component $e_{k,mn}$ through

$$d_{k,ij} = \sum_{mn} S_{ijmn} e_{k,mn} \tag{5.7}$$

where S_{ijmn} is the elastic compliance constant discussed in Section 3.2. In the case of zinc-blende crystals, the tensor $[d]$ can be expressed only by the component d_{14}:

$$d_{14} = S_{44} e_{14} \tag{5.8}$$

Like e_{14}, the calculated d_{14} of $In_{1-x}Ga_xAs_yP_{1-y}$ quaternary shows a weak nonlinearity with respect to the composition y (see Fig. 5.3).

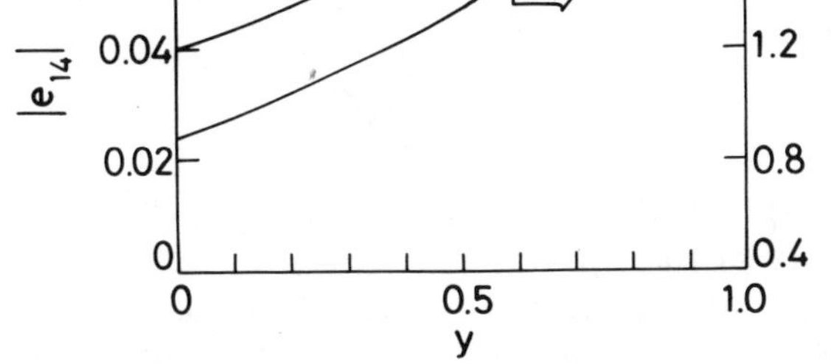

Figure 5.3 Piezoelectric stress ($|e_{14}|$) and strain constants ($|d_{14}|$) as a function of y for $In_{1-x}Ga_xAs_yP_{1-y}$ lattice-matched to InP.

TABLE 5.3 Piezoelectric Constants, $|e_{14}|$ and $|d_{14}|$, and Electromechanical Coupling Constant $K^2_{[110]}$ for InP, InAs, GaAs, GaP, $In_{1-x}Ga_xAs_yP_{1-y}$, and $In_{0.53}Ga_{0.47}As$

Material	$\lvert e_{14}\rvert$ (C/m²)	$\lvert d_{14}\rvert$ ($\times 10^{-12}$ m/V)	$10^3 K^2_{[110]}$
InP	0.040	0.88	0.32
InAs	0.045	1.14	0.40
GaAs	0.16	2.69	3.69
GaP	0.10	1.42	1.44
$In_{1-x}Ga_xAs_yP_{1-y}$	$0.040 + 0.033y$ $+0.026y^2$	$0.88 + 0.69y$ $+0.46y^2$	$0.32 + 0.30y$ $+1.00y^2$
$In_{0.53}Ga_{0.47}As$	0.099	2.03	1.62

5.2.2 Electromechanical Coupling Constant

It is known[34] that injected ultrasonic waves can be amplified in piezoelectric semiconductors by the application of a sufficiently high electric field because of the strong interaction with mobile electrons. Domains of intense acoustic flux, showing a broad band of frequencies in the low-GHz range, can also be produced in piezoelectric semiconductors, such as GaAs[35] and GaSb,[36] by acoustoelectric amplification of phonons from the thermal equilibrium spectrum. The gain of such acoustoelectric interactions can be explained in terms of the material parameter K^2, called the *electromechanical (piezoelectric) coupling constant*.[37,38] Carrier mobilities in semiconductors are also strongly affected by this parameter. The parameter value is also necessary in calculating figures of merit for piezoelectric (acoustoelectric) transducer devices.[39]

The coupling constant K^2 is a crystal-direction-dependent quantity. The maximum coupling of transverse acoustic waves is along the [110] direction for zinc-blende crystals, which is the reason why the majority of experiments have been carried out with crystals cut in this direction.[34-36] In this case, the coupling constant can be written as

$$K^2_{[110]} = \frac{e_{14}^2}{\epsilon_0 \epsilon_s C_{44}} \tag{5.9}$$

where e_{14} is the piezoelectric stress constant (see Table 5.3), ϵ_0 is the dielectric permittivity of free space, ϵ_s is the static dielectric constant (Table 5.1), and C_{44} is the elastic stiffness constant (Table 3.1). The calculated value of $K^2_{[110]}$ for $In_{1-x}Ga_xAs_yP_{1-y}$ quarternary, as a function of y, is given by a quadratic form:

$$K^2_{[110]} = (0.32 + 0.30y + 1.00y^2) \times 10^{-3} \tag{5.10}$$

5.3 FRÖHLICH COUPLING CONSTANT

The coupling between the electrons and longitudinal optical (LO) phonons cannot be neglected in the studies of transport and optical properties in polar semiconductors. A measure of the interaction between the electron and LO phonons can be represented by the well-known Fröhlich coupling constant:[40]

$$\alpha_F = \frac{1}{2}\frac{e^2/(\hbar/2m_c^\alpha\omega_{LO})^{1/2}}{\hbar\omega_{LO}}\left(\frac{1}{\epsilon_\infty} - \frac{1}{\epsilon_s}\right) \tag{5.11}$$

where e is the electronic charge and m_c^α is the effective electron mass (i.e., the conductivity effective mass) as will be discussed in Section 6.2. This constant depends strongly on the ionic polarization of the crystal, which is related to the dielectric constants ϵ_s and ϵ_∞. The high-frequency dielectric constant ϵ_∞ is determined by measuring the wavelength dependence of the refractive index. The Debye temperature of the LO phonon frequency is an important parameter in the study of the polaron mobility at finite temperatures. This temperature is defined by

$$\theta_{po} = \frac{\hbar\omega_{LO}}{k_B} \tag{5.12}$$

where k_B is the Boltzmann constant. The last parameter we need in order to consider the coupling constant α_F is the electron effective mass m_c^α. This mass corresponds to the conductivity effective band mass. [Note that the conductivity effective mass in the Γ minimum is equal to the density-of-states effective mass (see Section 6.2).]

Because of the lack of experimental data, we have to restrict ourselves in obtaining α_F of $In_{1-x}Ga_xAs_yP_{1-x}$ quaternary by the interpolation scheme. Figure 5.4 presents the results of this calculation. The end-binary data used are listed in Table 5.4. As seen in the figure, the composition dependence of α_F shows a relatively remarkable nonlinearity. To obtain accurate values of α_F in the alloy system, we must take into account the effects of multiple LO phonons present (see Section 3.3.3). The case of ternary alloys where two LO phonon modes exist has been considered by Nicholas et al.,[41] who defined a value of α_F for each LO phonon. Nash et al.[42] have also discussed the Fröhlich electron–phonon interaction and the Lyddane–Sachs–Teller splittings [see Eq. (5.3)] of optical phonon modes in alloys with two-mode behavior. They found that the electron–phonon coupling in $In_{0.53}Ga_{0.47}As$ is much weaker for the InAs-like modes than for the GaAs-like modes. The primary reason for this effect was considered to

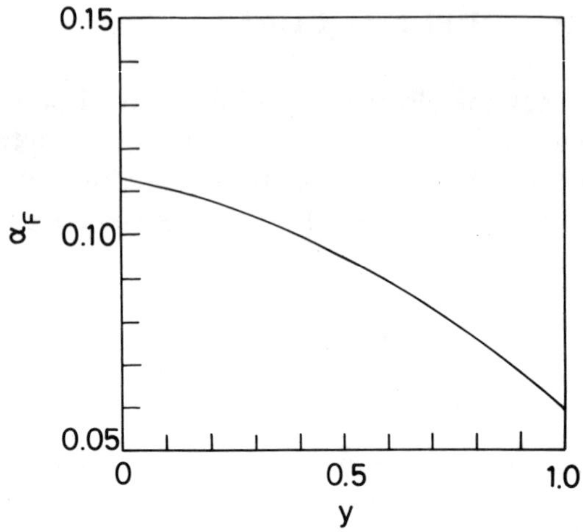

Figure 5.4 Fröhlich coupling constant α_F as a function of y for $In_{1-x}Ga_xAs_yP_{1-y}$ lattice-matched to InP.

TABLE 5.4 Fröhlich Coupling Constant α_F of InP, InAs, GaAs, GaP, $In_{1-x}Ga_xAs_yP_{1-y}$, and $In_{0.53}Ga_{0.47}As$

Material	α_F
InP[a]	0.113
InAs[a]	0.052
GaAs[a]	0.068
GaP[a]	0.201
$In_{1-x}Ga_xAs_yP_{1-y}$	$0.113 - 0.022y - 0.032y^2$
$In_{0.53}Ga_{0.47}As$	0.059

[a]J. T. Devreese, *Polarons in Ionic Crystals and Polar Semiconductors*, North-Holland, Amsterdam, 1972.

be coupling of the LO distortions of the alloy by the macroscopic electric field.

REFERENCES

1. See, for instance, S. M. Sze, *Physics of Semiconductor Devices*, Wiley, New York, 1981.
2. J. F. Nye, *Physical Properties of Crystals*, Clarendon, Oxford, 1972.
3. R. E. , Neidert, S. C. Binari, and T. Weng, *Electron. Lett.* **18,** 987 (1982).
4. L. G. Meiners, *J. Appl. Phys.* **59,** 1611 (1986).

5. K. Seeger, *Appl. Phys. Lett.* **54,** 1268 (1989).
6. J. Q. Howell, *IEEE Tran. Microwave Theory Tech.* **MTT-21,** 142 (1973).
7. A. N. Pikhtin and A. D. Yas'kov, *Sov. Phys. Semicond.* **12,** 622 (1978).
8. R. Newman, *Phys. Rev.* **111,** 1518 (1958).
9. S. Adachi, *J. Appl. Phys.* **53,** 8775 (1982).
10. E. Anastassakis, Y. S. Raptis, M. Hünermann, W. Richter, and M. Cardona, *Phys. Rev. B* **38,** 7702 (1988).
11. Y. T. Cherng, D. H. Jaw, M. J. Jou, and G. B. Stringfellow, *J. Appl. Phys.* **65,** 3285 (1989).
12. K. S. Champlin and G. H. Glover, *Appl. Phys. Lett.* **12,** 231 (1968).
13. I. Strzalkowski, S. Joshi, and C. R. Crowell, *Appl. Phys. Lett.* **28,** 350 (1976).
14. G. A. Samara, *Phys. Rev. B* **27,** 3494 (1983).
15. A. S. Barker, Jr., *Phys. Rev.* **165,** 917 (1968).
16. L. Patrick and P. J. Dean, *Phys. Rev.* **188,** 1254 (1969).
17. A. R. Goñi, K. Syassen, K. Strössner, and M. Cardona, *Semicond. Sci. Technol.* **4,** 246 (1989).
18. J. W. Harrison and J. R. Hauser, *J. Appl. Phys.* **47,** 292 (1976).
19. B. B. Pal, *Solid-State Electron.* **28,** 1235 (1985).
20. G. Arlt and P. Quadflieg, *Phys. Status Solidi* **25,** 323 (1968).
21. W. F. Boyle and R. J. Sladek, *Solid State Commun.* **16,** 323 (1975).
22. D. N. Nichols and R. J. Sladek, *Phys. Rev. B* **24,** 6161 (1981).
23. J. C. Phillips and J. A. Van Vechten, *Phys. Rev. Lett.* **23,** 1115 (1969).
24. R. M. Martin, *Phys. Rev. B* **5,** 1607 (1972).
25. K. Hübner, *Phys. Status Solidi B* **57,** 627 (1973).
26. M. Miura, H. Murata, Y. Shiro, and K. Iishi, *J. Phys. Chem. Solids* **42,** 931 (1981).
27. R. M. Martin and K. Kunc, *Phys. Rev. B* **24,** 2081 (1981).
28. J. B. McKitterick, *Phys. Rev. B* **28,** 7384 (1983).
29. M. R. Press and D. E. Ellis, *Phys. Rev. B* **38,** 3102 (1988).
30. S. D. Gironcoli, S. Baroni, and R. Resta, *Phys. Rev. Lett.* **62,** 2853 (1989).
31. D. L. Rode, *Phys. Rev. B* **2,** 1012 (1970).
32. J. C. Phillips, *Bonds and Bands in Semiconductors*, Academic, New York, 1973.
33. D. F. Nelson and E. H. Turner, *J. Appl. Phys.* **39,** 3337 (1968).
34. N. I. Meyer and M. H. Jørgensen, in *Advances in Solid State Physics*, O. Madelung, ed., Pergamon, Vieweg, 1970, p. 21.
35. D. L. Spears, *Phys. Rev. B* **2,** 1931 (1970).
36. P. O. Sliva and R. Bray, *Phys. Rev. Lett.* **14,** 372 (1965).
37. R. Klein, *Solid State Commun.* **7,** 917 (1969).
38. V. M. Bright and W. D. Hunt, *J. Appl. Phys.* **68,** 1985 (1990).

39. H. Jaffe and D. A. Berlincourt, *Proc. IEEE* **53,** 1372 (1965).
40. J. T. Devreese, *Polarons in Ionic Crystals and Polar Semiconductors*, North-Holland, Amsterdam, 1972.
41. R. J. Nicholas, R. A. Stradling, J. C. Portal, and S. Askenazy, *J. Phys. C* **12,** 1653 (1979).
42. K. J. Nash, M. S. Skolnick, and S. J. Bass, *Semicond. Sci. Technol.* **2,** 329 (1987).

6

ELECTRONIC ENERGY-BAND STRUCTURE

6.1 FUNDAMENTAL ABSORPTION EDGE AND OPTICAL TRANSITION ENERGIES

The $In_{1-x}Ga_xAs_yP_{1-y}$/InP heterojunction system offers, for device applications, the unique feature that the energy band gap can be varied by varying its alloy composition. Experimental data on the variation of the fundamental absorption edge, higher-energy direct gaps, conduction- and valence-band effective masses, and band-gap discontinuities are now available for this alloy system. Such band-structure related material parameters are basic to the description of radiative and non-radiative (Auger) recombinations, carrier mobility, impact ionization, carrier confinement efficiency, and other parameters.

The electronic energy-band structure of $In_{1-x}Ga_xAs_yP_{1-y}$ alloy has been studied experimentally in detail by a number of groups.[1-16] A wide variety of band-structure calculations have given fundamental understanding of the subject.[17-22] Some interpolation schemes based on a general principle of simplicity have also been proposed to predict the band-gap variations of this alloy system.[23-32]

In Fig. 6.1 we produce the energy-band structure of InP as calculated by an empirical nonlocal pseudopotential method by Chelikowsky and Cohen.[33] The fundamental absorption edge of InP corresponds to direct transitions from the highest valence band to the lowest conduction at the Γ point in the Brillouin zone (BZ) [i.e., $\Gamma_{15}^{v} \rightarrow \Gamma_{1}^{c}$; single-group notation]. The spin–orbit interaction splits the Γ_{15}^{v} valence band into Γ_{8}^{v} and Γ_{7}^{v} (dou-

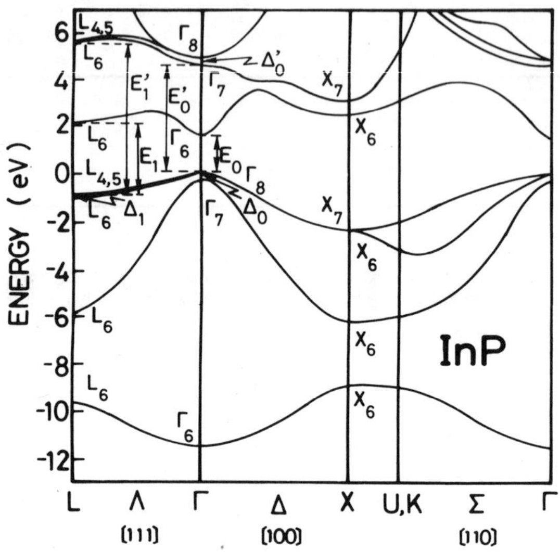

Figure 6.1 Electronic energy-band structure of InP along several lines of high-symmetry direction. (From Chelikowsky and Cohen.[33])

ble-group notation; splitting energy Δ_0), and the Γ_{15}^c conduction band into Γ_7^c and Γ_8^c (splitting energy Δ_0'). The corresponding optical transitions at or near $\boldsymbol{k} = 0$ (Γ) are, respectively, labeled E_0 [$\Gamma_8^v(\Gamma_{15}^v) \rightarrow \Gamma_6^c(\Gamma_1^c)$], $E_0 + \Delta_0[\Gamma_7^v(\Gamma_{15}^v) \rightarrow \Gamma_6^c(\Gamma_1^c)]$, $E_0'[\Gamma_8^v(\Gamma_{15}^v) \rightarrow \Gamma_7^c(\Gamma_{15}^c)]$, $E_0' + \Delta_0[\Gamma_7^v(\Gamma_{15}^v) \rightarrow \Gamma_7^c(\Gamma_{15}^c)$; dipole-forbidden], $E_0' + \Delta_0'[\Gamma_8^v(\Gamma_{15}^v) \rightarrow \Gamma_8^c(\Gamma_{15}^c)]$, and $E_0' + \Delta_0' + \Delta_0[\Gamma_7^v(\Gamma_{15}^v) \rightarrow \Gamma_8^c(\Gamma_{15}^c)]$. A further E_0' transition, found in the calculation,[33] is located along [100] (Δ) about 20% of the way to $X[E_0'(\Delta)$; $\Delta_5^v \rightarrow \Delta_5^c]$. The spin–orbit interaction also splits the $L_3^v(\Lambda_3^v)$ valence band into $L_{4,5}^v(\Lambda_{4,5}^v)$ and $L_6^v(\Lambda_6^v)$. The corresponding transitions are, respectively, labeled E_1 [$L_{4,5}^v(L_3^v) \rightarrow L_6^c(L_1^c)$ or $\Lambda_{4,5}^v(\Lambda_3^v) \rightarrow \Lambda_6^c(\Lambda_1^c)$] and $E_1 + \Delta_1$ [$L_6^v(L_3^v) \rightarrow L_6^c(L_1^c)$ or $\Lambda_6^v(\Lambda_3) \rightarrow \Lambda_6^c(\Lambda_1^c)$]. The E_2 transitions are expected to take place along the [110] (Σ) or near X, and occur in InP for energies close to the E_0' and $E_0' + \Delta_0'$ critical points (CPs). The E_1' transitions take place near the L point [$L_{4,5}^v(L_3^v) \rightarrow L_6^c(L_3^c)$]. The lowest indirect absorption edges correspond to transitions from the highest valence band at the Γ point to the lowest conduction band at the $L(X)$ point [i.e., $E_g^L(\Gamma_8^v \rightarrow L_6^c)$ and $E_g^X(\Gamma_8^v \rightarrow X_6^c)$].

We list in Table 6.1 the energies of CP's in InP measured at a few temperatures by several authors. We also list in Table 6.2 the energy eigenvalues at the Γ, X, and L points for the valence bands and first few conduction bands for InP as calculated by Chelikowsky and Cohen.[33] Figure 6.2 shows the calculated electronic density of states (DOS) for InP.[33] The dashed line is the experimentally obtained DOS for the valence bands

TABLE 6.1 Energies of CPs in InP at a Few Temperatures (in eV)

Temperature (K)	E_0	$E_0 + \Delta_0$	E_1	$E_1 + \Delta_1$	E_0' Range	E_2 Range
2	1.4241[a]					
	1.4182[b]					
	1.4183[c]					
	1.4185[c]					
	1.4185[d]					
	1.4243[e]					
4.2	1.423[f]	1.531[f]				
5	1.4182[g]	1.5263[g]	3.24[i]	3.38[i]	4.78[i]	5.10[i]
						5.77[i]
6	1.4205[h]					

[a] J. U. Fischbach et al., *Solid State Commun.* **11,** 721 (1972).
[b] A. M. White et al., *J. Phys. C* **5,** 1727 (1972).
[c] F. Evangelisti et al., *Phys. Rev. B* **9,** 1516 (1974).
[d] W. Rühle and W. Klingenstein, *Phys. Rev. B* **18,** 7011 (1978).
[e] M. A. Abdullaev et al., *Sov. Phys. Semicond.* **23,** 724 (1989).
[f] P. Rochon and E. Fortin, *Phys. Rev. B* **12,** 5803 (1975).
[g] J. Camassel et al., *Phys. Rev. B* **22,** 2020 (1980).
[h] W. J. Turner et al., *Phys. Rev. B* **136,** A1467 (1964).
[i] C. V. de Alvarez et al., *Phys. Rev. B* **6,** 1412 (1972).

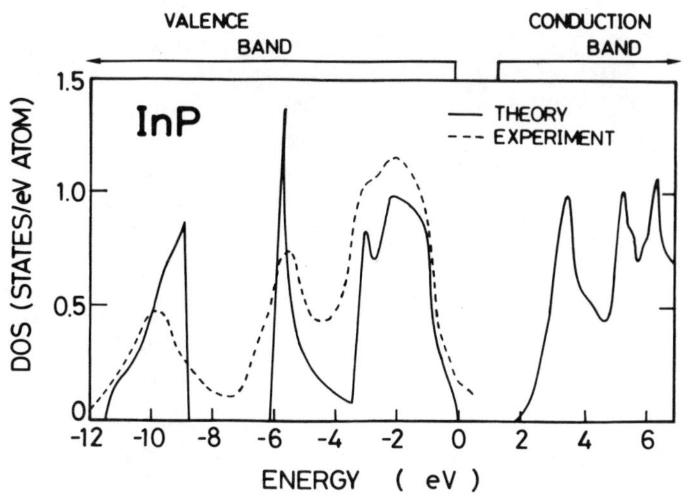

Figure 6.2 Calculated electronic density of states (DOS) for InP. The dashed line is the experimental result for the valence bands obtained by X-ray photoemission spectroscopy.[34] (From Chelikowsky and Cohen.[33])

by Ley et al. (X-ray photoemission data).[34] The DOS curve in the valence band shows a maximum near 2 eV associated with the $L_{4,5}$, L_6, X_6, and X_7 valence bands (see Table 6.2). The strong peak near 6.0 eV is due to the X_6 and L_6 valence bands. Chelikowsky and Cohen[33] pointed out that the overall agreement between the nonlocal result (solid curve) and experiment (dashed curve) are a considerable improvement over the local pseudopotential result.

A lot of work has been done to study the conduction-band structure of InP,[35–37] but the location of the subsidiary minima is controversial even today. Zollner et al.[38] have recently performed an *ab initio* linear muffin-tin-orbital band-structure calculation to obtain the valence bands with a high degree of accuracy. They then combined the calculated valence-band energies with published optical data and confirmed the interconduction-band separation energies $\Delta E_{\Gamma L} = 0.86 \pm 0.02$ eV ($\Gamma^c - L^c$) and $\Delta E_{\Gamma X}$

TABLE 6.2 Energy Eigenvalues at the Γ, *X*, and *L* Points for the Valence and First Few Conduction Bands of InP at 0 K (in eV)

Point	Level	Eigenvalue[a]
Γ	Γ_6^v	−11.42 (−11.48)
	Γ_7^v	−0.21
	Γ_8^v	0.00 (0.00)
	Γ_6^c	1.50 (1.51)
	Γ_7^c	4.64 (3.39)
	Γ_8^c	4.92
X	X_6^v	−8.91 (−9.30)
	X_6^v	−6.01 (−5.10)
	X_6^v	−2.09
	X_7^v	−2.06 (−2.17)
	X_6^c	2.44 (2.70)
	X_7^c	2.97 (4.58)
L	L_6^v	−9.67 (−10.15)
	L_6^v	−5.84 (−4.50)
	L_6^v	−1.09
	$L_{4,5}^v$	−0.94 (−1.17)
	L_6^c	2.19 (1.57)
	L_7^c	5.58 (6.45)
	$L_{4,5}^c$	5.70

[a]J. R. Chelikowsky and M. L. Cohen, *Phys. Rev. B* **14,** 556 (1976) (an empirical nonlocal pseudopotential calculation). Results of S. N. Sahu et al. [*Phys. Status Solidi B* **122,** 661 (1984)] are also shown in parentheses.

$= 0.96 \pm 0.02$ eV ($\Gamma^c - X^c$). Knowledge of such values is required for calculations of the velocity–electric field curve for electrons and for analyses of high-field transistor operations (see Chapter 10).

The only detailed theoretical studies of the electronic properties of semiconducting alloys have been coherent potential approximation (CPA) calculations for $In_{1-x}Ga_xAs_yP_{1-y}$ lattice-matched to InP. These have been carried out for the valence bands by Chen and Sher,[17] and for both the valence and conduction bands by Gera et al.[21] and Ekpenuma et al.[22]

Cheng et al.[39] studied experimentally the valence band of $In_{0.53}Ga_{0.47}As$ by ultraviolet photoemission using photon energies of 11.7, 16.8, and 21.2 eV. The large density of states associated with the valence band at L_3 was clearly visible in their spectra about 1.4 eV below the valence band edge. They concluded that the valence band of this material has a well-defined structure similar to that of III–V binary compounds.

A great deal of attention has been paid to the electronic energy-band-gap energies, especially the E_0 gap energy, of $In_{1-x}Ga_xAs_yP_{1-y}$ lattice-matched to InP by reasons of optoelectronic device applications. In the following subsections, we summarize the composition dependence of each CP (critical point) energy and indirect band-gap energies of this quaternary system.

6.1.1 E_0 and $E_0 + \Delta_0$ Gaps

The E_0 and $E_0 + \Delta_0$ transitions take part in the center of the BZ and are of three-dimensional M_0 CPs. Much effort has been made to evaluate the bowing parameter of these gaps. The expression of Eq. (2.2) suggests that most of the material parameter values Q for the $In_{1-x}Ga_xAs_yP_{1-y}$ quaternary can be written in good accuracy by the quadratic function with respect to the alloy composition y as

$$Q(x, y) = Q(y) = a + by + cy^2 \tag{6.1}$$

where a and b are determined by the end-point materials of the alloy series, and c, the so-called *bowing parameter*, measures the deviation from linearity. In general,[40] the bowing coefficient c composes of a periodic component c_i (intrinsic term), which is related to the change in bond length, and an aperiodic component c_e (extrinsic term) which is the root-mean-square (RMS) fluctuation in the potential from its periodic amplitude. Since the lattice constant in the $In_{1-x}Ga_xAs_yP_{1-y}$ quaternary system can be kept fixed as the composition is varied, the periodic term c_i is zero for the alloys lattice-matched to InP.[29,30]

Figure 6.3 compares the values of E_0 for $In_{1-x}Ga_xAs_yP_{1-y}$ lattice-

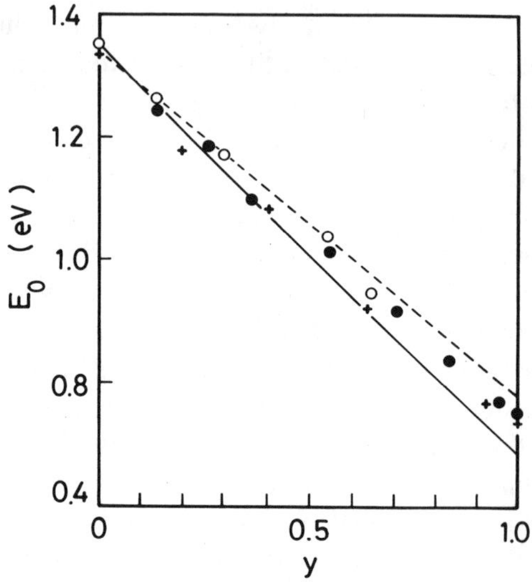

Figure 6.3 Values of the E_0 gap energy for $In_{1-x}Ga_xAs_yP_{1-y}$ lattice-matched to InP, obtained from CPA (coherent potential approximation) calculations (solid line[21] and dashed line[22]) along with the electroreflectance data taken from Refs. 12 (crosses) and 15 (open circles) and the photoluminescence data taken from Ref. 5 (solid circles).

matched to InP, obtained from CPA calculations (Ref. 21, solid line; Ref. 22, dashed line), with the electroreflectance data of Laufer et al.[12] and Lahtinen and Tuomi[15] and the photoluminescence data of Nakajima et al.[5] There is good agreement between the CPA calculations and experiments. In Table 6.3 we also list the expressions for the E_0 gap energy of $In_{1-x}Ga_xAs_yP_{1-y}$ lattice-matched to InP. They are obtained from the photoluminescence measurements,[1,5] electroreflectance measurements,[10–13,15] and calculated by using some interpolation schemes based on a general principle of simplicity.[28–30] Only the data of Nakajima et al.[5] showed a linear dependence on y. Pearsall[29,30] plotted the measured E_0 data versus y obtained from several sources[1,4,5,8,10] and found that there is very little scatter in these data. The measured data were also found to be in excellent agreement with his quadratic relation. However, he reported different values of b in Refs. 29 (-0.778 eV) and 30 (-0.775 eV). This may merely be a typographic error.

Moon et al.[24] have calculated E_0 versus composition by extrapolating ternary data into the quaternary region. The expression they obtained is given by (in electronvolts)

$$E_0(x, y) = 1.35 + (0.758x + 0.642)x + (0.101y - 1.101)y - (0.28x - 0.109y + 0.159)xy \qquad (6.2)$$

TABLE 6.3 Room-Temperature Values of the E_0, $E_0 + \Delta_0$, and Δ_0 Gap Energies for $In_{1-x}Ga_xAs_yP_{1-y}$ Lattice-Matched to InP (in eV)

E_0	$E_0 + \Delta_0$	Δ_0
$1.35 - 0.72y + 0.12y^2$[a]	$1.466 - 0.557y + 0.129y^2$[c]	$0.119 + 0.300y - 0.107y^2$[c]
$1.35 - 1.3y$[b]	$1.46 - 0.321y$[d]	$0.11 + 0.417y - 0.138y^2$[d]
$1.35 - 0.883y + 0.250y^2$[c]	$1.449 - 0.47y + 0.11y^2$[e]	$0.114 + 0.26y - 0.02y^2$[e]
$1.35 - 0.738y + 0.138y^2$[d]	$1.472 - 0.468y + 0.092y^2$[f]	$0.123 + 0.173y + 0.054y^2$[f]
$1.337 - 0.73y + 0.13y^2$[e]		
$1.349 - 0.641y + 0.038y^2$[f]		
$1.343 - 0.709y + 0.11y^2$[g]		
$1.35 - 0.778y + 0.149y^2$[h]		
$1.35 - 0.775y + 0.149y^2$[i]		

[a]R. E. Nahory et al., *Appl. Phys. Lett.* **33,** 659 (1978).
[b]K. Nakajima et al., *J. Appl. Phys.* **49,** 5944 (1978).
[c]E. H. Perea et al., *Appl. Phys. Lett.* **36,** 978 (1980).
[d]Y. Yamazoe et al., *Jpn. J. Appl. Phys.* **19,** 1473 (1980); *IEEE J. Quantum Electron.* **QE-17,** 139 (1981).
[e]P. M. Laufer et al., *Solid State Commun.* **36,** 419 (1980).
[f]J. A. Lahtinen and T. Tuomi, *Phys. Status Solidi B* **130,** 637 (1985).
[g]C. Papuzza et al., in *Gallium Arsenide and Related Compounds*, Institute of Physics, Bristol, UK, 1982, p. 275.
[h]T. P. Pearsall and C. Hermann, in *Gallium Arsenide and Related Compounds*, Institute of Physics, Bristol, UK, 1982, p. 269.
[i]T. P. Pearsall, in *GaInAsP Alloy Semiconductors*, T. P. Pearsall, ed., Wiley, Chichester, UK, 1982, p. 295.

Hsieh[7] has also reported emission wavelength ($E_{g\lambda}$) versus composition in the $In_{1-x}Ga_xAs_yP_{1-y}$/InP double-heterostructure lasers at 300 K and found $E_{g\lambda}$ to be (in electronvolts)

$$E_{g\lambda}(y) = 1.307 - 0.60y + 0.03y^2 \tag{6.3}$$

It is well known that modulation spectroscopy,[41] such as the electroreflectance and thermoreflectance, is very powerful for precise determination of band-structure parameters of semiconductors. Table 6.3 lists the quadratic expressions for the $E_0 + \Delta_0$ and Δ_0 gaps of $In_{1-x}Ga_xAs_yP_{1-y}$ lattice-matched to InP determined by electroreflectance measurements.

All data excepting Yamazoe et al.[13] indicated a nonlinear dependence of the $E_0 + \Delta_0$ gap on y. In ternary alloys having the zinc-blende structure it was found that the bowing for the spin–orbit splitting energy Δ_0 is downward, that is, with a positive value of c.[42,43] However, the studies quoted in Table 6.3 show negative values of c except that of Lahtinen and Tuomi ($c = +0.054$ eV).[15] Parayanthal and Pollak[31] have discussed the composition dependence of the spin–orbit splitting energies Δ_0 and Δ_1 in the $In_{1-x}Ga_xAs_yP_{1-y}$ quaternary based on a generalized Van Vechten–Berolo–Woolley model.[42,43] They showed that the model successfully accounts for the upward bowing of Δ_0 (negative c) for this alloy system.

6.1.2 E_1 and $E_1 + \Delta_1$ Gaps

The E_1 and $E_1 + \Delta_1$ gaps in the $In_{1-x}Ga_xAs_yP_{1-y}$ system take place along the Λ direction or at the L point in the BZ. These gaps are of the three-dimensional M_1 type and give pronounced structures in optical spectra (see Section 8.3). Figure 6.4 compares the values of E_1 for $In_{1-x}Ga_xAs_yP_{1-y}$ lattice-matched to InP, obtained from a CPA calculation[21] (solid line), with the electroreflectance data of Perea et al.,[10] Laufer et al.,[12] and Lahtinen and Tuomi.[15] It can be seen that the results obtained by Perea et al.[10] are consistently low. In Table 6.4 we summarize the quadratic expressions for the E_1, $E_1 + \Delta_1$, and Δ_1 gap energies of $In_{1-x}Ga_xAs_yP_{1-y}$ lattice-matched to InP. These results were obtained from electroreflectance[10, 12, 15] and spectroscopic–ellipsometry measurements.[14] All these results give a negative or upward bowing for the Δ_1 gap, as in the case for Δ_0.

6.1.3 E_0' Triplet Region

The more pronounced structure found in optical spectra of the quaternary in the region higher in energy than E_1 and $E_1 + \Delta_1$ is labeled E_0'.[14,44] The E_0' transitions in zinc-blende-type semiconductors are believed to take place

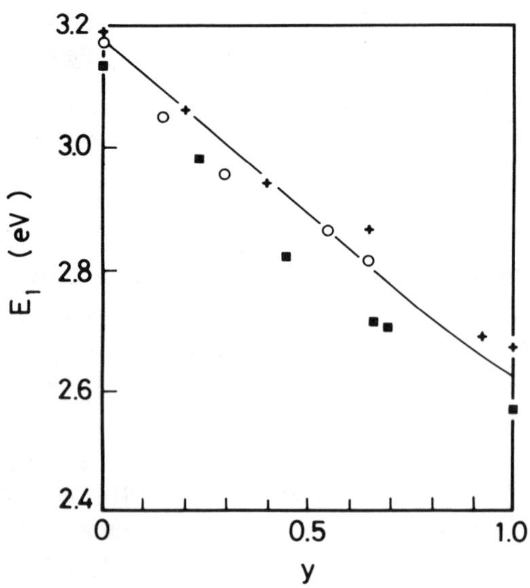

Figure 6.4 Values of the E_1 gap energy of $In_{1-x}Ga_xAs_yP_{1-y}$ lattice-matched to InP, obtained from a CPA calculation[21] (solid line) along with the electroreflectance data taken from Refs. 10 (solid squares), 12 (crosses), and 15 (open circles).

TABLE 6.4 Room-Temperature Values of the E_1, $E_1 + \Delta_1$, and Δ_1 Gap Energies for $In_{1-x}Ga_xAs_yP_{1-y}$ Lattice-Matched to InP (in eV)

E_1	$E_1 + \Delta_1$	Δ_1
$3.136 - 0.788y + 0.222y^2$ [a]	$3.281 - 0.615y + 0.158y^2$ [a]	$0.145 + 0.173y - 0.064y^2$ [a]
$3.122 - 0.80y + 0.21y^2$ [b]	$3.256 - 0.47y + 0.13y^2$ [b]	$0.135 + 0.33y - 0.20y^2$ [b]
$3.163 - 0.590y + 0.33y^2$ [c]	$3.296 - 0.466y + 0.26y^2$ [c]	$0.133 + 0.124y - 0.07y^2$ [c]
$3.181 - 0.787y + 0.184y^2$ [d]	$3.326 - 0.604y + 0.116y^2$ [d]	$0.146 + 0.183y - 0.067y^2$ [d]

[a] E. H. Perea et al., *Appl. Phys. Lett.* **36,** 978 (1980).
[b] P. M. Laufer et al., *Solid State Commun.* **36,** 419 (1980).
[c] S. M. Kelso et al., *Phys. Rev. B* **26,** 6669 (1982).
[d] J. A. Lahtinen and T. Tuomi, *Phys. Status Solidi B* **130,** 637 (1985).

at the Γ point or in the Δ direction near the Γ point in the BZ. As mentioned before, the spin–orbit interaction splits the *p*-like conduction band into two bands (splitting energy Δ_0') and the *p*-like valence band into two bands (Δ_0). The corresponding transitions are, respectively, labeled E_0', $E_0' + \Delta_0'$, and $E_0' + \Delta_0' + \Delta_0$. These are sometimes called the E_0' *triplet*. Although $E_0' + \Delta_0$ transitions are also possible, they are dipole-forbidden.

The band-structure calculation of Chelikowski and Cohen[33] for InP suggests various CPs in the E_0' triplet region. Three of them correspond to transitions at the Γ point (i.e., E_0' triplet). The $E_0' + \Delta_0' + \Delta_0$ transitions are thought to be too weak to be dominant in optical spectra. A further transition, found in the calculation in Ref. 33, is located along [100] about 20% of the way to *X* [i.e., $E_0'(\Delta)$].

The E_2 transitions are also expected to take place along the [110] direction or near *X*, and occur in InP for energies close to the E_0' and $E_0' + \Delta_0'$ CPs. However, the strength of the E_2 transitions in InP seems to be much weaker than those of the E_0' and $E_0' + \Delta_0'$ transitions. In an earlier work,[45] the $E_0' + \Delta_0'$ transitions in InP were attributed to the E_2 CP. However, from theoretical considerations about the spin–orbit splitting of the conduction band (Δ_0'),[14] the structure has more recently been assigned to the $E_0' + \Delta_0'$ transitions.

Since the nature of the E_0' triplet region is more complicated, it does not correspond to a single, well-defined CP. A few authors reported CP energies in the E_0' triplet region of $In_{1-x}Ga_xAs_yP_{1-y}$ lattice-matched to InP measured by the electroreflectance,[8,9,12] and spectroscopic-ellipsometry measurements.[14] But such data were not presented in graphical or tabular form. Only Kelso et al.[14] have reported the E_0' peak energy as a function of the alloy composition *y*, which can be written as (in electronvolts):

$$E_0'(y) = 4.72(\pm 0.01) - 0.31(\pm 0.02)y - 0.01(\pm 0.05)y^2 \quad (6.4)$$

6.1.4 Lowest-Indirect Band Gaps

Even though the basic $In_{1-x}Ga_xAs_yP_{1-y}$/InP heterostructure concepts are well understood at this time, practical device parameters in this system have been hampered by a lack of definite knowledge of many material parameters. To our knowledge, no detailed information is available about the indirect-band-gap energy E_g^{ID} in this alloy system. This necessitates the use of some sort of an interpolation scheme.[46] The indirect transitions play an important part in the spectral dependence of optical constants, such as ϵ_1, ϵ_2, and α (absorption coefficient), as discussed in Section 8.3.

The quaternary alloy, $In_{1-x}Ga_xAs_yP_{1-y}$, is thought to be constructed of four ternary compounds: $In_{1-x}Ga_xP$, $In_{1-x}Ga_xAs$, $GaAs_yP_{1-y}$, and $InAs_yP_{1-y}$. The material parameters in many ternary alloys can usually be approximated by a quadratic function [see Eq. (6.1)]. If relationships for the ternary (T) parameters are available, the quaternary parameter $Q(x, y)$ can be given by

$$Q(x, y) = \frac{1}{x(1-x) + y(1-y)} \{x(1-x)[yT_{ABC}(x) + (1-y)T_{ABD}(x)] + y(1-y)[xT_{ACD}(y) + (1-x)T_{BCD}(y)]\} \tag{6.5}$$

In Table 6.5 we list the indirect-band-gap energies E_g^X and E_g^L for the binaries of interest (InP, InAs, GaAs, and GaP) and the bowing parameters of E_g^X and E_g^L for some ternaries of interest ($In_{1-x}Ga_xP$, $In_{1-x}Ga_xAs$, $GaAs_yP_{1-y}$, and $InAs_yP_{1-y}$). The quaternary indirect-band-gap energies as a function of composition are obtained by numerically solving Eq. (6.5), using the relation of Eq. (2.1) and the numeric values listed in Table 6.5. In Fig. 6.5 we present the indirect-band-gap energies E_g^X and E_g^L as a function of y for $In_{1-x}Ga_xAs_yP_{1-y}$ lattice-matched to InP. The variations with composition of the E_0,[1] E_1,[14] and E_0' gaps[14] are also plotted in the figure. It is evident from the figure that the quaternary system has the direct band gap E_0 as the fundamental absorption edge in a full range of the y composition ($0 \leqq y \leqq 1.0$). It is also found that the E_g^X and E_g^L gaps lie in energies between the E_0 and E_1 gaps. The indirect-band-gap energy versus y plotted in Fig. 6.5 can now be expressed as (in electronvolts)

$$E_g^X(y) = 2.21 - 0.88y \tag{6.6a}$$

$$E_g^L(y) = 2.05 - 0.85y \tag{6.6b}$$

The E_g^X and E_g^L gap transitions thus occur after the onset of the E_0 transitions but before the predominance of the E_1 gap transitions. It is also worth not-

TABLE 6.5 Indirect-Band-Gap Energies, E_g^X and E_g^L, for the Binaries of Interest (InP, InAs, GaAs, and GaP) and the Bowing Parameters *c* for Some Ternaries of Interest ($In_{1-x}Ga_xP$, $In_{1-x}Ga_xAs$, $GaAs_yP_{1-y}$, and $InAs_yP_{1-y}$) (at 300 K; in eV)

Binary	E_g^X	E_g^L
InP	2.21	2.05
InAs	1.37	1.07
GaAs	1.91	1.73
GaP	2.26	2.63
	Bowing Parameter *c*	
Ternary	E_g^X	E_g^L
$In_{1-x}Ga_xP$	0.18[a]	0.43[a]
$In_{1-x}Ga_xAs$	1.40[b]	0.72[b]
$GaAs_yP_{1-y}$	0.21[c]	0.42[c]
$InAs_yP_{1-y}$	0.28[d]	0.27[d]

[a]A.-B. Chen and A. Sher, *Phys. Rev. B* **23,** 5360 (1981).
[b]W. Porod and D. K. Ferry, *Phys. Rev. B* **27,** 2587 (1983).
[c]H. J. Lee et al., *Phys. Rev. B* **21,** 659 (1980).
[d]Estimated [see S. Adachi, *J. Appl. Phys.* **61,** 4869 (1987)].

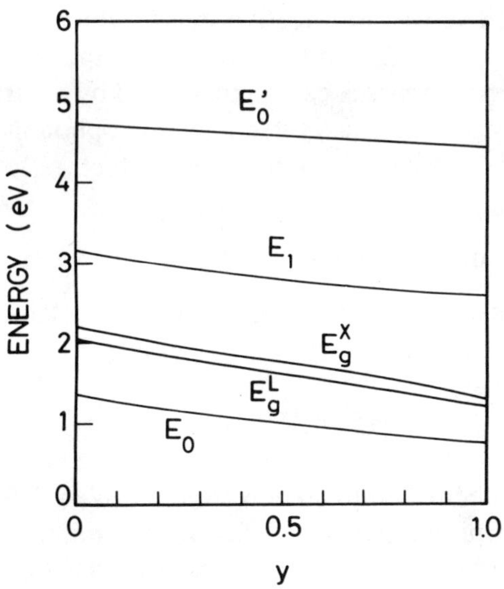

Figure 6.5 Variation with composition *y* of E_0 (Ref. 1), E_1 (Ref. 14), E_0' (Ref. 14), E_g^X, and E_g^L for $In_{1-x}Ga_xAs_yP_{1-y}$ lattice-matched to InP.

ing that the E_g^X gap lies above the E_g^L one. This leads to the assurance that the Γ_6–L_6–X_6 conduction-band order increases with energy holds for the entire range of compositions of this alloy system.

6.1.5 Energy Gaps in $In_{0.53}Ga_{0.47}As$

$In_{0.53}Ga_{0.47}As$ lattice-matched to InP ($y = 1.0$) is a semiconductor with important optoelectronic device applications. A wide variety of calculations[47–52] and experiments[39, 53–62] have given detailed information about the electronic energy-band structure of $In_{0.53}Ga_{0.47}As$ ternary. The material is a direct-band-gap semiconductor whose room-temperature band-gap energy is 0.75 eV.[58] Because of its low electron effective mass, $m_e^\Gamma/m_0 = 0.041$,[63] the material is also highly suited for high-speed transistor applications. The valence band of $Ga_{0.53}In_{0.47}As$ ternary has been studied by Cheng et al.[39] by means of ultraviolet photoemission. They used the observed spectra to determine the interconduction-band separation energy $\Delta E_{\Gamma L}$ of 0.55 eV at 300 K. This value for GaAs is 0.33 ± 0.04 eV.[64]

6.2 ELECTRON AND HOLE EFFECTIVE MASSES

To analyze a number of important semiconducting properties, a quite detailed knowledge of band-structure parameters at the principal band extrema is required. The carrier effective mass, which is strongly connected with the carrier mobility, is known to be one of the most important device parameters.[65] Effective masses can be measured by a variety of techniques, such as the Shubnikov–de Haas effect, magnetophonon resonance, cyclotron resonance, and interband magnetooptic effects.

6.2.1 Electron Effective Mass

The density-of-states (DOS) effective mass m_e^α for electrons in the conduction-band minima $\alpha (= \Gamma, X, \text{or } L)$ is given by

$$m_e^\alpha = N^{2/3} m_{t\alpha}^{2/3} m_{l\alpha}^{1/3} \tag{6.7}$$

where N is the number of equivalent α minima ($N = 1$ for the Γ minimum, $N = 3$ for the X minima, and $N = 4$ for the L minima) and $m_{t\alpha}$,$m_{l\alpha}$ are the transverse and longitudinal masses of the minima ($m_{t\alpha} = m_{l\alpha}$ for the Γ minimum), respectively. The DOS effective mass can be used for DOS calculations.

The conductivity effective mass m_c^α, which can be used for conductivity calculations, is also obtained from the equation

$$1/m_c^\alpha = \frac{1}{3}\left(\frac{2}{m_{t\alpha}} + \frac{1}{m_{l\alpha}}\right) \tag{6.8}$$

Since $m_{t\alpha} = m_{l\alpha}$ at the Γ minimum (Γ_6^c), we have the relation $m_e^\Gamma = m_c^\Gamma$.

The $\boldsymbol{k} \cdot \boldsymbol{p}$ method is a powerful procedure for evaluating the conduction-band effective-mass parameters in the vicinity of certain important points in $\boldsymbol{k}$ space.[43,50,66–69] The simple five-level $\boldsymbol{k} \cdot \boldsymbol{p}$ theory of Hermann and Weisbuch at $\boldsymbol{k} = 0$ can be shown to lead to the electron effective-mass ratio[50]:

$$\frac{m_0}{m_e^\Gamma} = 1 + \frac{P^2}{3}\left[\frac{2}{E_0} + \frac{1}{E_0 + \Delta_0}\right] - \frac{P'^2}{3}\left[\frac{2}{E(\Gamma_8^c) - E_0} + \frac{1}{E(\Gamma_7^c) - E_0}\right] + C \tag{6.9}$$

where m_0 is the free electron mass, P is the momentum matrix element connecting the p-like valence band with the s-like conduction band, P' is the momentum matrix element connecting the s-like conduction band with next-higher-lying p-like conduction bands, and C is a small correction for all higher-lying bands. For all III–V compounds, P'^2 is considerably smaller than P^2.[50] [The energy denominators in Eq. (6.9) for the P'^2 term are also much larger than those for P^2.] The effective mass is thus strongly connected with the lowest direct gaps E_0 and $E_0 + \Delta_0(P^2)$.

There are many investigations on the dependence on composition of the electron effective mass in the III–V ternary systems. Figure 6.6 shows, as an example, the nonlinear variation of the electron effective mass m_e^Γ with composition $x(y)$ for $In_{1-x}Ga_xAs$, $InAs_ySb_{1-y}$, and $In_{1-x}Ga_xSb$ alloy systems.[43] As discussed in Section 6.1, the energy gaps in these alloy systems show a nonlinearity. The electron effective mass also shows a nonlinearity with respect to the alloy composition. [This is especially the case for the $InAs_ySb_{1-y}$ alloy (see Fig. 6.6).] Berolo et al.[43] have analyzed the effects of alloy-disorder-induced valence-conduction band mixing on the effective mass. With a consistent set of assumptions for interband and intraband mixing and by using the standard $\boldsymbol{k} \cdot \boldsymbol{p}$ calculation in the virtual-crystal approximation, they found that it is possible to account for the nonlinear variation of the electron effective mass in such alloys.

The dependence on composition of the electron effective mass in the $In_{1-x}Ga_xAs_yP_{1-y}$/InP quaternary system has been studied experimentally

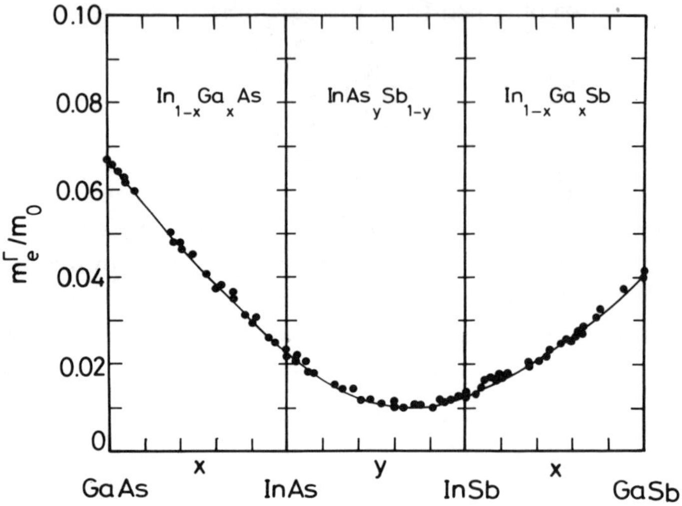

Figure 6.6 Variation of the electron effective mass m_e^Γ with composition $x(y)$ for $In_{1-x}Ga_xAs$, $InAs_ySb_{1-y}$, and $In_{1-x}Ga_xSb$ alloy systems. The solid line marks a possible explanation of the experimental findings.

by many workers.[70–77] The techniques used were the Shubnikov–de Haas oscillation,[70–72,74,75] magnetophonon resonance,[71,72,76] cyclotron resonance,[71–73,76] and shallow donor Zeeman spectroscopy.[77] These results are plotted in Fig. 6.7.

It can be seen that there is very large scatter in these data. Because the samples used for the Shubnikov–de Haas oscillations are heavily doped, the Fermi level lies above the band edge. For carrier levels above 10^{17} cm^{-3}, this effect becomes important, and the measured mass must be corrected for the band nonparabolicity of the conduction band. Such a nonparabolic-band correction can be given by[78]

$$\frac{1}{m_e^*} = \frac{1}{m_e^\Gamma}\left(1 - \frac{2\phi}{E_0}\right) \tag{6.10}$$

where m_e^* is the measured effective mass, $\phi = 0.8156(3\pi^2 n)^{2/3}(\hbar^2/2m_c^\Gamma)$ is in joules, and n is the free-electron concentration. The data plotted in Fig. 6.7 were corrected for this nonparabolicity. The cyclotron effective masses are determined from the resonance magnetic field B using $m_e^* = eB/\omega$, where ω is the measured electromagnetic (photon) frequency. Employing a wide range of photon frequencies, one can establish the electron effective mass at various energies above the band edge. In other words, in order to deduce the band edge mass from cyclotron

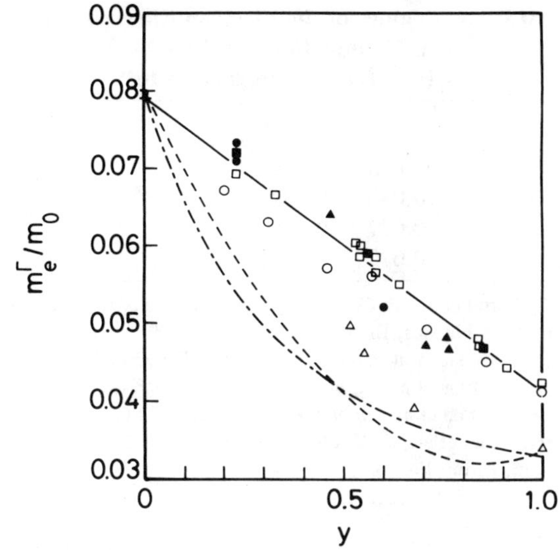

Figure 6.7 Electron effective mass m_e^Γ/m_0 as a function of y for $In_{1-x}Ga_xAs_yP_{1-y}$ lattice-matched to InP. The experimental data are taken from Refs. 72 (solid circles), 73 (open triangles), 74 (open circles), 75 (solid triangles), 76 (open squares), and 77 (solid squares). The dashed line is the calculated result of Eq. (6.13). The solid line is the result of the linear interpolation between the values of InP ($m_e^\Gamma/m_0 = 0.079$; see Ref. 82) and $In_{0.53}Ga_{0.47}As$ ($m_e^\Gamma/m_0 = 0.041$; see Ref. 63). The dashed–dotted line is also the result of Eq. (A.4) (in Appendix).

resonance measurements it is also necessary to correct for band nonparabolicity. Brendecke et al.[73] found an increase of $m_e^\Gamma \sim 15\%$ from $\hbar\omega = 5.96$ to 17.54 meV in the $In_{1-x}Ga_xAs_yP_{1-y}$ alloy of compositions $x = 0.312$ and $y = 0.674$.

It is also noted that the cyclotron resonance experiment gives the polaron mass rather than the band mass. The electron–LO-phonon (Fröhlich) coupling is known to modify the electron effective mass. In order to deduce the true "bare" mass from cyclotron resonance measurements it is thus necessary to correct for such a polaron enhancement. This correction can be written as[79]

$$m_e^\Gamma = \left(\frac{1 - 0.0008\alpha_F^2}{1 - (\alpha_F/6) + 0.0034\alpha_F^2}\right) m_e^* \qquad (6.11)$$

where α_F is the Fröhlich coupling constant (see Section 5.3) and m_e^* is the cyclotron mass. Typical values for this correction are 1–2% for $In_{1-x}Ga_xAs_yP_{1-y}$ lattice-matched to InP. These are less than the other experimental uncertainties.

TABLE 6.6 Values of the Electron Effective Mass m_e^Γ/m_0 at the Γ_6^c Conduction Band for $In_{1-x}Ga_xAs_yP_{1-y}$ Lattice-Matched to InP

m_e^Γ/m_0
$0.080 - 0.039y$[a,b,c]
$0.080 - 0.082y + 0.036y^2$[d]
$0.077 - 0.050y + 0.014y^2$[e]
$0.079 - 0.038y$[f]

[a]J. C. Portal et al., in *Gallium Arsenide and Related Compounds*, Institute of Physics, Bristol, UK, 1979, p. 829.
[b]R. J. Nicholas et al., *Appl. Phys. Lett.* **34,** 492 (1979).
[c]R. J. Nicholas et al., *Appl. Phys. Lett.* **37,** 178 (1980).
[d]H. Brendecke et al., *Appl. Phys. Lett.* **35,** 772 (1979).
[e]E. H. Perea et al., *J. Electron. Mater.* **9,** 459 (1980).
[f]Obtained from Fig. 6.7.

We summarize in Table 6.6 the expressions for m_e^Γ/m_0 as a function y for $In_{1-x}Ga_xAs_yP_{1-y}$ lattice-matched to InP obtained from various works. The linear dependence of m_e^Γ on alloy composition was suggested in Refs. 71, 72, 76, and 77, while Restorff et al.[70] reported a value 20% lower for one particular alloy composition $y = 0.22$. Brendecke et al.[73] and Perea et al.[74] also reported significant deviations of m_e^Γ from the linearly interpolated values. Alavi et al.[80] measured interband magnetooptic absorption for $y = 0.52$ which gives $m_e^\Gamma/m_0 = 0.061$ obeying the linear dependence well. It has also been shown[72,73] that the linear interpolation of P^2 and P'^2 between the end members of the quaternary (InP and $In_{0.53}Ga_{0.47}As$) leads to a slight bowing in the m_e^Γ versus composition curve.

The value of m_e^Γ/m_0 for InP is believed to be 0.080.[81] In a more recent report by Rochon and Fortin[82] interband-magnetooptic-transition results have been reported that give $m_e^\Gamma/m_0 = 0.079 \pm 0.001$. A value for the band-edge mass of $m_e^\Gamma/m_0 = 0.07927$ has also been reported by Hopkins et al.[83] who utilized the cyclotron resonance over a wide range of energies using the photoconductive detection technique. They also measured conduction-band nonparabolicity in InP and GaAs. The electron effective mass in $In_{1-x}Ga_xAs$ ternary alloys has also been studied.[43,50,63,84–86] A value of $m_e^\Gamma/m_0 = 0.041$ for $In_{0.53}Ga_{0.47}As$ lattice-matched to InP was reported by Pearsall et al.[63] from the Shubnikov–de Haas oscillation, magnetoresonance, and cyclotron resonance measurements.

Theoretical values for m_e^Γ are interpolated from the masses of related end binaries.[70] Starting with the effective masses of InP, InAs, GaAs, and GaP, we can obtain the effective mass for the ternary materials $In_{1-x}Ga_xAs$,

$In_{1-x}Ga_xP$, $GaAs_xP_{1-x}$, and $InAs_xP_{1-x}$ by the formula

$$\frac{1}{m_e^\Gamma(x)} = \frac{x}{m_1^\Gamma} + \frac{(1-x)}{m_2^\Gamma} \tag{6.12}$$

where m_1^Γ and m_2^Γ are the effective masses of the two end-point binaries. Justification for this formula is suggested by effective-mass transformations of conventional Schrödinger representations.[87] This interpolation formula leads to a mass variation with composition that is bowed downward with respect to a linear interpolation.

The ternary interpolations can be combined to obtain an interpolation for the quaternary. Restorff et al.[70] obtained $m_e^\Gamma(x, y)$ for the $In_{1-x}Ga_xAs_yP_{1-y}$ quaternary lattice-matched to InP by performing such a scheme:

$$\frac{m_e^\Gamma(x, y)}{m_0} = 0.08 - 0.116y + 0.026x - 0.059xy + (0.064 - 0.02y)x^2 + (0.06 + 0.032x)y^2 \tag{6.13}$$

The values of the binary masses used for this calculation are obtained from Lawaetz.[69] The dashed line in Fig. 6.7 represents the calculated result of Eq. (6.13). Discrepancy between the calculation and experiments is rather large.

The solid line is the result of the linear interpolation between the values of InP (0.079)[82] and $In_{0.53}Ga_{0.47}As$ (0.041):[63]

$$\frac{m_e^\Gamma(y)}{m_0} = 0.079 - 0.038y \tag{6.14}$$

We can recognize that this simple interpolation agrees with the experimentally determined m_e^Γ well.

6.2.2 Hole Effective Mass

The electron effective mass is rather accurately known from numerous experiments, but valence-band masses must essentially be calculated from band theory. This approach was made for InP by different authors,[66,69,88,89,90] but their results span a large range of effective mass ratios (see Table 6.7).

The effective masses of the light (m_{lh}) and heavy holes (m_{hh}) in p-type InP determined by Leotin et al.[91] using a cyclotron resonance were m_{lh}/m_0

TABLE 6.7 Γ-Point Light- and Heavy-Hole Masses, m_{lh} and m_{hh}, for $In_{1-x}Ga_xAs_yP_{1-y}$ Lattice-Matched to InP [Masses (m_{lh}, m_{hh}, m_{so}, m_c^h, m_d^h) Estimated in this Present Study Listed in Lower Part of Table]

	m_{lh}/m_0	
y	Experimental	Calculated
0.0	0.12 ± 0.01[a]	0.112[f]
	0.12 ± 0.01 (⟨111⟩)[b]	0.086[g]
	0.12 ± 0.01 (⟨100⟩)[b]	0.089[h]
		0.1149[c]
		0.13[i]
		0.10[i]
0.33	0.09 ± 0.01[c]	0.0872[c]
0.34	0.09 ± 0.01[c]	0.0872[c]
0.52	0.078[d]	
0.60	0.072 ± 0.01[c]	0.07072[c]
0.61	0.072 ± 0.01[c]	0.07072[c]
0.84	0.062 ± 0.007[c]	0.0575[c]
1.0	0.051 ± 0.003[c]	0.0505[c]
	0.050[e]	

	m_{hh}/m_0	
y	Experimental	Calculated
0.0	0.45 ± 0.05[a]	0.50[g]
	0.60 ± 0.02 (⟨111⟩)[b]	0.85[h]
	0.56 ± 0.02 (⟨100⟩)[b]	0.66[i]
		0.52[i]
0.52	0.45[d]	
1.0	0.47[e]	

Mass	InP	$In_{1-x}Ga_xAs_yP_{1-y}$	$In_{0.53}Ga_{0.47}As$
m_{lh}/m_0	0.120	$0.120 - 0.099y + 0.030y^2$	0.051
m_{hh}/m_0	0.46	0.46	0.46
m_{so}/m_0	0.21	$0.21 - 0.01y - 0.05y^2$	0.15
$(m_c^h/m_0)^j$	0.50	$0.50 - 0.03y$	0.47
$(m_d^h/m_0)^k$	0.42	$0.42 + 0.03y$	0.45

[a]P. Rochon and E. Fortin, *Phys. Rev. B* **12,** 5803 (1975).
[b]J. Leotin et al., *Solid State Commun.* **15,** 693 (1974).
[c]C. Hermann and T. P. Pearsall, *Appl. Phys. Lett.* **38,** 450 (1981).
[d]K. Alavi et al., *J. Magnet. Mag. Mater.* **11,** 136 (1979).
[e]K. Alavi et al., *Phys. Rev. B* **21,** 1311 (1980).
[f]R. Braunstein and E. O. Kane, *J. Phys. Chem. Solids* **23,** 1423 (1962).
[g]M. Cardona, *J. Phys. Chem. Solids* **24,** 1543 (1963).
[h]P. Lawaetz, *Phys. Rev. B* **4,** 3460 (1971).
[i]M. Cardona et al., *Phys. Rev. B* **38,** 1806 (1988).
[j]Conductivity effective mass at the valence-band maximum [see Eq. (10.18)].
[k]Density-of-states effective mass at the valence-band maximum [see Eq. (10.19)].

$= 0.12 \pm 0.01$, $m_{hh}/m_0 = 0.60 \pm 0.02$ in the $\langle 111 \rangle$ direction and $m_{lh}/m_0 = 0.12 \pm 0.01$, $m_{hh}/m_0 = 0.56 \pm 0.02$ in the $\langle 100 \rangle$ direction. Rochon and Fortin[82] have made a study of interband magnetooptic transitions in n-type InP and obtained the following masses (m values) and the Luttinger's valence-band parameters (γ values): $m_{lh}/m_0 = 0.12 \pm 0.01$; $m_{hh}/m_0 = 0.45 \pm 0.05$; $\gamma_1^L = 5.15 \pm 0.05$; $\gamma_2^L = 0.94 \pm 0.03$; and $\gamma_3^L = 1.62 \pm 0.03$.

Hermann and Pearsall[92] have carried out determinations of the Γ-point light-hole mass as a function of alloy composition for $In_{1-x}Ga_xAs_yP_{1-y}$ lattice-matched to InP by means of a conduction-electron spin-polarized photoluminescence analysis. They showed that the results are in good agreement with masses calculated using the four-band approximation of Kane's model (the $\boldsymbol{k} \cdot \boldsymbol{p}$ theory).[93] Alavi et al.[80] have determined m_{lh} and m_{hh} for $y = 0.52$ by the method of interband magnetooptic absorption. They also determined several interband parameters, such as the band-gap energy and Luttinger parameters, for $In_{0.53}Ga_{0.47}As$ using the same method.[94] In Table 6.7 we list the experimental and calculated light- and heavy-hole masses, m_{lh}/m_0 and m_{hh}/m_0, versus composition y for $In_{1-x}Ga_xAs_yP_{1-y}$ lattice-matched to InP.

In Figs. 6.8 and 6.9 we also plot the experimentally determined masses

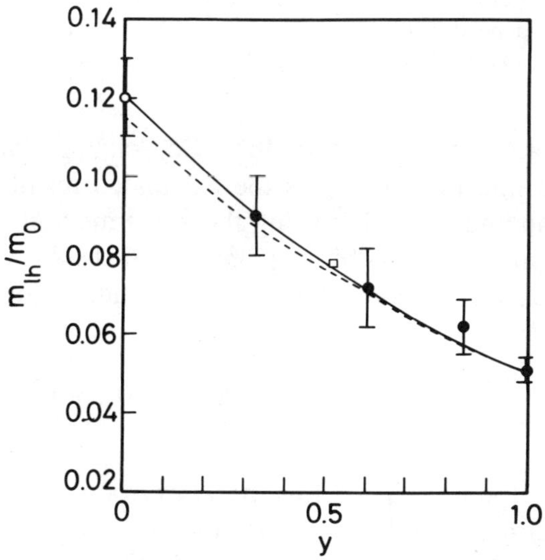

Figure 6.8 Effective mass of the light holes m_{lh}/m_0 as a function of y for $In_{1-x}Ga_xAs_yP_{1-y}$ lattice-matched to InP. The experimental data are taken from Refs. 82 and 91 (open circle), 92 (solid circles), and 80 (open square). The dashed line is the calculated result of Hermann and Pearsall using the $\boldsymbol{k} \cdot \boldsymbol{p}$ method.[92] The solid line marks a possible explanation of the experimental findings (see text).

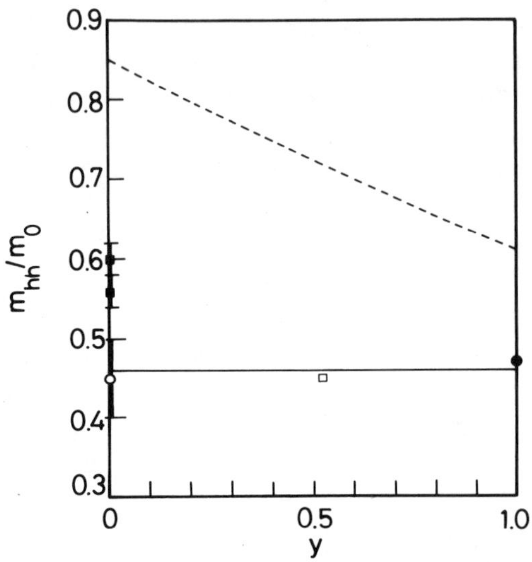

Figure 6.9 Effective mass of the heavy holes m_{hh}/m_0 as a function of y for $In_{1-x}Ga_xAs_yP_{1-y}$ lattice-matched to InP. The experimental data are taken from Refs. 82 (open circle), 91 (solid squares), 80 (open square), and 94 (solid circle). Note that the data of Leotin et al.[91] (solid squares) indicate an anisotropy in the heavy-hole masses between the two principal directions ⟨100⟩ (0.56 ± 0.02) and ⟨111⟩ (0.60 ± 0.02). The dashed line is the interpolated result of Eq. (2.3) using the binary masses of Lawaetz.[69] Some experimental data (i.e., the open circle, open square, and solid circle) suggest that the heavy-hole mass is independent of the composition y. The solid line in the figure indicates such a trend (i.e., $m_{hh}/m_0 \simeq 0.46$).

m_{lh} and m_{hh}, as a function of y, for $In_{1-x}Ga_xAs_yP_{1-y}$ lattice-matched to InP. The dashed line in Fig. 6.8 is the calculated result of the effective mass by Hermann and Pearsall[92] using the $\boldsymbol{k} \cdot \boldsymbol{p}$ method. The solid curve in Fig. 6.8 also marks a possible explanation of these experimental findings; thus, the curve is calculated assuming a quadratic dependence on y:

$$\frac{m_{lh}(y)}{m_0} = 0.120 - 0.099y + 0.030y^2 \tag{6.15}$$

Leotin et al.[91] found that for InP the anisotropy of m_{lh} is zero within experimental accuracy, while the heavy-hole mass m_{hh} shows a 7% anisotropy between the two principal directions ⟨100⟩ and ⟨111⟩. In view of the extreme anisotropy of the heavy-hole bands in InP, spherically averaged bands and resulting effective masses, which can be calculated from the Dresselhaus band parameters presented in Ref. 91, may be questionable. The dashed line in Fig. 6.9 is the interpolated result of Eq. (2.3) using the binary masses of Lawaetz [the averaged (DOS) masses].[69] It is noted that the interpolated values are considerably larger than the experimental findings. The data of Refs. 82 (InP), 80 ($y = 0.52$), and 94 ($y = 1.0$) suggest that the heavy-hole mass may be independent of the composition y. The solid line in the figure indicates such a trend, specifically, $m_{hh}/m_0 \simeq 0.46$.

To our knowledge, no experimental or theoretical value has been reported for the composition dependence of the spin–orbit splitoff band effective mass m_{so} in the $In_{1-x}Ga_xAs_yP_{1-y}$ quaternary. There only exist reports for InP[66,69,82,88,89] that give $m_{so}/m_0 = 0.21 \pm 0.01$ (the interband magnetooptic absorption measurements),[82] 0.310,[66] 0.18,[88] 0.17,[69] 0.20 (the $\boldsymbol{k} \cdot \boldsymbol{p}$ calculations),[89] and 0.18 (the linear muffin-tin-orbital-method calculation).[89]

In Fig. 6.10 we plot the interpolated m_{so} [Eq. (2.3)] as a function of y for $In_{1-x}Ga_xAs_yP_{1-y}$ lattice-matched to InP. The solid circle represents the experimental data taken from Ref. 82 (InP). The dashed line is calculated from the binary masses of Lawaetz,[69] while the solid line is obtained from Eq. (2.3) using the experimentally determined binary masses: $m_{so}/m_0 = 0.21 \pm 0.01$ for InP (interband magnetooptic transitions),[82] 0.14 ± 0.01

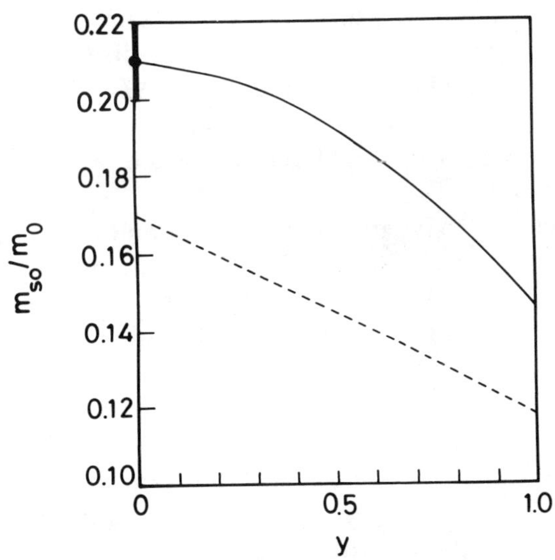

Figure 6.10 Effective mass of the spin–orbit splitoff band m_{so}/m_0 as a function of y for $In_{1-x}Ga_xAs_yP_{1-y}$ lattice-matched to InP. The experimental data are taken from Ref. 82 (solid circle). The dashed line is the interpolated result of Eq. (2.3) using the binary masses of Lawaetz.[69] The solid line is also the interpolated result of Eq. (2.3) using the experimentally determined binary masses (see text).

for InAs (interband magnetooptic transitions),[95] 0.154 ± 0.010 for GaAs (interband magnetooptic transitions),[96] and 0.34 for GaP (indirect interband optical absorption).[97] The dashed line indicates that the mass varies almost linearly with composition y. The solid line, on the other hand, gives an upward bowing of the m_{so} variation with respect to the composition y:

$$\frac{m_{so}(y)}{m_0} = 0.21 - 0.01y - 0.05y^2 \tag{6.16}$$

6.3 CONDUCTION- AND VALENCE-BAND OFFSETS

6.3.1 $In_{1-x}Ga_xAs_yP_{1-y}$/InP Heterojunction System

One of the most important parameters for the design and analysis of heterojunction and quantum-well electronic and optoelectronic devices is the heterojunction band offset. There have been several reports on band offsets in the $In_{1-x}Ga_xAs_yP_{1-y}$/InP heterojunction system.[98–108]

Chin et al.[98,99] have determined the band-gap discontinuity between the $y = 0.37$ quaternary and InP by measuring the photoluminescence spectrum of quantum-well layers. (Note that the $y = 0.37$ composition is a corrected value, and refers to the "$y = 0.29$" samples of Refs. 98 and 99; see Ref. 103.) They reported the valence-band offset to be $\Delta E_v = 80$ meV and the conduction-band offset to be $\Delta E_c = 2\Delta E_v$. The valence- and conduction-band offsets are thus approximately $E_v = 0.33\Delta E_0$ and $\Delta E_c = 0.67\Delta E_0$, where ΔE_0 is the band-gap difference in E_0 between these heterojunction materials.

Brunemeier et al.[103] have measured photoluminescence spectra of single-well, quantum-well $In_{1-x}Ga_xAs_yP_{1-y}$/InP heterostructures ($y = 0.25$, 0.37, 0.39, 0.91). The ΔE_v determination indicated that $\Delta E_v = 0.35\Delta E_0$ and $\Delta E_c = 0.65\Delta E_0$ for all $In_{1-x}Ga_xAs_yP_{1-y}$ compositions lattice-matched to InP. Forrest et al.,[102,104] on the other hand, found a conduction-band offset of $\Delta E_c = (0.39 \pm 0.01)\Delta E_0$ ($\Delta E_v = 0.61\Delta E_0$) for several compositions of $In_{1-x}Ga_xAs_yP_{1-y}$ grown lattice-matched on InP using capacitance–voltage techniques. Montie et al.[105] and Soucail et al.[107,108] also found the ratio of the conduction- to valence-band discontinuities to be 35 : 65 and 43 : 57, respectively. There is thus a considerable disagreement in the band offsets determined by these authors.

Results of the above-mentioned extensive measurements obtained at 77 K are summarized in Fig. 6.11. The solid line is a recent proposal of Langer et al.,[109] who used a fact that transition-metal impurity levels in semi-

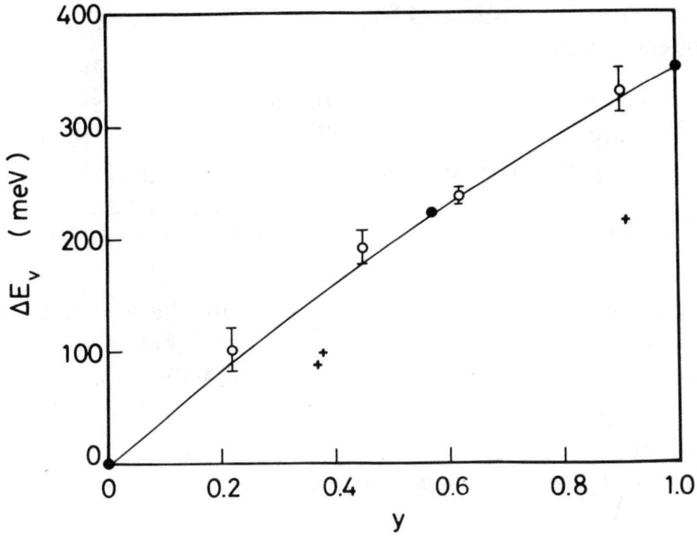

Figure 6.11 Valence-band offsets ΔE_v in $In_{1-x}Ga_xAs_yP_{1-y}$/InP heterojunction at 77 K. The experimental data are taken from Refs. 102 (open circles) and 103 (crosses). The solid circles are the Fe acceptor-level energies in $In_{1-x}Ga_xAs_yP_{1-y}$ relative to InP (see Ref. 109). The solid line is the calculated result of Eq. (6.17a).

conductors may serve as a reference in band-gap alignment in semiconductor heterojunctions.[110] We find that their result can be written approximately as (in millielectronvolts)

$$\Delta E_v(y) = 502y - 152y^2 \tag{6.17a}$$

The corresponding ΔE_c can now be written, with the use of the 77-K $E_0(y) - y$ data in Ref. 111, as (in millielectronvolts)

$$\Delta E_c(y) = 268y + 3y^2 \tag{6.17b}$$

Equation (6.17b) suggests that the conduction-band discontinuity ΔE_c varies almost linearly with respect to the alloy composition y. It is reasonable to consider that the ratio of the conduction- to valence-band discontinuities is not dependent on temperature. It is also noted that the energy difference ΔE_0 is almost independent of temperature in the range between 4 and 300 K. These arguments support the belief that the discontinuities derived from Eq. (6.17) are valid not only at 77 K but also, at least, in the range 4–300 K.

6.3.2 $In_{0.53}Ga_{0.47}As/InP$ and $In_{0.53}Ga_{0.47}As/In_{0.52}Al_{0.48}As$ Heterojunction Systems

The band offsets in the $In_{0.53}Ga_{0.47}As/InP$ heterojunction system have been reported by many authors.[106,112–127] In Table 6.8 we show the band offsets, ΔE_c and ΔE_v, for this heterojunction system obtained by different authors.[106,112,115,117,119–124] The ratios of the ΔE_c to ΔE_v spread from 36:64,[106,124] 40:60,[115,127] 42:58,[117] 45:55,[118] to 100:0.[116] Further study is, therefore, needed.

Like $In_{0.53}Ga_{0.47}As$, $In_{0.52}Al_{0.48}As$ can be grown lattice-matched on InP. The band-gap discontinuity $\Delta E_0(= \Delta E_c + \Delta E_v)$ between $In_{0.53}Ga_{0.47}As$ ($E_g = 0.75$ eV) and $In_{0.52}Al_{0.48}As$ ($E_g = 1.47$ eV) is about 0.72 eV. This large band-gap discontinuity makes the heterojunction system $In_{0.53}Ga_{0.47}/In_{0.52}Al_{0.48}As/InP$ attractive for possible fabrications of optoelectronic and hot-electron-injection devices. The ΔE_c between

TABLE 6.8 Conduction- and Valence-Band Discontinuities, ΔE_c and ΔE_v, in Heterojunction System between $In_{0.53}Ga_{0.47}As$ and InP (IV = Current–Voltage Measurements; CV = Capacitance–Voltage Measurements; PR = Photoresponse Characteristics; PL = Photoluminescence Measurements; AS = Admittance Spectroscopy; OA = Optical Absorption Measurements)

ΔE_c (meV)	ΔE_v (meV)	Technique
220 ± 20^a		IV, CV, PR
230^b	380^b	PL, PR
250 ± 10^c	346 ± 10^c	AS
235 ± 20^d	380 ± 30^d	OA
220^e		CV
210 ± 20^f		AS
203 ± 15^g		PR
219^h	390^h	PR
217 ± 5^i		CV
185 ± 5^j		AS

[a] S. R. Forrest and O. K. Kim, *J. Appl. Phys.* **52,** 5838 (1981).
[b] M. S. Skolnick et al., *Semicond. Sci. Technol.* **1,** 29 (1986).
[c] D. V. Lang et al., *Appl. Phys. Lett.* **50,** 736 (1987); *J. Vacuum Sci. Technol. B* **5,** 1215 (1987).
[d] M. S. Skolnick et al., *Appl. Phys. Lett.* **51,** 24 (1987).
[e] L. Y. Leu and S. R. Forrest, *J. Appl. Phys.* **64,** 5030 (1988).
[f] R. E. Cavicchi et al., *Appl. Phys. Lett.* **54,** 739 (1989).
[g] M. A. Haase et al., *Appl. Phys. Lett.* **54,** 1457 (1989).
[h] M. Zachau et al., *Superlatt. Microstruct.* **5,** 19 (1989).
[i] M. T. Furtado et al., *Superlatt. Microstruct.* **5,** 507 (1989).
[j] A. W. Higgs et al., *Semicond. Sci. Technol.* **5,** 581 (1990).

TABLE 6.9 Conduction-Band Discontinuity, ΔE_c, in Heterojunction System between $In_{0.53}Ga_{0.47}As$ and $In_{0.52}Al_{0.48}As$ (IV = Current–Voltage Measurements; PL = Photoluminescence Measurements)

ΔE_c (eV)	Technique
0.52[a]	IV
0.50 ± 0.05[b]	IV
0.52[c]	PL
0.51 ± 0.04[d]	IV
0.553 ± 0.020[e]	IV
0.51 ± 0.02[f]	PL

[a]D. V. Morgan et al., *Phys. Status Solidi* **72,** 251 (1982).
[b]R. People et al., *Appl. Phys. Lett.* **43,** 118 (1983).
[c]D. F. Welch et al., *J. Appl. Phys.* **55,** 3176 (1984).
[d]C. K. Peng et al., *J. Appl. Phys.* **60,** 1709 (1986).
[e]Y. Sugiyama et al., *Jpn. J. Appl. Phys.* **25,** L648 (1986).
[f]A. Sandhu et al., *Jpn. J. Appl. Phys.* **26,** 1709 (1987).

$In_{0.53}Ga_{0.47}As$ and $In_{0.52}Al_{0.48}As$ have been reported by several authors.[128–133] These results are summarized in Table 6.9.

The $In_{0.52}Al_{0.48}Al$/InP interface is the least well characterized.[125,134,135] The staggered alignment is generally agreed on.[125] Two experiments provide values for the offsets: low-temperature luminescence from a series of superlattices with varying period implied the range of values $\Delta E_v = 0.255 - 0.325$ eV and $\Delta E_c = 0.345 - 0.415$ eV;[134] photoemission gave $\Delta E_v = 0.16 - 0.05$ eV.[135]

6.3.3 Lattice-Mismatched Heterojunctions Based on $In_{1-x}Ga_xAs$

Some lattice-mismatched heterojunction systems like $In_{1-x}Ga_xAs$/GaAs are important material systems for high-speed and optoelectronic device applications. The band offset depends not only on the heterojunction semiconductors involved but also on the amount of mismatch strain at the interface. The strain field in the strained-heterojunction wells and barriers can, therefore, lead to several plausible configurations of the quantizing superpotentials.[136] The large tetragonal elastic strains in the strained layers can produce marked effects on the electronic properties through the deformation–potential interactions. The hydrostatic component of the strain causes shifts in the bulk energy levels of the layers, and the shear component causes splittings of certain degenerate valence-band levels. The energy levels of the layers are therefore determined including the strain modification of the layer structure. There are reports of the band-offset values in the $In_{1-x}Ga_xAs$-based, strained-heterojunction systems, such as $In_{1-x}Ga_xAs/GaAs$,[137–154] $In_{1-x}Ga_xAs/InP$,[155,156] and $In_{1-x}Ga_xAs/Al_xGa_{1-x}As$.[149,157–162]

6.4 EXTERNAL PERTURBATION EFFECTS ON BAND PARAMETERS

6.4.1 Effects of Temperature

The temperature-induced change in the band-gap energy E_g can be commonly given in terms of the α and β coefficients of the Varshni equation[163]

$$E_g(T) = E_g(0) - \frac{\alpha T^2}{\beta + T} \tag{6.18}$$

where $E_g(0)$ is the band-gap energy at 0 K, α is in electronvolts per degree kelvin, and β is proportional to the Debye temperature (in kelvins).

Besides the changes of band structure induced by thermal expansion of the lattice, the temperature dependence of E_g is mainly due to the electron–phonon interactions [Debye–Waller term (two-phonon process and Fan term (one-phonon process)].[164] Lautenschlager et al.[44] have, thus, recently proposed a temperature dependence with an equation containing the Bose–Einstein occupation factor for phonons:

$$E_g(T) = E_B - a_B\left(1 + \frac{2}{e^{\theta/T} - 1}\right) \tag{6.19}$$

where the parameter θ describes the mean frequency of the phonons involved and a_B is the strength of the interaction.

O'Donnell and Chen[165] have also advocated the three-parameter fit equation

$$E_g(T) = E_g(0) - S\langle\hbar\omega\rangle\left[\coth\left(\frac{\langle\hbar\omega\rangle}{2k_BT}\right) - 1\right] \tag{6.20}$$

as a direct replacement of the Varshni equation, where S is a dimensionless coupling constant and $\langle\hbar\omega\rangle$ is an average phonon energy. It is noted that the term within the brackets in Eq. (6.19) readily reduces to coth $(\theta/2T)$ so that Eq. (6.19) can be written in a form similar to Eq. (6.20). The expression Eq. (6.20) has shown to be compatible with reasonable assumptions about the influence of phonons on the band-gap energy. They also used Eq. (6.20) to derive expressions for the thermodynamic functions, such as the Gibbs energy, enthalpy, and entropy of formation of electron–hole pairs in semiconductors.

Figure 6.12 shows the temperature variation of E_0 and $E_0 + \Delta_0$ for InP. The data below 30 K (solid circles) are taken from a tabulation in Ref. 44. Using spectroscopic ellipsometry, Lautenschlager et al.[44] measured the energies of E_0 and $E_0 + \Delta_0$ from 30 to 420 K. They also measured the E_1, E_0', and $E_0' + \Delta_0'$ gap variation from 30 to 750 K. The solid lines are the results of their measurements [best fitted to Eq. (6.19)]. Hang et al.[166] have also measured the temperature variation of E_0 and $E_0 + \Delta_0$ from 77 K to 600°C using photoreflectance modulation spectroscopy. The dashed lines indicate least-squares-fitted results of Hang's measurements to Eq. (6.18).

The temperature dependence of the broadening function of interband electronic transitions can also yield information about the electron–phonon interactions, excitonic effects, and similar phenomena. The broadening function Γ can be, in general, expressed by a sum of two different contributions:[44]

$$\Gamma(T) = \Gamma_0 + \Gamma_1 \left(1 + \frac{2}{e^{\theta/T} - 1}\right) \tag{6.21}$$

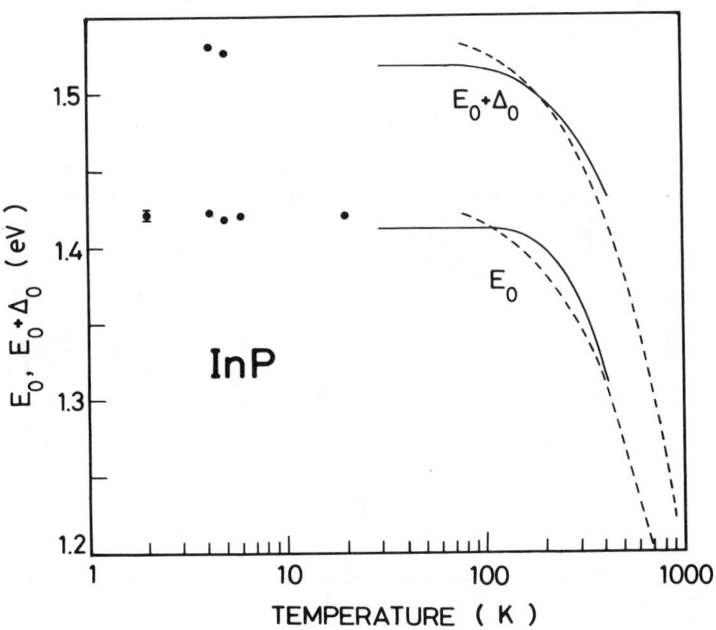

Figure 6.12 Temperature variation of the E_0 and $E_0 + \Delta_0$ gaps for InP. The data below 30 K are taken from Ref. 44. The solid and dashed lines are the experimental results of Lautenschlager et al.[44] and Hang et al.,[166] respectively.

where Γ_0 is independent of temperature, arising mainly from crystalline imperfections, and Γ_1 is a contribution through emission and absorption of phonons of the average frequency θ. This expression ensures a relatively constant Γ value from low T up to, in many cases, $T \simeq 100$ K, at which point a component Γ_1 becomes discernible, and then increases nearly linearly with T for high T. We show as an example, in Fig. 6.13, the temperature dependence of the broadening parameter Γ for E_0 of InP. The solid circles are the experimental values estimated from photoreflectance modulation spectroscopy.[166] The solid line is the least-squares fit to Eq. (6.21) with $\Gamma_0 = 26$ meV, $\Gamma_1 = 33$ meV, and $\theta = 670$ K. As pointed out in Ref. 44, the temperature θ obtained when fitting the temperature dependence of Γ is not necessarily the same as that for the shift of the corresponding energy gap [i.e., Eq. 6.19)]. For the energy shift due to the electron–phonon interaction, acoustic as well as optical phonons contribute (while for the broadening parameter mainly optical phonons are responsible). The dashed line in Fig. 6.12 for E_0 represents a fit with $\theta = 259$ K.

The temperature coefficient of the band gap (dE_g/dT) is known to be almost linear for temperatures higher than 150 K, as in most semiconductors. Turner et al.[167] have obtained a value for dE_0/dT in InP of -2.9×10^{-4} eV/K from optical absorption measurements. Spectroscopic-ellipsometry data of Lautenschlager et al.[44] provide $dE_0/dT = -3.3 \times 10^{-4}$ eV/K, while from the data of Hang et al.[166] we can calculate the coefficient to be -4.0×10^{-4} eV/K.

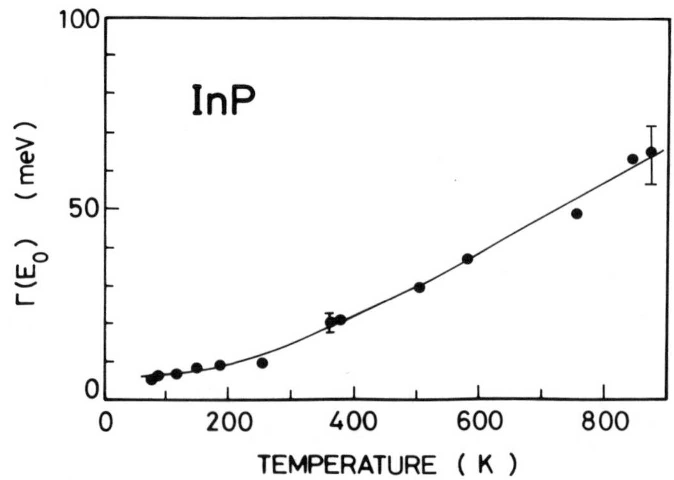

Figure 6.13 Temperature variation of the broadening parameter Γ for E_0 of InP. The solid line is a least-squares fit to Eq. (6.21). (From Hang et al.[166])

Satzke et al.[16] have determined the Varshni parameters $\alpha = -0.426$ meV/K and $\beta = 224$ K for E_0 of $In_{1-x}Ga_xAs_yP_{1-y}$ from the temperature dependence of electroabsorption spectra where the alloy used was tuned to the emission of 1.33 μm at room temperature (i.e., $y \simeq 0.6$). These parameters provide $dE_0/dT = -3.5 \times 10^{-4}$ eV/K near 300 K. Madelon and Dore[168] ascertained that $\alpha = 0.75$ meV/K, $\beta = 611$ K, and $dE_0/dT = -3.87 \times 10^{-4}$ eV/K in the range 150–300 K for $In_{1-x}Ga_xAs_yP_{1-y}$/InP with $y = 0.59$ from derivative transmission measurements. Temkin et al.[169] also reported from photoluminescence measurements values for the Varshni coefficients for $In_{1-x}Ga_xAs_yP_{1-y}$/InP quaternary of $\alpha = 0.49$ meV/K and $\beta = 327$ K, independent of composition. Yamazoe et al.[13] found that the temperature coefficient of E_0 in $In_{1-x}Ga_xAs_yP_{1-y}$/InP near 300 K is about -0.4 meV/K over the whole composition range of this quaternary.

Pearsall[170] estimated the temperature variation of E_0 for $In_{0.53}Ga_{0.47}As$ ternary as given by (in electronvolts)

$$E_0(T) = 0.812 - 3.26 \times 10^{-4}T + 3.31 \times 10^{-7}T^2 \qquad (6.22)$$

that is, a linear temperature coefficient of -3.26×10^{-4} eV/K. Results of these measurements are summarized in Fig. 6.14. The solid line is the linearly interpolated result of Eq. (2.3) using the binary data listed in Table 6.10.

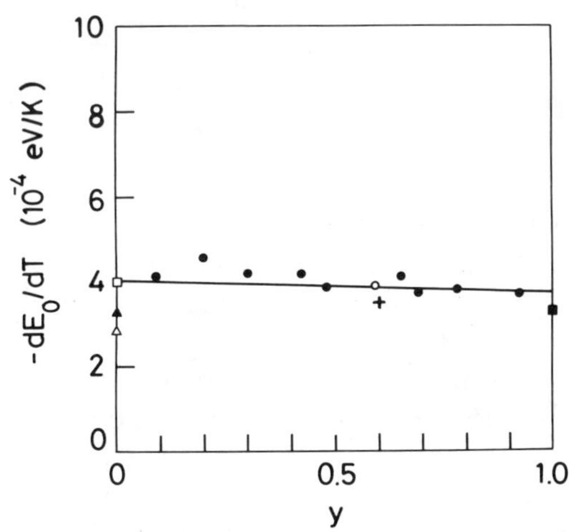

Figure 6.14 Linear temperature coefficient $-dE_0/dT$ as a function of y for $In_{1-x}Ga_xAs_yP_{1-y}$ lattice-matched to InP. The experimental data are taken from Refs. 44 (solid triangle), 166 (open square), 167 (open triangle), 13 (solid circles), 16 (cross), 168 (open circle), and 170 (solid square). The solid line is the interpolated result of Eq. (2.3) using the binary data listed in Table 6.10.

TABLE 6.10 Linear Temperature Coefficient dE_0/dT and Pressure Coefficient of Direct-Gap Energy E_0 for InP, InAs, GaAs, GaP, $In_{1-x}Ga_xAs_yP_{1-y}$, and $In_{0.53}Ga_{0.47}As$

		$E_0(p) = E_0(0) + ap + bp^2$ (eV)	
Material	dE_0/dT ($\times 10^{-4}$ eV/K)	a ($\times 10^{-2}$ eV/GPa)	b ($\times 10^{-4}$ eV^2/GPa^2)
InP	-4.0[a]	7.5 ± 0.2[e]	-12 ± 5[e]
InAs	-3.5[b]	10.0[f]	
GaAs	-3.95[c]	10.8 ± 0.3[g]	-14 ± 2[g]
GaP	-4.6[d]	9.7 ± 0.8[h]	-35 ± 6[h]
$In_{1-x}Ga_xAs_yP_{1-y}$	$-4.0 + 0.3y$[i]	$7.5 + 2.9y$[i]	
$In_{0.53}Ga_{0.47}As$	-3.7[i]	10.4[i]	

[a] Z. Hang et al., *Solid State Commun.* **73,** 15 (1990).
[b] F. Matossi and F. Stern, *Phys. Rev.* **111,** 472 (1958).
[c] M. Zvára, *Phys. Status Solidi* **27,** K157 (1968).
[d] R. Zallen and W. Paul, *Phys. Rev.* **134,** A1628 (1964).
[e] C. S. Menoni et al., *Phys. Rev. B* **33,** 5896 (1986).
[f] R. Zallen and W. Paul, *Phys. Rev.* **155,** 703 (1967).
[g] A. R. Goñi et al., *Phys. Rev. B* **36,** 1581 (1987).
[h] S. Ves et al., *Solid State Commun.* **55,** 327 (1985).
[i] Linearly interpolated value using Eq. (2.3).

Paul et al.[171] also proposed an empirical expression for both parameters as a function of alloy composition x and temperature for $In_{1-x}Ga_xAs$ ternary. Their derived expression can be given by (in electronvolts)

$$E_g(x, T) = 0.42 + 0.625x - \left(\frac{5.8}{T + 300} - \frac{4.19}{T + 273}\right)10^{-4}T^2x - \frac{4.19 \times 10^{-4}}{T + 271}T^2 + 0.475x^2 \qquad (6.23)$$

The $E_g(0.47, T)$ values calculated from Eq. (6.23) give values larger at low T and smaller at high T than those calculated from Eq. (6.22). This difference, however, seemed to be less than other experimental uncertainties.

The effective band masses are also known to be functions of temperature.[172–175] Figure 6.15 shows the temperature dependence of the Γ-band electron effective masses m_e^Γ for InP, InAs, and GaAs. The data are taken for InP from Ref. 174 and InAs and GaAs, from Ref. 175. Hamaguchi and coworkers[174] have determined the temperature dependence of the electron effective masses in some III–V semiconductors (InP, GaAs, and InSb) from the magnetophonon resonance experiments in the temperature range between 77 and 300 K. The effective masses in these semiconductors de-

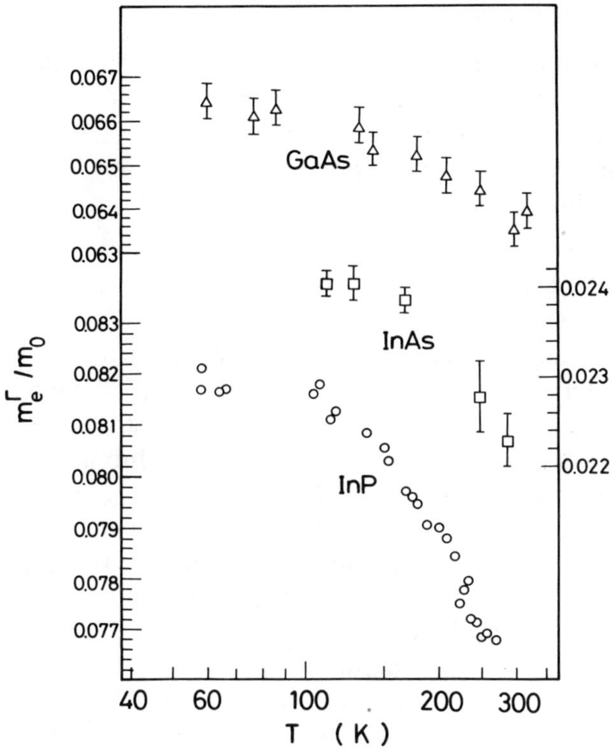

Figure 6.15 Temperature variation of the Γ-band electron effective mass m_e^Γ/m_0 for InP, InAs, and GaAs. The experimental data are taken for InP from Ref. 174 and InAs and GaAs, from Ref. 175.

creased with increasing temperature. Stradling and Wood[175] have also studied the band-edge effective masses in *n*-type InAs, InSb, and GaAs over a wide range of temperatures using the magnetophonon resonance effect. They deduced the temperature dependence of the effective mass in each semiconductor and compared their results with that predicted from the changes of dilatational component in the band gap with temperature. It was found that the agreement is excellent for InSb but, with InAs and GaAs, the observed change is greater than that predicted although less than that obtained by substitution of the change of the optical energy gap with temperature. A similar comparison was also carried out in Ref. 174 and reasonable agreement was achieved for GaAs, but the agreement was not as good for InP and InSb. No detailed reports have been, however, reported on the temperature dependence of the effective mass in the $In_{1-x}Ga_xAs_yP_{1-y}$/InP alloy system at the present time.

6.4.2 Effects of Pressure

The band-structure parameters, such as band gaps and effective masses, are strongly dependent on pressure. The pressure effects on the band parameters have been studied for InP and $In_{1-x}Ga_xAs_yP_{1-y}$ by many authors.[176–193] Increasing the hydrostatic pressure p usually increases the band-gap energy E_g in the following manner:

$$E_g(p) = E_g(0) + ap + bp^2 \tag{6.24}$$

The pressure dependence of the energy-band gaps for InP has been studied both theoretically[177] and experimentally.[176,178–182] Gorczyca et al.[177] used self-consistent linear muffin-tin-orbital band-structure calculations to investigate the optical and structural properties of InP under pressure. The calculated linear and nonlinear coefficients, a and b, of the energy-band gaps at high-symmetry critical points agreed with published experimental data well. Their calculated pressure at the crossover from the direct to indirect gap is 10.4 ± 0.2 GPa (104 ± 2 kbar) in zinc-blende structure. Müller et al.[178] studied the experimental hydrostatic pressure dependence of the fundamental absorption edge of InP at 300 K. They obtained the crossing of the Γ and X conduction-band minima at 10.4 ± 0.1 GPa. This value agrees with that estimated by Gorczyca et al.[177] Similar measurements, but at 20 K (photoluminescence), have been performed by Menoni et al.[180] Their results showed only the direct-band-gap transitions ($\Gamma_1^c - \Gamma_{15}^v$) at pressures up to 12 GPa. Recent low-temperature photoluminescence measurements of n-InP by Leroux[182] also showed no evidence for the Γ − X crossing up to the phase transition pressure (~10 GPa). On the other hand, Kobayashi et al.[181] have suggested the Γ − X crossing in InP for p ~ 8 Gpa.

People et al.[183] have studied the pressure dependence of E_0 gap of $In_{0.53}Ga_{0.47}As$ and found that the gap varies sublinearly with pressure for $p > 10$ kbar, having an initial slope 12.44 meV/kbar. They also determined the hydrostatic-pressure deformation potential for the Γ-band gap [−(7.79 ± 0.4) eV]. The pressure coefficient dE_0/dp for $In_{0.53}Ga_{0.47}As$ has also been measured[192] by photoluminescence in a diamond anvil cell at room temperature and at 80 K. The dE_0/dp was found to be independent of temperature between 300 and 80 K (10.95 ± 0.1 meV/kbar).

The composition dependence of the pressure coefficient dE_0/dp for $In_{1-x}Ga_xAs_yP_{1-y}$ lattice-matched to InP has been studied[184] by measuring photoconductive band-edge shifts. Wiesenfeld[185] has also determined the pressure coefficient for the $y = 0.66$ quaternary ($dE_0/dp = 8.9 \pm 1.3$

meV/kbar) from time-resolved transmission and reflection changes photoexcited by subpicosecond optical pulses.

The experimental dE_0/dp values for InP and $In_{1-x}Ga_xAs_yP_{1-y}$ are plotted in Fig. 6.16. Table 6.10 also summarizes the dE_0/dp for InP, InAs, GaAs, GaP, $In_{1-x}Ga_xAs_yP_{1-y}$, and $In_{0.53}Ga_{0.47}As$. The solid line in Fig. 6.16 is obtained from the linear interpolation scheme [Eq. (2.3)] by using the end-point binary data. This scheme provides nearly linear variation of dE_0/dp with composition y (see Table 6.10).

Increasing the pressure increases the effective band mass[186-191] as in the case for the energy-band gaps. The magnetophonon effects in n-type InP and GaAs have been observed up to 15 kbar at 295 K by Pitt et al.[186] using a nonmagnetic high-pressure apparatus. The pressure coefficient dm_e^Γ/dp was determined as 0.62 and 0.70%/kbar for InP and GaAs, respectively. The influence of the hydrostatic pressure on the conduction-band effective mass m_e^Γ in $In_{1-x}Ga_xAs_yP_{1-y}$ (y = 0.55, and 1.0) has been studied by Shantharama et al.[188] and Adams et al.[189] using a magnetophonon resonance. They found an anomalously large increase in m_e^Γ, which is thought to be due to a breakdown of the approximations used in $\mathbf{k} \cdot \mathbf{p}$ theory [Eq. (6.9)] for alloys (E_0 was determined in their work by photoconductive techniques). The dm_e^Γ/dE_0 values they found are $(0.084 \pm 0.005)m_0$ eV^{-1} for $y = 0.55$ and $(0.074 \pm 0.005)m_0$ eV^{-1} for $y = 1.0$.

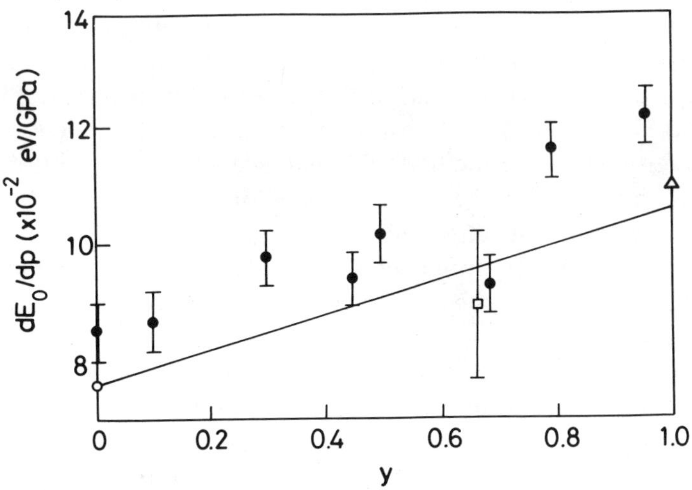

Figure 6.16 Pressure coefficient dE_0/dp as a function of y for $In_{1-x}Ga_xAs_yP_{1-y}$ lattice-matched to InP. The open circle represents the experimental data taken from Ref. 180. The solid circles are taken from Ref. 184. The open square is taken from Ref. 185. The open triangle is also taken from Ref. 192.

Adams and Shantharama[190] have also studied the pressure dependence of hole mobility in InP, GaAs, and $In_{1-x}Ga_xAs_yP_{1-y}$/InP ($y = 0.65$). The mobility of holes in InP and GaAs was observed to increase with pressure at 0.18 and 0.31%/kbar, respectively. By contrast, the mobility in the quaternary was observed to decrease with pressure. They considered that the result indicates the presence of alloy scattering in hole-scattering mechanisms. They also estimated the logarithmic pressure coefficients of the heavy-hole mass $d \ln (m_{hh}/m_0)/dp$ as $+8.0 \times 10^{-4}$ kbar^{-1} for InP, $+19 \times 10^{-4}$ kbar^{-1} for the quaternary, and -1.0×10^{-4} kbar^{-1} for GaAs.

The effective mass in $In_{0.53}Ga_{0.47}As$/InP heterojunction (two-dimensional electron gas) under hydrostatic pressure up to 15 kbar was studied by Gauthier et al.[191] Cyclotron magnetophonon resonance measurements were used to determine the coefficients of the effective masses in the samples. The pressure coefficient was found to be dependent on the carrier concentration of the measured samples. A mass increase in the highest carrier concentration sample was 1 ± 0.1%/kbar ($dm_e^{\Gamma}/dF_0 = 0.044m_0$ eV^{-1}). This value is two times smaller than that found experimentally in InGaAs bulk material ($dm_e^{\Gamma}/dE_0 = 0.074m_0$ eV^{-1}; see Refs. 188 and 189), but is closer to the values found in binary III–V compounds [$dm_e^{\Gamma}/dE_0 = 0.049m_0$ eV^{-1}; see Ref. 189 (InP)].

The high doping concentration of semiconductors contributes carriers to the conduction band in an n-type crystal or to the valence band in a p-type one. The shift of absorption edge due to band filling is called the *Moss–Burstein shift*. Accurate knowledge of the shifts in conduction- and valence-band edges due to heavy doping effects is crucial in modeling semiconductor device structures that utilize heavily doped layers. The shift of the E_0 and $E_0 + \Delta_0$ gaps in InP has been studied by Lahtinen[194] by means of the electroreflectance method. The valence-band-edge shift due to doping in p^+-GaAs has also been studied by Silberman et al.[195] by means of X-ray photoemission spectroscopy.

The band mass is also influenced by the doping concentration of impurities. Usually, the lower the doping concentration, the smaller the effective mass; for instance, $m_e^{\Gamma}/m_0 = 0.040$ in lightly doped ($N_d = 2 \times 10^{17}$ cm^{-3}) n-type $In_{0.53}Ga_{0.47}As$ and 0.074 in degenerate sample ($N_d = 6 \times 10^{18}$ cm^{-3}).[63] Such a change in m_e^{Γ} is due mainly to the nonparabolicity in the conduction band.[83,196] An increase in m_e^{Γ} beyond the value expected from the nonparabolicity of the conduction band in GaAs has also been observed.[197] The electron mass at the conduction-band minimum changes from $0.0636m_0$ for pure GaAs to $0.73m_0$ for $n = 8.0 \times 10^{18}$ cm^{-1}. This change in m_e^{Γ} is interpreted as a distortion of the conduction band produced by the impurities. It has also been reported[198] that the electron effective mass in InP determined from Zeeman splitting of the donor level indicates a strong magnetic field dependence. This Zeeman effective mass is approximately 45% smaller than the cyclotron mass at low magnetic fields. Such an anomaly is qualitatively explained on the basis of level mixing resulting from random static electric fields present throughout the crystal due to ionized impurity centers.

The effective mass of the two-dimensional electron gas at the InGaAs(P)/InP interface is significantly larger than the electron effective mass at the band edge in bulk material; for example, $m_e^\Gamma/m_0 = 0.047 \pm 0.001$ in a $In_{0.53}Ga_{0.47}As$/InP heterojunction.[199,200] It is noted that the nonparabolicity in the band mass strongly influences the electronic subband structure of heterojunctions and quantum wells.[201]

REFERENCES

1. R. E. Nahory, M. A. Pollack, W. D. Johnston, Jr., and R. L. Barns, *Appl. Phys. Lett.* **33,** 659 (1978).
2. G. A. Antypas and R. L. Moon, *J. Electrochem. Soc.* **120,** 1574 (1973).
3. M. A. Pollack, R. E. Nahory, J. C. DeWinter, and A. A. Ballman, *Appl. Phys. Lett.* **33,** 314 (1978).
4. T. Nishino, Y. Yamazoe, and Y. Hamakawa, *Appl. Phys. Lett.* **33,** 861 (1978).
5. K. Nakajima, A. Yamaguchi, K. Akita, and T. Kotani, *J. Appl. Phys.* **49,** 5944 (1978).
6. Y. Yamazoe, H. Takakura, T. Nishino, Y. Hamakawa, and T. Kariya, *J. Cryst. Growth.* **45,** 454 (1978).
7. J. J. Hsieh, *J. Electron. Mater.* **7,** 31 (1978).
8. H. H. Caspers and H. H. Wieder, *Solid State Commun.* **29,** 403 (1979).
9. S. N. Grinyaev, M. A. Il'in, A. I. Lukomskiĭ, V. A. Chaldyshev, and V. M. Chupakhina, *Sov. Phys. Semicond.* **14,** 446 (1980).
10. E. H. Perea, E. E. Mendez, and C. G. Fonstad, *Appl. Phys. Lett.* **36,** 978 (1980).
11. Y. Yamazoe, T. Nishino, Y. Hamakawa, and T. Kariya, *Jpn. J. Appl. Phys.* **19,** 1473 (1980).
12. P. M. Laufer, F. H. Pollak, R. E. Nahory, and M. A. Pollack, *Solid State Commun.* **36,** 419 (1980).
13. Y. Yamazoe, T. Nishino, and Y. Hamakawa, *IEEE J. Quantum Electron.* **QE-17,** 139 (1981).
14. S. M. Kelso, D. E. Aspnes, M. A. Pollack, and R. E. Nahory, *Phys. Rev. B* **26,** 6669 (1982).
15. J. A. Lahtinen and T. Tuomi, *Phys. Status Solidi B* **130,** 637 (1985).

16. K. Satzke, G. Weiser, R. Höger, and W. Thulke, *J. Appl. Phys.* **63,** 5485 (1988).
17. A.-B. Chen and A. Sher, *Phys. Rev. B* **19,** 3057 (1979).
18. L. Fen and Z. Keiming, *Chinese J. Semicond.* **1,** 257 (1980).
19. V. L. Panyutin, B. É. Ponedel'nikov, A. É. Rozenson, and V. I. Chizhikov, *Sov. Phys. Semicond.* **14,** 594 (1980).
20. W. Porod and D. K. Ferry, *Phys. Rev. B* **27,** 2587 (1983).
21. V. B. Gera, R. Gupta, and K. P. Jain, *Phys. Rev. B* **36,** 9657 (1987).
22. S. N. Ekpenuma, C. W. Myles, and J. R. Gregg, *Phys. Rev. B* **41,** 3582 (1990).
23. S. Adachi, *J. Appl. Phys.* **53,** 8775 (1982).
24. R. L. Moon, G. A. Antypas, and L. W. James, *J. Elecctron. Mater.* **3,** 635 (1974).
25. T. H. Glisson, J. R. Hauser, M. A. Littlejohn, and C. K. Williams, *J. Electron. Mater.* **7,** 1 (1978).
26. G. H. Olsen, T. J. Zamerowski, R. T. Smith, and E. P. Bertin, *J. Electron. Mater.* **9,** 977 (1980).
27. G. B. Stringfellow, *J. Electron. Mater.* **10,** 919 (1981).
28. C. Papuzza, D. Campi, L. Benassi, and F. Taiariol, in *Gallium Arsenide and Related Compounds*, Institute of Physics, Bristol, UK, 1982, p. 275.
29. T. P. Pearsall and C. Hermann, in *Gallium Arsenide and Related Compounds*, Institute of Physics, Bristol, UK, 1982, p. 269.
30. T. P. Pearsall, in *GaInAsP Alloy Semiconductors*, T. P. Pearsall, ed., Wiley, New York, 1982, p. 295.
31. P. Paraynathal and F. H. Pollak, *Phys. Rev. B* **28,** 3632 (1983).
32. G. F. Smirnova, *Inorg. Mater.* (USA) **18,** 1252 (1983).
33. J. R. Chelikowsky and M. L. Cohen, *Phys. Rev. B* **14,** 556 (1976).
34. L. Ley, R. A. Pollak, F. R. McFeely, S. P. Kowalczyk, and D. A. Shirley, *Phys. Rev. B* **9,** 600 (1974).
35. O. Wada, A. Majerfeld, and A. N. M. M. Choudhury, *J. Appl. Phys.* **51,** 423 (1980).
36. G. P. Williams, F. Cerrina, G. J. Lapeyre, J. R. Anderson, R. J. Smith, and J. Hermanson, *Phys. Rev. B* **34,** 5548 (1986).
37. L. Sorba, V. Hinkel, H. U. Middelmann, and K. Horn, *Phys. Rev. B* **36,** 8075 (1987).
38. S. Zollner, U. Schmid, N. E. Christensen, and M. Cardona, *Appl. Phys. Lett.* **57,** 2339 (1990).
39. K. Y. Cheng, A. Y. Cho, S. B. Christman, T. P. Pearsall, and J. E. Rowe, *Appl. Phys. Lett.* **40,** 423 (1982).
40. J. A. Van Vechten and T. K. Bergstresser, *Phys. Rev. B* **1,** 3351 (1970).
41. See, for instance, R. K. Willardson and A. C. Beer, *Semiconductors and Semimetals*, Vol. 9, Academic, New York, 1972.

42. J. A. Van Vechten, O. Berolo, and J. C. Woolley, *Phys. Rev. Lett.* **29,** 1400 (1972).
43. O. Berolo, J. C. Woolley, and J. A. Van Vechten, *Phys. Rev. B* **8,** 3794 (1973).
44. P. Lautenschlager, M. Garriga, and M. Cardona, *Phys. Rev. B* **36,** 4813 (1987).
45. M. Cardona, K. L. Shaklee, and F. H. Pollak, *Phys. Rev.* **154,** 696 (1967).
46. S. Adachi, *Phys. Rev. B* **39,** 12612 (1989).
47. R. Hill, *J. Phys. C* **7,** 516 (1974).
48. R. Hill, *J. Phys. C* **7,** 521 (1974).
49. K.-R. Schulze, H. Neumann, and K. Unger, *Phys. Status Solidi B* **75,** 493 (1976).
50. C. Hermann and C. Weisbuch, *Phys. Rev. B* **15,** 823 (1977).
51. A.-B. Chen and A. Sher, *Phys. Rev. B* **22,** 3886 (1980).
52. R. Gupta, V. B. Gera, and K. P. Jain, *Solid State Commun.* **61,** 253 (1987).
53. A. G. Thompson and J. C. Woolley, *Can. J. Phys.* **45,** 255 (1967).
54. E. W. Williams and V. Rehn, *Phys. Rev.* **172,** 798 (1968).
55. J. C. Woolley, M. B. Thomas, and A. G. Thompson, *Can. J. Phys.* **46,** 157 (1968).
56. W. M. Coderre and J. C. Woolley, *Can. J. Phys.* **48,** 463 (1970).
57. R. E. Nahory, M. A. Pollack, and J. C. DeWinter, *J. Appl. Phys.* **46,** 775 (1975).
58. Y. Takeda, A. Sasaki, Y. Imamura, and T. Takagi, *J. Appl. Phys.* **47,** 5405 (1976).
59. Y.-T. Leu, F. A. Thiel, H. Scheiber, Jr., J. J. Rubin, B. I. Miller, and K. J. Bachmann, *J. Electron. Mater.* **8,** 663 (1979).
60. J. E. Rowe, T. P. Pearsall, and R. A. Logan, *Physica* **117 & 118B,** 347 (1983).
61. J. M. Wrobel, J. L. Aubel, U. K. Reddy, S. Sundaram, J. P. Salerno, and J. V. Gormley, *J. Appl. Phys.* **59,** 266 (1986).
62. J. Hwang, P. Pianetta, Y.-C. Pao, C. K. Shih, Z.-X. Shen, P. A. P. Lindberg, and R. Chow, *Phys. Rev. Lett.* **61,** 877 (1988).
63. T. P. Pearsall, R. Bisaro, P. Merenda, G. Laurencin, R. Ansel, J. C. Portal, C. Houlbert, and M. Quillec, in *Gallium Arsenide and Related Compounds*, Institute of Physics, Bristol, UK, 1979, p. 94.
64. D. E. Aspnes and M. Cardona, *Phys. Rev. B* **17,** 741 (1978).
65. See, for instance, R. K. Willardson and A. C. Beer, *Semiconductors and Semimetals*, Vol. 10, Academic, New York, 1975.
66. R. Braunstein and E. O. Kane, *J. Phys. Chem. Solids* **23,** 1423 (1962).
67. F. H. Pollak, C. W. Higginbotham, and M. Cardona, *J. Phys. Soc. Jpn.* **21,** (suppl.), 20 (1966).
68. R. L. Bowers and G. D. Mahan, *Phys. Rev.* **185,** 1073 (1969).

69. P. Lawaetz, *Phys. Rev. B* **4,** 3460 (1971).
70. J. B. Restorff, B. Houston, J. R. Burke, and R. E. Hayes, *Appl. Phys. Lett.* **32,** 189 (1978).
71. J. C. Portal, P. Perrier, M. A. Renucci, S. Askenazy, R. J. Nicholas, and T. P. Pearsall, in *Gallium Arsenide and Related Compounds*, Institute of Physics, Bristol, UK, 1979, p. 829.
72. R. J. Nicholas, J. C. Portal, C. Houlbert, P. Perrier, and T. P. Pearsall, *Appl. Phys. Lett.* **34,** 492 (1979).
73. H. Brendecke, H. L. Störmer, and R. J. Nelson, *Appl. Phys. Lett.* **35,** 772 (1979).
74. E. H. Perea, E. E. Mendez, and C. G. Fonstad, *J. Electron. Mater.* **9,** 459 (1980).
75. J. B. Restorff, B. Houston, R. S. Allgaier, M. A. Littlejohn, and S. B. Phatak, *J. Appl. Phys.* **51,** 2277 (1980).
76. R. J. Nicholas, S. J. Sessions, and J. C. Portal, *Appl. Phys. Lett.* **37,** 178 (1980).
77. J. M. Chamberlain, A. A. Reeder, R. J. Turner, E. Kuphal, and J. L. Benchimol, *Solid-State Electron.* **30,** 217 (1987).
78. A. Raymond, J. L. Robert, and B. Pistoulet, in *Gallium Arsenide and Related Compounds*, Institute of Physics, Bristol, UK, 1977, p. 105.
79. D. C. Langreth, *Phys. Rev.* **159,** 717 (1967).
80. K. Alavi, R. L. Aggarwal, and S. H. Groves, *J. Magnet. Mag. Mater.* **11,** 136 (1979).
81. J. M. Chamberlain, P. E. Simmonds, R. A. Stradling, and C. C. Bradley, in *Proc. 11th Int. Conf. Phys. Semicond.*, Warsaw, 1972, p. 1016.
82. P. Rochon and E. Fortin, *Phys. Rev. B* **12,** 5803 (1975).
83. M. A. Hopkins, R. J. Nicholas, P. Pfeffer, W. Zawadzki, D. Gauthier, J. C. Portal, and M. A. DiForte-Poisson, *Semicond. Sci. Technol.* **2,** 568 (1987).
84. M. B. Thomas and J. C. Woolley, *Can. J. Phys.* **49,** 2052 (1971).
85. H. Fetterman, J. Waldman, and C. M. Wolfe, *Solid State Commun.* **11,** 375 (1972).
86. L. Hrivnak, *Phys. Status Solidi A* **116,** K73 (1989).
87. J. W. Harrison and J. R. Hauser, *J. Appl. Phys.* **47,** 292 (1976).
88. M. Cardona, *J. Phys. Chem. Solids* **24,** 1543 (1963).
89. M. Cardona, N. E. Christensen, and G. Fasol, *Phys. Rev. B* **38,** 1806 (1988).
90. N. P. Belov, V. T. Prokopenko, and A. D. Yas'kov, *Sov. Phys. Semicond.* **23,** 1296 (1989).
91. J. Leotin, R. Barbaste, S. Askenazy, M. S. Skolnick, R. A. Stradling, and J. Tuchendler, *Solid State Commun.* **15,** 693 (1974).
92. C. Hermann and T. P. Pearsall, *Appl. Phys. Lett.* **38,** 450 (1981).

93. E. O. Kane, *J. Phys. Chem. Solids* **1,** 249 (1957).
94. K. Alavi, R. L. Aggarwal, and S. H. Groves, *Phys. Rev. B* **21,** 1311 (1980).
95. C. R. Pidgeon, S. H. Groves, and J. Feinleib, *Solid State Commun.* **5,** 677 (1967).
96. M. Reine, R. L. Aggarwal, B. Lax, and C. M. Wolfe, *Phys. Rev. B* **2,** 458 (1970).
97. W. P. Dumke, M. R. Lorenz, and G. D. Pettit, *Phys. Rev. B* **5,** 2978 (1972).
98. R. Chin, N. Holonyak, Jr., S. W. Kirchoefer, R. M. Kolbas, and E. A. Rezek, *Appl. Phys. Lett.* **34,** 862 (1979).
99. E. A. Rezek, R. Chin, N. Holonyak, Jr., S. W. Kirchoefer, and R. M. Kolbas, *J. Electron. Mater.* **9,** 1 (1980).
100. Y. Takanashi and Y. Horikoshi, in *Gallium Arsenide and Related Compounds*, Institute of Physics, Bristol, UK, 1981, p. 263.
101. M. Nakao, S. Yoshida, and S. Gonda, *Solid State Commun.* **49,** 663 (1984).
102. S. R. Forrest, P. H. Schmidt, R. B. Wilson, and M. L. Kaplan, *Appl. Phys. Lett.* **45,** 1199 (1984).
103. P. E. Brunemeier, D. G. Deppe, and N. Holonyak, Jr., *Appl. Phys. Lett.* **46,** 755 (1985).
104. S. R. Forrest, P. H. Schmidt, R. B. Wilson, and M. L. Kaplan, *J. Vacuum Sci. Technol. B* **4,** 37 (1986).
105. E. A. Montie, P. J. A. Thijs, and G. W. 'tHooft, *Appl. Phys. Lett.* **53,** 1611 (1988).
106. M. Zachau, P. Helgesen, A. Kux, F. Kock, D. Grützmacher, R. Meyer, H. Jürgensen, and P. Balk, *Superlatt. Microstruct.* **5,** 19 (1989).
107. B. Soucail, P. Voisin, M. Voos, D. Rondi, J. Nagle, and B. de Crémoux, *Superlatt. Microstruct.* **8,** 279 (1990).
108. B. Soucail, P. Voisin, M. Voos, D. Rondi, J. Nagle, and B. de Crémoux, *Semicond. Sci. Technol.* **5,** 918 (1990).
109. J. M. Langer, C. Delerue, M. Lannoo, and H. Heinrich, *Phys. Rev. B* **38,** 7723 (1988).
110. J. Tersoff, *Phys. Rev. Lett.* **52,** 465 (1984); *Phys. Rev. B* **30,** 4874 (1984).
111. T. P. Pearsall, L. Eaves, and J. C. Portal, *Appl. Phys. Lett.* **54,** 1037 (1983).
112. S. R. Forrest and O. K. Kim, *J. Appl. Phys.* **52,** 5838 (1981).
113. M. Ogura, M. Mizuta, K. Onaka, and H. Kukimoto, *Jpn. J. Appl. Phys.* **22,** 1502 (1983).
114. H. Temkin, M. B. Panish, P. M. Petroff, R. A. Hamm, J. M. Vandenberg, and S. Sumski, *Appl. Phys. Lett.* **47,** 394 (1985).
115. M. S. Skolnick, P. R. Tapster, S. J. Bass, A. D. Pitt, N. Apsley, and S. P. Aldred, *Semicond. Sci. Technol.* **1,** 29 (1986).
116. K. Steiner, R. Schmitt, R. Zuleeg, L. M. F. Kaufmann, K. Heime, E. Kuphal, and J. Wolter, *Surface Sci.* **174,** 331 (1986).

117. D. V. Lang, M. B. Panish, F. Capasso, J. Allam, R. A. Hamm, A. M. Sergent, and W. T. Tsang, *Appl. Phys. Lett.* **50,** 736 (1987); *J. Vacuum Sci. Technol. B* **5,** 1215 (1987).

118. D. J. Westland, A. M. Fox, A. C. Maciel, J. F. Ryan, M. D. Scott, J. I. Davies, and J. R. Riffat, *Appl. Phys. Lett.* **50,** 839 (1987).

119. M. S. Skolnick, L. L. Taylor, S. J. Bass, A. D. Pitt, D. J. Mowbray, A. G. Cullis, and N. G. Chew, *Appl. Phys. Lett.* **51,** 24 (1987).

120. L. Y. Leu and S. R. Forrest, *J. Appl. Phys.* **64,** 5030 (1988).

121. R. E. Cavicchi, D. V. Lang, D. Gershoni, A. M. Sergent, J. M. Vandenberg, S. N. G. Chu, and M. B. Panish, *Appl. Phys. Lett.* **54,** 739 (1989).

122. M. A. Haase, N. Pan, and G. E. Stillman, *Appl. Phys. Lett.* **54,** 1457 (1989).

123. A. W. Higgs, H. J. Hutchinson, L. L. Taylor, N. Apsley, and S. J. Bass, *Semicond. Sci. Technol.* **5,** 581 (1990).

124. M. T. Furtado, M. S. S. Loural, A. C. Sachs, and P. J. Shieh, *Superlatt. Microstruct.* **5,** 507 (1989).

125. M. S. Hybertsen, *Appl. Phys. Lett.* **58,** 1759 (1991).

126. M. S. Hybertsen, *J. Vacuum Sci. Technol. B* **8,** 773 (1990).

127. B. R. Nag and S. Mukhopadhyay, *Appl. Phys. Lett.* **58,** 1056 (1991).

128. D. V. Morgan, K. Board, C. E. C. Wood, and L. F. Eastman, *Phys. Status Solidi* **72,** 251 (1982).

129. R. People, K. W. Wecht, K. Alavi, and A. Y. Cho, *Appl. Phys. Lett.* **43,** 118 (1983).

130. D. F. Welch, G. W. Wicks, and L. F. Eastman, *J. Appl. Phys.* **55,** 3176 (1984).

131. C. K. Peng, A. Ketterson, H. Morkoç, and P. M. Solomon, *J. Appl. Phys.* **60,** 1709 (1986).

132. Y. Sugiyama, T. Inata, T. Fujii, Y. Nakata, S. Muto, and S. Hiyamizu, *Jpn. J. Appl. Phys.* **25,** L648 (1986).

133. A. Sandhu, Y. Nakata, S. Sasa, K. Kodama, and S. Hiyamizu, *Jpn. J. Appl. Phys.* **26,** 1709 (1987).

134. L. Aina, M. Mattingly, and L. Stecker, *Appl. Phys. Lett.* **53,** 1620 (1988).

135. J. R. Waldrop, E. A. Kraut, C. W. Farley, and R. W. Grant, *J. Vacuum Sci. Technol. B* **8,** 768 (1990).

136. J. Y. Marzin, M. N. Charasse, and B. Sermage, *Phys. Rev. B* **31,** 8298 (1985).

137. P. K. Bhattacharya, H.-J. Bühlmann, M. Ilegems, P. Schmid, and H. Melchior, *Appl. Phys. Lett.* **41,** 449 (1982).

138. L. P. Ramberg, P. M. Enquist, Y.-K. Chen, F. E. Najjar, L. F. Eastman, E. A. Fitzgerald, and K. L. Kavanagh, *J. Appl. Phys.* **61,** 1234 (1987).

139. G. Ji, D. Huang, U. K. Reddy, H. Unlu, T. S. Henderson, and H. Morkoç, *J. Vacuum Sci. Technol. B* **5,** 1346 (1987).

140. G. Ji, D. Huang, U. K. Reddy, T. S. Henderson, R. Houdré, and H. Morkoç, *J. Appl. Phys.* **62,** 3366 (1987).

141. J. Hwang and P. Pianetta, *Appl. Phys. Lett.* **51,** 1632 (1987).

142. J. Meńendez, A. Pinczuk, D. J. Werder, S. K. Sputz, R. C. Miller, D. L. Sivco, and A. Y. Cho, *Phys. Rev. B* **36,** 8165 (1987).

143. R. L. S. Devine and W. T. Moore, *Solid State Commun.* **65,** 177 (1988).

144. T. G. Andersson, Z. G. Chen, V. D. Kulakovskii, A. Uddin, and J. T. Vallin, *Phys. Rev. B* **37,** 4032 (1988).

145. J. Menéndez and A. Pinczuk, *IEEE J. Quantum Electron.* **QE-24,** 1698 (1988).

146. C. Priester, G. Allan, and M. Lannoo, *Phys. Rev. B* **38,** 9870 (1988).

147. W. Xinghua and R. Laiho, *Superlatt. Microstruct.* **5,** 79 (1989).

148. S. Niki, C. L. Lin, W. S. C. Chang, and H. H. Wieder, *Appl. Phys. Lett.* **55,** 1339 (1989).

149. T. G. Andersson, V. D. Kulakovski, Z. G. Chen, A. Uddin, J. T. Vallin, and J. Westin, *J. Phys. Colloq.* (suppl.) **48,** C5-163 (1987).

150. K. F. Huang, K. Tai, S. N. G. Chu, and A. Y. Cho, *Appl. Phys. Lett.* **54,** 2026 (1989).

151. U. K. Reddy, G. Ji, T. S. Henderson, D. Huang, R. Houdré, and H. Morkoç, *J. Vacuum Sci. Technol. B* **7,** 1106 (1989).

152. X. Marie, J. J. Barrau, B. Brousseau, Th. Amand, M. Brousseau, E. V. K. Rao, and F. Alexandre, *J. Appl. Phys.* **69,** 812 (1991).

153. Y. Zou, P. Grodzinski, E. P. Menu, W. G. Jeong, P. D. Dapkus, J. J. Alwan, and J. J. Coleman, *Appl. Phys. Lett.* **58,** 601 (1991).

154. X. Letartre and D. Stievenard, *Appl. Phys. Lett.* **58,** 1047 (1991).

155. R. People, *J. Appl. Phys.* **62,** 2551 (1987).

156. D. Gershoni, H. Temkin, J. M. Vandenberg, S. N. G. Chu, R. A. Hamm, and M. B. Panish, *Phys. Rev. Lett.* **60,** 448 (1988).

157. T. G. Andersson, Z. G. Chen, V. D. Kulakovskii, A. Uddin, and J. T. Vallin, *Solid State Commun.* **64,** 379 (1987).

158. M. A. Reed and J. W. Lee, *Appl. Phys. Lett.* **50,** 845 (1987).

159. M. J. Joyce, M. J. Johnson, M. Gal, and B. F. Usher, *Phys. Rev. B* **38,** 10978 (1988).

160. N. Debbar, D. Biswas, and P. Bhattacharya, *Phys. Rev. B* **40,** 1058 (1989).

161. S. Y. Lin, D. C. Tsui, H. Lee, and D. Ackley, *Appl. Phys. Lett.* **55,** 2211 (1989).

162. D. J. Arent, *Phys. Rev. B* **41,** 9843 (1990).

163. Y. P. Varshni, *Physica* **34,** 149 (1967).

164. S. Zollner, S. Gopalan, and M. Cardona, *Solid State Commun.* **77,** 485 (1991).

165. K. P. O'Donnell and X. Chen, *Appl. Phys. Lett.* **58,** 2924 (1991).

166. Z. Hang, H. Shen, and F. H. Pollak, *Solid State Commun.* **73,** 15 (1990).
167. W. J. Turner, W. E. Reese, and G. D. Pettit, *Phys. Rev.* **136,** A1467 (1964).
168. R. Madelon and M. Dore, *Solid State Commun.* **39,** 639 (1981).
169. H. Temkin, V. G. Keramidas, M. A. Pollack, and W. R. Wagner, *J. Appl. Phys.* **52,** 1574 (1981).
170. T. P. Pearsall, *IEEE J. Quantum Electron.* **QE-16,** 709 (1980).
171. S. Paul, J. B. Roy, and P. K. Basu, *J. Appl. Phys.* **69,** 827 (1991).
172. A. C. Sharma, N. M. Ravindra, S. Auluck, and V. K. Srivastava, *Phys. Status Solidi B* **120,** 715 (1983).
173. G. Grégoris, J. Beerens, S. B. Amor, L. Dmowski, J. C. Portal, F. Alexandre, D. L. Sivco, and A. Y. Cho, *Phys. Rev. B* **37,** 1262 (1988).
174. H. Hazama, T. Sugimasa, T. Imachi, and C. Hamaguchi, *J. Phys. Soc. Jpn.* **55,** 1282 (1985).
175. R. A. Stradling and R. A. Wood, *J. Phys. C* **3,** L94 (1970).
176. D. Tatel, C. S. Menoni, and I. L. Spain, *J. Appl. Phys.* **66,** 1658 (1989).
177. I. Gorczyca, N. E. Christensen, and M. Alouani, *Phys. Rev. B* **39,** 7705 (1989).
178. H. Müller, R. Trommer, M. Cardona, and P. Vogl, *Phys. Rev. B* **21,** 4879 (1980).
179. T. Kobayashi, T. Tei, K. Aoki, K. Yamamoto, and K. Abe, in *Physics of Solids under High Pressure*, J. S. Schilling and R. N. Shelton, eds., North-Holland, Amsterdam, 1981, p. 141.
180. C. S. Menoni, H. D. Hochheimer, and I. L. Spain, *Phys. Rev. B* **33,** 5896 (1986).
181. T. Kobayashi, K. Aoki, and K. Yamamoto, *Physica* **139 & 140B,** 537 (1986).
182. M. Leroux, *Semicond. Sci. Technol.* **4,** 231 (1989).
183. R. People, A. Jayaraman, K. W. Wecht, D. L. Sivco, and A. Y. Cho, *Appl. Phys. Lett.* **52,** 2124 (1988).
184. J. R. Hayes, D. Patel, A. R. Adams, and P. D. Greene, *J. Electron. Mater.* **11,** 155 (1982).
185. J. M. Wiesenfeld, *J. Appl. Phys.* **47,** 143 (1985).
186. G. D. Pitt, J. Lees, R. A. Hoult, and R. A. Stradling, *J. Phys. C* **6,** 3282 (1973).
187. L. G. Shantharama, A. R. Adams, C. N. Ahmad, and R. J. Nicholas, *J. Phys. C* **17,** 4429 (1984).
188. L. G. Shantharama, R. J. Nicholas, A. R. Adams, and C. K. Sarkar, *J. Phys. C* **18,** L443 (1985).
189. A. R. Adams, L. G. Shantharama, R. J. Nicholas, and C. K. Sarkar, *Physica* **139 & 140B,** 401 (1986).
190. A. R. Adams and L. G. Shantharama, *Physica* **139 & 140B,** 419 (1986).

191. D. Gauthier, L. Dmowski, J. C. Portal, D. Leadley, M. A. Hopkins, M. A. Brummell, R. J. Nicholas, M. Razeghi, and P. Maurel, *Superlatt. Microstruct.* **4,** 201 (1988).

192. J. D. Lambkin and D. J. Dunstan, *Solid State Commun.* **67,** 827 (1988).

193. S. Massidda, A. Continenza, A. J. Freeman, T. M. de Pascale, F. Meloni, and M. Serra, *Phys. Rev. B* **41,** 12079 (1990).

194. J. A. Lahtinen, *Phys. Rev. B* **33,** 2550 (1986).

195. J. A. Silberman, T. J. de Lyon, and J. M. Woodall, *Appl. Phys. Lett.* **58,** 2126 (1991).

196. C. K. Sarkar, R. J. Nicholas, J. C. Portal, M. Razeghi, J. Chevrier, and J. Massies, *J. Phys. C* **18,** 2667 (1985).

197. D. M. Szmyd, P. Porro, A. Majerfeld, and S. Lagomarsino, *J. Appl. Phys.* **68,** 2367 (1990).

198. D. C. Reynolds, K. K. Bajaj, and T. C. Collins, *J. Appl. Phys.* **69,** 6478 (1991).

199. Y. Guldner, J. P. Vieren, P. Voisin, M. Voos, M. Razeghi, and M. A. Poisson, *Appl. Phys. Lett.* **40,** 877 (1982).

200. P. Maurel, M. Razeghi, Y. Guldner, and J. P. Vieren, *Semicond. Sci. Technol.* **2,** 695 (1987).

201. S. Yamada, A. Taguchi, and A. Sugimura, *Appl. Phys. Lett.* **46,** 675 (1985).

7

ELECTRON AND HOLE DEFORMATION POTENTIALS

7.1 DEFORMATION POTENTIALS FOR HOLES AT THE Γ POINT

The deformation potentials of the electronic states at the Brillouin zone (BZ) center ($\boldsymbol{k} = 0$) play an important role in many physical phenomena. The lattice mobility of holes in III–V compounds is limited primarily by acoustic and nonpolar optical phonon scattering. The strengths of these scattering mechanisms are determined essentially by the valence-band deformation potentials [a, b, and d (Pikus–Bir notation)].[1] Fundamental piezooptical properties, such as Raman and Brillouin scattering[2,3] and piezobirefringence (see Section 9.1), are also strongly affected by these potentials. A review of the importance of these quantities and their definition[4] and methods of calculation[5] was given by Blacha et al.[6]

The orbital-strain and stress-dependent spin–orbit interaction Hamiltonians H_{ev1} and H_{ev2} for the valence bands at $\boldsymbol{k} = 0$ in the zinc-blende semiconductors can be written as[1]

$$H_{ev1} = -a_1(e_{xx} + e_{yy} + e_{zz}) - 3b_1[(L_x^2 - \tfrac{1}{3}\boldsymbol{L}^2)e_{xx} + \text{cp}] - \sqrt{3}\, d_1[(L_xL_y + L_yL_x)e_{xy} + \text{cp}] \tag{7.1a}$$

$$H_{ev2} = -a_2(e_{xx} + e_{yy} + e_{zz})(\boldsymbol{L} \cdot \boldsymbol{s}) - 3b_2[(L_xs_x - \tfrac{1}{3}\boldsymbol{L} \cdot \boldsymbol{s})e_{xx} + \text{cp}] - \sqrt{3}\, d_2[(L_xs_y + L_ys_x)e_{xy} + \text{cp}] \tag{7.1b}$$

where e_{ij} denotes the components of the strain tensor, $\boldsymbol{L}$ is the angular momentum operator, $\boldsymbol{s}$ is the Pauli matrix, a_i is hydrostatic deformation potential, b_i and d_i are shear deformation potentials (b_1 and d_1 describe orbit–strain interaction, and b_2 and d_2 describe stress-dependent spin–orbit interaction), and cp denotes cyclic permutations with respect to the rectangular coordinates x, y, z.

The application of a uniaxial stress to a semiconductor produces a strain that reduces the symmetry of the material and results in significant change in the electronic energy bands. The uniaxial strain effect on the direct band edge at $\boldsymbol{k} = 0$ (Γ) for a zinc-blende-type semiconductor is shown schematically in Fig. 7.1. The remarkable effect of the compressive or tensile uniaxial strain (stress) on the electronic energy bands is to split the heavy hole ($J = 3/2$, $m_j = \pm 3/2$ in spherical notation) and light hole ($J = 3/2$, $m_j = \pm 1/2$) degeneracy at the Γ point. If we introduce the strain components [Eq. (3.27)] into Eq. (7.1), we obtain the changes in the band-gap energies to first order in stress[7]

$$\Delta(E_{v1} - E_{v2}) = \delta E \tag{7.2a}$$

$$\Delta(E_{v2} - E_{v3}) = \Delta_0 - \tfrac{1}{2}\,\delta E \tag{7.2b}$$

where $\delta E = 2b(S_{11} - S_{12})X$ for the [100] stress ($b = b_1 + 2b_2$) and $\delta E = (d/\sqrt{3})S_{44}X$ for the [111] stress ($d = d_1 + 2d_2$); thus, the splitting energies are proportional to the shear deformation potential b or d.

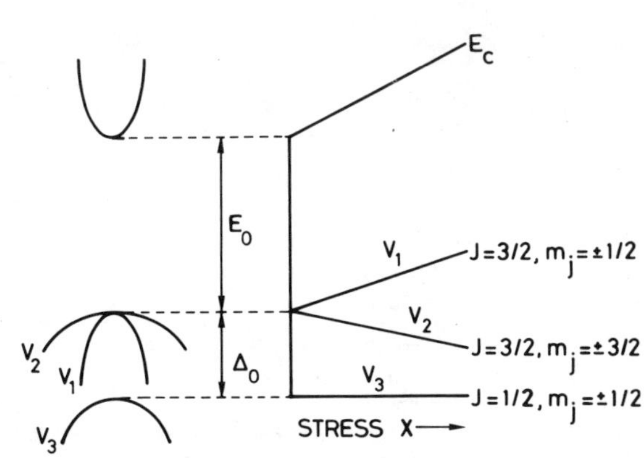

Figure 7.1 Band structure of InP for the lowest conduction band and the highest valence band and its uniaxial compressive stress dependence. The $|J = \frac{3}{2}, m_j = \pm\frac{1}{2}\rangle$, $|J = \frac{3}{2}, m_j = \pm\frac{3}{2}\rangle$, and $|J = \frac{1}{2}, m_j = \pm\frac{1}{2}\rangle$ states correspond to the light-hole, heavy-hole, and spin–orbit splitoff valence bands, respectively. Optical polarization selection rules indicate the higher-lying valence band being the light-hole band for an uniaxial compressive stress (see Refs. 7 and 25).

TABLE 7.1 Calculated and Experimental Valence-Band Deformation Potentials, *a*, *b*, and *d*, for InP, InAs, GaAs, and GaP (in eV)

InP	InAs	GaAs	GaP	Comment
		a		
−5.1[a]	−5.2[a]	−4.8[a]	−6.8[a]	Calc.
−0.4[b]	−0.6[b]	−1.6[b]	−1.5[b]	Calc.
1.27[c]	1.00[c]	1.16[c]	1.70[c]	Calc.
		1.1[d]		Calc.
−0.6[e]		−0.7 ± 1.0[e]		Exper.
		−1.0[f]		Exper.
		b		
−1.7[a]	−1.0[a]	−2.2[a]	−0.1[a]	Calc.
	−1.55[c]	−1.90[c]		Calc.
−3.3[g]	−3.2[g]	−3.9[g]	−4.1[g]	Calc.
−1.8[h]	−1.7[h]	−2.3[h]	−2.3[h]	Calc.
−2.11[i]	−2.33[i]	−2.79[i]	−2.79[i]	Calc.
−1.0[j]	−1.1[j]	−1.3[j]	−1.4[j]	Calc.
		−1.43[k]		Calc.
−2.0 ± 0.5[l]	−1.8[p]	−1.75[n]	−1.5 ± 0.2[u]	Exper.
−1.55[m]		−1.7 ± 0.2[q]	−1.8 ± 0.3[v]	Exper.
−1.55[n]		−1.96 ± 0.1[r]	−1.8[w]	Exper.
−1.5 ± 1.5[o]		−2.0 ± 0.2[s]		Exper.
		−1.66 ± 0.1[t]		Exper.
		−2.00 ± 0.2[x]		Exper.
		d		
−7.1[a]	−5.8[a]	−4.2[a]	−6.5[a]	Calc.
	−3.10[c]	−4.23[c]		Calc.
−4.7[g]	−4.4[g]	−4.6[g]	−4.7[g]	Calc.
−2.5[h]	−2.4[h]	−2.7[h]	−2.7[h]	Calc.
−3.54[i]	−3.83[i]	−4.77[i]	−4.75[i]	Calc.
		−4.54[k]		Calc.
		−4.18[k]		Calc.
−5.0 ± 0.5[l]	−3.6[p]	−5.55[n]	−4.6 ± 0.2[u]	Exper.
−4.2[m]		−4.4 ± 0.6[q]	−4.0 ± 0.5[v]	Exper.
−4.4[n]		−5.4 ± 0.3[r]	−4.3[w]	Exper.
−2.8 ± 1.2[o]		−6.0 ± 0.4[s]		Exper.
		−4.52 ± 0.25[t]		Exper.
		−4.43 ± 0.6[x]		Exper.

[a]Empirical pseudopotential calculation [A. Blacha et al., *Phys. Status Solidi B* **126,** 11 (1984)].
[b]Linear muffin-tin-orbital method [M. Cardona and N. E. Christensen, *Phys. Rev. B* **35,** 6182 (1987); erratum, ibid. **36,** 2906 (1987)].
[c]*Ab initio* pseudopotential calculation [C. G. Van de Walle, *Phys. Rev. B* **39,** 1871 (1989)].
[d]*Ab initio* pseudopotential calculation [C. G. Van de Walle and R. M. Martin, *Phys. Rev. Lett.* **62,** 2028 (1989)].
[e]D. D. Nolte et al., *Phys. Rev. Lett.* **59,** 501 (1987).
[f]W. Walukiewicz, *J. Appl. Phys.* **59,** 3577 (1986).

Although the shear deformation potentials b and d have been measured for many materials, it is difficult to obtain values for the hydrostatic potential a [$= a_1 + a_2$ (Γ_8^v)] since most experiments measure changes in energy gaps and their related effects rather than absolute shifts of the band edges. Lawaetz (cited in Ref. 8) has, however, proposed a theoretical expression based on the dielectric band theory of Phillips[9,10] that allows a for various semiconductors to be estimated with reasonable accuracy. His theory leads to the following expression for a:

$$a = -0.4\frac{E_{\text{vh}}^2}{E_{\text{v}}} - 0.7\frac{C^2}{E_{\text{v}}} \tag{7.3}$$

In Eq. (7.3) a, E_v, E_{vh}, and C are in electronvolts. The parameter C is Phillips's ionic energy gap, and E_v and E_{vh} are functions of a_0 (lattice constant) and C (see Ref. 8). A similar calculation has also recently been made by Blacha et al.[6] on the basis of the empirical pseudopotential method (EPM) and a linear combination of the atomic orbitals calculation (LCAO) by Cardona and Christensen[11] based on the linear muffin-tin-orbital method (LMTO), and by Van de Walle[12] and Van de Walle and Martin[13] based on the *ab initio* pseudopotential calculation (AIPC). These authors also calculated the shear deformation potentials b and d for several semiconductors (see Table 7.1). The shear potentials b and d were also calculated by Pries-

[g]Linear combination of atomic orbitals based on a simpler sp^3 hybrid model [A. Blacha et al., *Phys. Status Solidi B* **126,** 11 (1984)].
[h]Linear combination of atomic orbitals based on the so-called sp^3s^* hybrid model in which one excited state s^* per atom is added to the basis of the sp^3 model [A. Blacha et al., *Phys. Status Solidi B* **126,** 11 (1984)].
[i]Tight-binding approximation [C. Priester et al., *Phys. Rev. B* **37,** 8519 (1988)].
[j]Tight-binding approximation [E. P. O'Reilly, *Semicond. Sci. Technol.* **1,** 128 (1986)].
[k]Linear muffin-tin-orbital method [N. E. Christensen, *Phys. Rev. B* **30,** 5753 (1984)].
[l]J. Camassel et al., *Phys. Rev. B* **22,** 2020 (1980).
[m]G. Weber and W. Rühle, *Phys. Status Solidi B* **92,** 425 (1979).
[n]A. Gavini and M. Cardona, *Phys. Rev. B* **1,** 672 (1970).
[o]N. Suzuki and K. Tada, *Jpn. J. Appl. Phys.* **22,** 441 (1983).
[p]P. Y. Yu et al., *Phys. Rev. B* **3,** 340 (1971).
[q]I. Balslev, *Solid State Commun.* **5,** 315 (1967).
[r]R. N. Bhargava and M. I. Nathan, *Phys. Rev.* **161,** 695 (1967).
[s]F. H. Pollak and M. Cardona, *Phys. Rev.* **172,** 816 (1968).
[t]M. Chandrasekhar and F. H. Pollak, *Phys. Rev. B* **15,** 2127 (1977).
[u]H. Mathieu et al., *Phys. Rev. B* **19,** 2209 (1979).
[v]L. N. Glurdzhidze et al., *Sov. Phys. Semicond.* **7,** 305 (1973).
[w]R. G. Humphreys et al., *Phys. Rev. B* **18,** 5590 (1978).
[x]H. Qiang et al., *Solid State Commun.* **76,** 1087 (1990).

ter et al.[14] by the method of the tight-binding approximation (TBA) and by Christensen[15] by the LMTO.

As mentioned previously, experimental determination of the hydrostatic deformation potential a is quite difficult. Nolte et al.[16] reported experimental values for the conduction- and valence-band-edge hydrostatic deformation potentials of InP and GaAs based on the use of transition-metal impurity levels as reference levels in band-structure lineups. They found for the valence-band-edge deformation potential values of $a = -0.6$ eV for InP and -0.7 ± 1.0 eV for GaAs from uniaxial-stress, deep-level transient spectroscopy on defect levels of Ti and V. Walukiewicz[17] has also determined a value of the deformation potential $a \simeq -1.0$ eV for GaAs from an analysis of the temperature-dependent hole mobilities in a p-type modulation-doped GaAs/AlGaAs heterostructure.

There have been many experimental data for the shear deformation potentials b and d for various semiconductors, including for InP,[18–21] InAs,[22] GaAs,[20,23–26] and GaP.[27–29] These data were obtained by means of optical techniques, such as modulation spectroscopy (wavelength-modulated reflectivity,[18] electroreflectance,[7,25] piezoreflectance,[20,23] wavelength-modulated absorption,[29] and photoreflectance[26]), fundamental reflectivity,[19] piezobirefringence,[22,28] photoluminescence,[24] and optical absorption,[27] under uniaxial stress conditions or estimated from Dixon-Cohen's elastooptic method.[21] In Table 7.1 we compare these experimental results with the theoretical values. Experimentally the b values are close to -2.0 eV and d values generally lie in the range -4 to -6 eV for these semiconductors. Using these binary data[18,22,25,27] and Eq. (2.3), we estimate the composition dependence of the shear deformation potentials for $In_{1-x}Ga_xAs_yP_{1-y}$ lattice-matched to InP (in electrovolts):

$$b(y) = -2.0 + 0.2y \tag{7.4a}$$

$$d(y) = -5.0 + 0.5y \tag{7.4b}$$

Let us now proceed to find some general trend of the shear deformation potentials of semiconductors. In Fig. 7.2 we plot the values of $\eta = d/\sqrt{3}\,b$ as a function of the Phillips ionicity f_i for measured covalent and zinc-blende semiconductors. The quantity η is the ratio of the splitting for a shear strain along [111] to the splitting for a shear strain of the same magnitude along [100]. The bold line in the figure is the locus for $In_{1-x}Ga_xAs_yP_{1-y}$ lattice-matched to InP obtained from Eq. (7.4). The dashed line also represents a reasonable fit to the experimental findings. We can recognize from the figure that η increases with increasing f_i. Such a trend may be interpreted by a point-ion model proposed by Gavini and Cardona.[20,30]

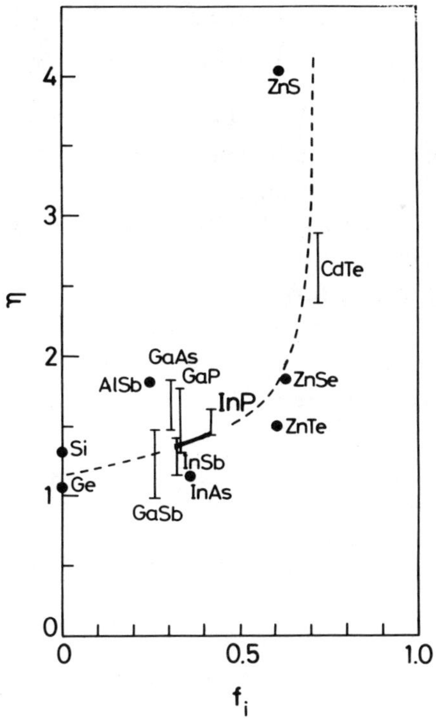

Figure 7.2 Plots of the measured $\eta = d/\sqrt{3}b$ versus Phillips's ionicity f_i for tetragonal semiconductors. The bold line is the locus for $In_{1-x}Ga_xAs_yP_{1-y}$ lattice-matched to InP obtained from Eq. (7.4). The dashed line represents an average of the experimental findings.

The phonons that usually dominate in the scattering probability are the long-wavelength optical phonons that produce a short-range potential in the crystal that shifts the electronic band states. In polar semiconductors, like InP, the phonons are also accompanied by a long-range macroscopic electric field that produces additional scattering. The shifts of the electronic band states per unit ionic displacement associated with long-wavelength optical phonons are called the *optical-phonon deformation potential* d_o.

The deformation potential d_o is defined by the splitting of the Γ_{15} state produced, in the absence of spin-orbit interaction, by a phonon along [111][6]

$$\delta E = d_o \frac{u}{a_0} \tag{7.5}$$

where δE is the shift of the singlet component of Γ_{15} polarized along [111], a_0 is the lattice constant, and u is the change of the bond length in the [111] direction due to the displacement of the two sublattices. The deformation potentials d and d_o are related linearly through the internal strain parameter ζ:[6]

$$d = d' - \tfrac{1}{4}\zeta d_o \tag{7.6}$$

A direct measurement of ζ is, however, rather difficult since it requires the measurement of the intensity of an X-ray reflection forbidden in the unstrained crystal.

The experimental deduction of d_o from either low-field transport[31,32] or Raman data[33] is quite involved. Several theoretical calculations of d_o for the covalent and zinc-blende semiconductors were carried out by the methods of the EPM,[5,6] LCAO,[5,6] SCBO (self-consistent Born–Oppenheimer approximation),[34] LMTO,[15,35,36] and AIPC.[37,38] In Table 7.2 we list theoretical and experimental results for d_o for InP, InAs, GaAs, and GaP.

Lawaetz[39] has determined an effective acoustic-mode deformation potential, Ξ_{eff}, appropriate for low-field transport in p-type materials with Ge-like (zinc-blende) valence bands. This parameter is written as

$$\Xi_{\text{eff}}^2 = a^2 + \frac{C_l}{C_t}\left(b^2 + \tfrac{1}{2}d^2\right) \tag{7.7}$$

where C_l and C_t are spherically averaged elastic coefficients given by

$$C_l = \tfrac{1}{5}(3C_{11} + 2C_{12} + 4C_{44}) \tag{7.8a}$$

$$C_t = \tfrac{1}{5}(C_{11} - C_{12} + 3C_{44}) \tag{7.8b}$$

The calculated composition dependence of Ξ_{eff} versus y for $In_{1-x}Ga_xAs_yP_{1-y}$ lattice-matched to InP can be written as (in electronvolts)

$$\Xi_{\text{eff}}(y) = 7.4 - 4.3y + 2.2y^2 \tag{7.9}$$

where the elastic constants used are taken from Table 3.1. This expression gives $\Xi_{\text{eff}}(\text{InP}) = 7.4$ eV, which compares very favorably with the value 6.6 eV reported in Ref. 16.

Although the deformation constant Ξ_{eff} is a somewhat phenomenologic average over the longitudinal and transverse acoustic phonon modes, it is widely used to model the performance of semiconductor devices and is particularly useful for electrical engineers. The deformation constant Ξ_{eff} is also related to the phenomenologic acoustic deformation potential, E_{ac}, proposed by Wiley and DiDomenico[31,40] by the equation

$$E_{\text{ac}}^2 = \frac{\beta + 2}{6\beta}\,\Xi_{\text{eff}}^2 \tag{7.10}$$

where $\beta = C_l/C_t$. We find that the calculated dependence of E_{ac} on y for $In_{1-x}Ga_xAs_yP_{1-y}$ quaternary is also a linear relationship given by (in elec-

TABLE 7.2 Calculated and Experimental Optical-Phonon Deformation Potentials d_o for InP, InAs, GaAs, and GaP

Optical-Phonon Deformation Potential d_o (eV)				
InP	InAs	GaAs	GaP	Comment
35.6[a]	33.3[a]	36.4[a]	28.9[a]	Calc.
30.5[b]	29.5[b]	36.3[b]	38.1[b]	Calc.
16.3[c]	16.1[c]	20.9[c]	21.5[c]	Calc.
31.7[d]	28.5[d]	29.1[d]	26.5[d]	Calc.
32.0[e]	31.0[e]	37.0[e]	39.4[e]	Calc.
		25.9[f]		Calc.
		16.5[g]		Calc.
14.1[h]	14.3[h]	17.5[h]		Calc.
20.1[i]	20.8[i]	25.0[i]	24.3[i]	Calc.
		34.0[j]	31.5[j]	Calc.
22.6[k]	25.3[k]	37.1[k]	33.1[k]	Calc.
42[l]	42[l]	41[l]	44[l]	Exper.
35[m]		41[m]	47[m]	Exper.
		48[n]		Exper.

[a]Empirical pseudopotential calculation [A. Blacha et al., *Phys. Status Solidi B* **126,** 11 (1984)].
[b]Linear combination of atomic orbitals based on a simpler sp^3 hybrid model [A. Blacha et al., *Phys. Status Solidi B* **126,** 11 (1984)].
[c]Linear combination of atomic orbitals based on the so-called sp^3s^* hybrid model in which one excited state s^* per atom is added to the basis of the sp^3 model [A. Blacha et al., *Phys. Status Solidi B* **126,** 11 (1984)].
[d]Empirical pseudopotential calculation [W. Pötz and P. Vogl, *Phys. Rev. B* **24,** 2025 (1981)].
[e]Linear combination of atomic orbitals calculation [W. Pötz and P. Vogl, *Phys. Rev. B* **24,** 2025 (1981)].
[f]Self-consistent Born–Oppenheimer approximation [A. J. Hernández-Cabrera et al., *J. Phys. C* **16,** 2251 (1983)].
[g]Linear muffin-tin-orbital method [N. E. Christensen, *Phys. Rev. B* **30,** 5753 (1984)].
[h]Linear muffin-tin-orbital method [N. E. Christensen et al., *Phys. Rev. B* **36,** 1032 (1987)].
[i]Linear muffin-tin-orbital method [L. Brey et al., *Phys. Rev. B* **36,** 2638 (1987)].
[j]*Ab initio* pseudopotential calculation [B.-S. Wang et al., *Phys. Rev. B* **39,** 12789 (1989); see also comment by M. Cardona and N. E. Christensen, ibid. **41,** 5407 (1990)].
[k]*Ab initio* pseudopotential calculation [Z.-Q. Gu et al., *Phys. Rev. B* **41,** 8333 (1990)].
[l]J. D. Wiley, in *Semiconductors and Semimetals*, Vol. 10, R. K. Willardson and A. C. Beer, eds., Academic, New York, 1975, p. 91.
[m]See, for instance, W. Pötz and P. Vogl, *Phys. Rev. B* **24,** 2025 (1981).
[n]M. H. Grimsditch et al., *Phys. Rev. B* **20,** 1758 (1979).

tronvolts)

$$E_{ac}(y) = 8.3 - 5.4y + 2.8y^2 \qquad (7.11)$$

Wiley and DiDomenico reviewed[40] the circumstances under which polar mode scattering is important in the III–V compounds and concluded that it is the dominant intrinsic scattering mechanism only in the n-type direct-band-gap materials. For p-type III–V materials, they found that the dominant hole-scattering mechanism is a combination of acoustic and nonpolar optical mode scattering. Their analyses of the hole mobility data yielded the values of $E_{ac} = 6.5 \pm 1.5$ eV for InP, $E_{ac} = 5.3$ eV for GaAs, and $E_{ac} = 5.7$ eV for GaP. An estimate in Ref. 31 finds $E_{ac} = 4$ eV for GaAs, which was recently confirmed with ultrafast laser cooling experiments by Rühle et al ($E_{ac} = 4.8 \pm 1.0$ eV).[41]

7.2 DEFORMATION POTENTIALS FOR ELECTRONS

If one neglects stress-induced coupling between the *s*-like conduction band Γ_6^c and the *p*-like valence band $\Gamma_8^v(M_j = \pm\frac{1}{2})$, the energy displacement of the Γ_6^c minimum is linear in strain and its orbit-strain Hamiltonian may be written as

$$H_{ec} = C_1(e_{xx} + e_{yy} + e_{zz}) \tag{7.12}$$

where $e_{ii}(i = x, y, \text{ or } z)$ is the diagonal component of the strain tensor.

Let us now consider the case of the scattering of electrons due to the strain caused by acoustic waves, that is, the intravalley (acoustic) deformation potential scattering. If the strains involved are small as in the usual case, the electronic energy shifts caused by them may be described adequately with linear terms in the strain. By symmetry, for spherical constant energy surfaces and acoustic-mode scattering, one may write for the shift of the Γ_6 minimum ΔE_c^Γ as

$$\Delta E_c^\Gamma = E_1(e_{xx} + e_{yy} + e_{zz}) \tag{7.13}$$

where E_1 is the so-called *intravalley (acoustic) deformation potential*. Equation (7.13) is based on the fact that the matrix element of Eq. (7.12) is practically equal to that obtained by replacing H_{ec} by ΔE_c^Γ. The deformation potential $E_1(C_1)$ can now be written in a phenomenologic form as[42]

$$E_1 = -B_u \frac{dE_c^\Gamma}{dp} \tag{7.14a}$$

or[13]

$$E_1 = \frac{dE_c^\Gamma}{d \ln V} \tag{7.14b}$$

where B_u is the bulk modulus (see Section 3.2), dE_c^Γ/dp is the hydrostatic pressure coefficient of the Γ_6^c minimum, and $d \ln V = dV/V$ is the fractional volume change of the crystal.

There are various theoretical models for finding values of conduction-band-edge deformation potentials for tetrahedral semiconductors.[6, 11–13, 15, 43] Vergés et al.[43] used self-consistent LMTO calculations for bulk solids to derive the deformation potentials. In this method the solid is broken up into atom-like spheres and all potentials are referred to a reference level that is chosen so that the Hartree potential of a single atomic sphere is zero at infinity. The solid can be terminated at any sphere while leaving the electronic charge distribution in this sphere equal to what it would have in the bulk. An attempt was made to evaluate in this manner the electron–phonon interaction constants relevant to longitudinal acoustic phonons. Recently, it was argued that screening needs to be included in the previous approach, and Cardona and Christensen[11] proposed a model to calculate these effects. Van de Walle and Martin[13] have also reported the AIPC determination of the deformation potentials, which takes into account long-range electrostatic effects in their theory. In Table 7.3 we summarize the calculated conduction-band deformation potential E_1 for InP, InAs, GaAs, and GaP. Since GaP is an indirect-band-gap semiconductor, we list the results for not only the lowest direct minimum (Γ_6^c) but also the lowest conduction minima (X_6^c).

Experimental determinations of the conduction-band deformation potentials have been quite controversial. This is because those measurements are indirect and require a significant amount of analysis, interpretation, and assumption. For example, reported values for InP range widely from $|E_1| = 3.4$ to 21 eV.[44–54] Hamilton[52] pointed out that the estimated values 18.5 and 21 eV may be perhaps too high because of the use of the approximate Matthiessen's mobility rule ("$1/\mu$"; see Chapter 10). Electron mobility analysis by Takeda and Sasaki[45] yielded $|E_1| = 3.4$ eV. Taguchi and Yamada[46] found $|E_1| = 6.5$ eV from electron mobility analysis for high-purity n-type InP by taking into account the compensation ratio. This value (6.5 eV) is in acceptable agreement with the one recently calculated by Cardona and Christensen [5.9 eV (screened value)].[11]

The value $|E_1| = 5.8$ eV for InAs[47] was determined from an analysis of electron transport data. Free-carrier absorption data for InAs also provided a value of $|E_1| = 10$ eV.[55]

Many low-field transport and free-electron optical absorption data are available for GaAs,[47, 48, 56–66] and their analyses yield the $|E_1|$ values ranging from 7 to 16.5 eV. Recently, Nolte et al.[16] have found a value of $E_1 = -9.3 \pm 1.0$ eV for GaAs based on the use of transition-metal impurity levels as reference levels in band-structure lineups. This value is suffi-

TABLE 7.3 Calculated and Experimental Conduction-Band Deformation Potentials E_1 for InP, InAs, GaAs, and GaP (in eV)

Conduction-Band Deformation Potential E_1				
InP	InAs	GaAs	GaP	Comment
−13.7[a]	−11.7[a]	−18.3[a]	−16.1[a,b]	Calc.
		−17.1[c]		Calc.
−5.9[d]	−7.3[d]	−8.8[d]	+0.8[d,e]	Calc.
−5.04[f]	−5.08[f]	−7.17[f]	−7.14,[b,f] + 3.26[e,f]	Calc.
21[g]	11.5[h]	8.6[k]	13.0[e,k,ee]	Exper.
14.5[h,i]	5.8[k]	9.0[q]	~10[e,ff]	Exper.
18[j]	10[r]	−6.3[s,t]	9 ± 1[e,gg]	Exper.
6.8[k]		16.0 ± 0.5[u]		Exper.
14.2 ± 0.7[l]		−15.7[v]		Exper.
3.6[m]		13.5[w,x]		Exper.
3.4[n]		7[y,z]		Exper.
14[o]		12[aa]		Exper.
6.5[p]		11 ± 1[bb]		Exper.
11.5[q]		−9.3 ± 1.0[cc]		Exper.
		11 − 12[dd]		Exper.

[a]Empirical pseudopotential calculation [A. Blacha et al., *Phys. Status Solidi B* **126,** 11 (1984)].
[b]Γ_6^c conduction-band minimum.
[c]Linear muffin-tin-orbital model [N. E. Christensen, *Phys. Rev. B* **30,** 5753 (1984)].
[d]Linear muffin-tin-orbital method [M. Cardona and N. E. Christensen, *Phys. Rev. B* **35,** 6182 (1987); erratum, ibid. **36,** 2906 (1987)].
[e]X_6^c conduction-band minima.
[f]*Ab initio* pseudopotential calculation [C. G. Van de Walle, *Phys. Rev. B* **39,** 1871 (1989)].
[g]V. V. Galavanov and N. V. Siukaev, *Phys. Status Solidi* **38,** 523 (1970).
[h]D. L. Rode, *Phys. Rev. B* **3,** 3287 (1971).
[i]D. K. Hamilton, *Solid-State Electron.* **24,** 317 (1981).
[j]B. R. Nag and G. M. Dutta, *J. Phys. C* **11,** 119 (1978).
[k]D. L. Rode, in *Semiconductors and Semimetals*, Vol. 10, R. K. Willardson and A. C. Beer, eds., Academic, New York, 1975, p. 1.
[l]H. J. Lee et al., *Can. J. Phys.* **58,** 923 (1980).
[m]J. R. Hayes et al., *J. Electron. Mater.* **11,** 155 (1982).
[n]Y. Takeda and A. Sasaki, *Solid-State Electron.* **27,** 1127 (1984).
[o]D. A. Anderson and N. Apsley, *Semicond. Sci. Technol.* **1,** 187 (1986).
[p]A. Taguchi and S. Yamada, *J. Appl. Phys.* **61,** 2412 (1987).
[q]D. Lancefield et al., *J. Appl. Phys.* **62,** 2342 (1987).
[r]E. Haga and H. Kimura, *J. Phys. Soc. Jpn.* **19,** 471 (1964).
[s]S. Perkowitz, *J. Appl. Phys.* **40,** 3751 (1969).
[t]K. Osamura and Y. Murakami, *Jpn. J. Appl. Phys.* **11,** 365 (1972).
[u]H. J. Lee et al., *Can. J. Phys.* **57,** 233 (1979).
[v]P. Pfeffer et al., *Solid State Commun.* **51,** 179 (1984).
[w]E. E. Mendez et al., *Appl. Phys. Lett.* **45,** 294 (1984).
[x]P. J. Price, *Phys. Rev. B* **32,** 2643 (1985).
[y]K. Lee et al., *J. Appl. Phys.* **54,** 6432 (1983).
[z]W. Walukiewicz et al., *Phys. Rev. B* **30,** 4571 (1984); **32,** 2645 (1985).
[aa]B. Vinter, *Phys. Rev. B* **33,** 5904 (1986).
[bb]K. Hirakawa and H. Sakaki, *Appl. Phys. Lett.* **49,** 889 (1986).
[cc]D. D. Nolte et al., *Phys. Rev. Lett.* **59,** 501 (1987).
[dd]T. Kawamura and S. Das Sarma, *Phys. Rev. B* **42,** 3725 (1990).
[ee]D. L. Rode, *Phys. Status Solidi B* **53,** 245 (1972).
[ff]J. D. Wiley and M. DiDomenico, Jr., *Phys. Rev. B* **1,** 1655 (1970).
[gg]S. Kocsis, *Phys. Status Solidi A* **28,** 133 (1975).

ciently close to $E_1 = -8.8$ eV predicted theoretically by Cardona and Christensen.[11]

For GaP, values of $|E_1| = 13$ eV and 9 ± 1 eV have been obtained by Rode[47,67] and Kocsis,[68] respectively, from the analyses of electron-transport data. Wiley and DiDomenico[69] have also obtained $|E_1| \sim 10$ eV from free-carrier absorption in *n*-type GaP. These values are the deformation constants at the indirect-conduction minima (X_6^c) but not at the Γ_6^c minimum. Note that the reported experimental values are much higher than the theoretically predicted one [$E_1 = +0.8$ eV (screened); see Ref. 11]. In Table 7.3 we list the experimentally determined deformation potential values E_1 for the III–V binaries (InP, InAs, GaAs, and GaP).

The deformation potential E_1 for $In_{0.53}Ga_{0.47}As$ ternary has been determined by several authors from analyses of electron-transport data.[70–72] The values obtained are $|E_1| = 5.89$ eV[18,70] and 9.4 eV.[72] To our knowledge, however, no experimental data has been reported for $In_{1-x}Ga_xAs_yP_{1-y}$ quaternary.

It is generally accepted that the Gunn effect arises from a negative conductance mechanism in which the electrons are transferred from a low-mass central valley (Γ) to higher-lying large mass satellite valley (*L* and/or *X*) such as exist in the conduction band of GaAs. The strength of this electron-transfer mechanism can be represented by the coupling constant D_{ij}.[73,74] The constant D_{ij} $(i, j = x, y, \text{ or } z)$ is the so-called *intervalley deformation potential constant* (in electronvolts per centimeter), where $i = j$ corresponds to the equivalent intervalley scattering and $i \neq j$ corresponds to the nonequivalent one.

Figure 7.3 shows the conduction band in the Γ–*X* and the Γ–*L* directions for InP, including a schematic representation of the intervalley deformation potential scattering. Although the scattering processes play an important role in the analysis of electron-transfer properties, there seems to be no reliable experimental data to date. The experimental deformation potential constants reported in the literature largely scatter for many semiconductors (see, e.g., Refs. 75 and 76).

The intervalley deformation constants D_{ij} between the lowest Γ-, *X*-, and *L*-conduction-band valleys in zinc-blende semiconductors have recently been calculated by Zollner et al.[77,78] using empirical pseudopotentials. The matrix element for the intervalley scattering of a carrier $|\boldsymbol{k}\rangle$ by absorption or emission of a phonon $|\boldsymbol{q}\rangle$ can be simply written as $\langle \boldsymbol{k} \pm \boldsymbol{q}|H_{iv}|\boldsymbol{k}\rangle$, where H_{iv} is the intervalley electron–phonon interaction Hamiltonian. As H_{iv} has the symmetry of the phonon involved, the scattering is forbidden unless the representation belonging to the state $|\boldsymbol{k} \pm \boldsymbol{q}\rangle$ is contained in the product of the representations belonging to $|\boldsymbol{q}\rangle$ and $|\boldsymbol{k}\rangle$. Their calculated D_{ij} values for InP, InAs, GaAs, and GaP are listed in Table 7.4.

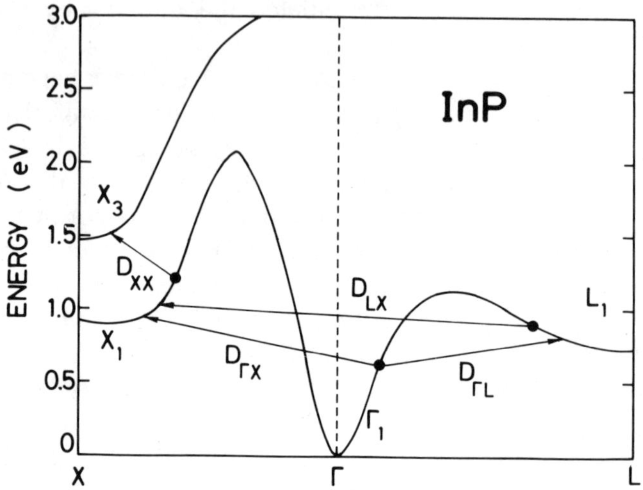

Figure 7.3 Conduction band in the Γ–X and the Γ–L directions for InP, including a schematic representation of the intervalley deformation potential scattering.

Zollner et al.[77] pointed out that Γ–X_1 scattering by LO phonons and Γ–X_3 scattering by LA phonons are not allowed for InP, InAs, and GaP; conversely, Γ–X_1 scattering by LA phonons and Γ–X_3 scattering by LO phonons are not allowed for GaAs. This can be explained as follows.[77] In zinc-blende materials the symmetries for the two lowest conduction band states at X and X_1 and X_3 (single group), with the anion at the origin. The state X_1 is lower in energy than X_3 for semiconductors of interest here. The transverse phonons have X_5 symmetry and do not contribute to Γ–X scattering. The LA and LO phonons have X_1 and X_3 symmetry. If the cation is

TABLE 7.4 Intervalley Deformation Potential Constant D_{ij} for InP, InAs, GaAs, and GaP (in 10^8 eV/cm)

D_{ij}	Phonon	InP	InAs	GaAs	GaP
$D_{\Gamma L}$	LA	1.4	1.7	3.0	0.8
	LO	1.8	1.0	0.4	0.7
	LA + LO	2.3	2.0	3.0	1.1
$D_{\Gamma X}(\Gamma–X_1)$	LA/LO	1.6	2.2	2.9	1.1
$D_{\Gamma X}(\Gamma–X_3)$	LA/LO	2.6	2.0	3.3	0.8
D_{LL}	LA + LO	0.9	1.1	1.2	0.6
$D_{XX}(X_1–X_1)$	LA/LO	3.1	2.5	4.9	3.0
$D_{LX}(L–X_1)$	TA	0.7	0.6	0.8	0.4
	LA	0.5	0.6	0.4	0.7
	LO	2.8	1.9	1.8	1.6
	TO	1.2	1.1	1.8	0.7

Source: S. Zollner et al., *J. Appl. Phys.* **68,** 1682 (1990).

heavier than the anion, as in the case of InP, InAs, and GaP, then the X_3 state has higher energy (LO) than the X_1 (LA) state. In the reverse case (such as GaSb, AlAs, and AlSb), the LO (LA) phonon has $X_1(X_3)$ symmetry. The latter case is also true for GaAs, where the two masses are similar.

The intervalley deformation constants D_{ij} in III–V compounds have also been calculated using a method of nonlocal pseudopotential by Grinyaev et al.[79] Zollner et al.[80] have employed the rigid pseudoion method to calculate the $\boldsymbol{q}$ dependence of D_{ij} values for GaP from the Γ point to the X valley. They found that not only the LA and LO phonons are allowed to couple away from X, but the fast TA phonon also has to be considered along certain directions. Using spectroscopic ellipsometry, they also obtained an intervalley scattering time of 40 fs, which yields an intervalley deformation potential of about 4×10^8 eV/cm, whereas the rigid pseudoion theory gives a value of only about 1×10^8 eV/cm.

REFERENCES

1. G. E. Pikus and G. L. Bir, *Sov. Phys. Solid State* **1,** 136 (1959).
2. M. Cardona, *Light Scattering in Solids*, Springer, Berlin, 1975.
3. K. Ando and C. Hamaguchi, *Phys. Rev. B* **11,** 3876 (1975); S. Adachi and C. Hamaguchi, *Phys. Rev. B* **19,** 938 (1979).
4. G. L. Bir and G. E. Pikus, *Symmetry and Strain-Induced Effects in Semiconductors*, Wiley, New York, 1974.
5. W. Pötz and P. Vogl, *Phys. Rev. B* **24,** 2025 (1981).
6. A. Blacha, H. Presting, and M. Cardona, *Phys. Status Solidi B* **126,** 11 (1984).
7. F. H. Pollak and M. Cardona, *Phys. Rev.* **172,** 816 (1968).
8. J. D. Wiley, *Solid State Commun.* **8,** 1865 (1970).
9. J. C. Phillips, *Bonds and Bands in Semiconductors*, Academic, New York, 1973.
10. J. C. Phillips, *Phys. Rev. Lett.* **20,** 550 (1968).
11. M. Cardona and N. E. Christensen, *Phys. Rev. B* **35,** 6182 (1987).
12. C. G. Van de Walle, *Phys. Rev. B* **39,** 1871 (1989).
13. C. G. Van de Walle and R. M. Martin, *Phys. Rev. Lett.* **62,** 2028 (1989).
14. C. Priester, G. Allan, and M. Lannoo, *Phys. Rev. B* **37,** 8519 (1988).
15. N. E. Christensen, *Phys. Rev. B* **30,** 5753 (1984).
16. D. D. Nolte, W. Walukiewicz, and E. E. Haller, *Phys. Rev. Lett.* **59,** 501 (1987).
17. W. Walukiewicz, *J. Appl. Phys.* **59,** 3577 (1986).

18. J. Camassel, P. Merle, L. Bayo, and H. Mathieu, *Phys. Rev. B* **22,** 2020 (1980).
19. G. Weber and W. Rühle, *Phys. Status Solidi B* **92,** 425 (1979).
20. A. Gavini and M. Cardona, *Phys. Rev. B* **1,** 672 (1970).
21. N. Suzuki and K. Tada, *Jpn. J. Appl. Phys.* **22,** 441 (1983).
22. P. Y. Yu, M. Cardona, and F. H. Pollak, *Phys. Rev. B* **3,** 340 (1971).
23. I. Balslev, *Solid State Commun.* **5,** 315 (1967).
24. R. N. Bhargava and M. I. Nathan, *Phys. Rev.* **161,** 695 (1967).
25. M. Chandrasekhar and F. H. Pollak, *Phys. Rev. B* **15,** 2127 (1977).
26. H. Qiang, F. H. Pollak, and G. Hickman, *Solid State Commun.* **76,** 1087 (1990).
27. H. Mathieu, P. Merle, E. L. Ameziane, B. Archilla, J. Camassel, and G. Poiblaud, *Phys. Rev. B* **19,** 2209 (1979).
28. L. N. Glurdzhidze, A. P. Izergin, Z. N. Kopylova, and A. D. Remenyuk, *Sov. Phys. Semicond.* **7,** 305 (1973).
29. R. G. Humphreys, U. Rössler, and M. Cardona, *Phys. Rev. B* **18,** 5590 (1978).
30. A. Gavini and M. Cardona, *Phys. Rev.* **177,** 1351 (1969).
31. J. D. Wiley, in *Semiconductors and Semimetals*, Vol. 10, R. K. Willardson and A. C. Beer, eds., Academic, New York, 1975, p. 91.
32. See, for instance, W. Pötz and P. Vogl, *Phys. Rev. B* **24,** 2025 (1981).
33. M. H. Grimsditch, D. Olego, and M. Cardona, *Phys. Rev. B* **20,** 1758 (1979).
34. A. J. Hernández-Cabrera, J. Sanchez-Dehesa, and C. Tejedor, *J. Phys. C* **16,** 2251 (1983).
35. L. Brey, N. E. Christensen, and M. Cardona, *Phys. Rev. B* **36,** 2638 (1987).
36. N. E. Christensen, S. Satpathy, and Z. Pawlowska, *Phys. Rev. B* **36,** 1032 (1987).
37. B.-S. Wang, Z.-Q. Gu, J.-Q. Wang, and M.-F. Li, *Phys. Rev. B* **39,** 12789 (1989); see also comment by M. Cardona and N. E. Christensen, ibid. **41,** 5407 (1990).
38. Z.-Q. Gu, M.-F. Li, J.-Q. Wang, and B.-S. Wang, *Phys. Rev. B* **41,** 8333 (1990).
39. P. Lawaetz, *Phys. Rev.* **166,** 763 (1968); **174,** 867 (1968); **183,** 730 (1969).
40. J. D. Wiley and M. DiDomenico, Jr., *Phys. Rev. B* **2,** 427 (1970).
41. W. W. Rühle, K. Leo, and E. Bauser, *Phys. Rev. B* **40,** 1756 (1989).
42. H. Ehrenreich, *Phys. Rev.* **120,** 1951 (1960).
43. J. A. Vergés, D. Glötzel, M. Cardona, and O. K. Andersen, *Phys. Status Solidi B* **113,** 519 (1982).
44. J. R. Hayes, D. Patel, A. R. Adams, and P. D. Greene, *J. Electron. Mater.* **11,** 155 (1982).

45. Y. Takeda and A. Sasaki, *Solid-State Electron.* **27,** 1127 (1984).
46. A. Taguchi and S. Yamada, *J. Appl. Phys.* **61,** 2412 (1987).
47. D. L. Rode, in *Semiconductors and Semimetals*, Vol. 10, R. K. Willardson and A. C. Beer, eds., Academic, New York, 1975, p. 1.
48. D. Lancefield, A. R. Adams, and M. A. Fisher, *J. Appl. Phys.* **62,** 2342 (1987).
49. D. A. Anderson and N. Apsley, *Semicond. Sci. Technol.* **1,** 187 (1986).
50. H. J. Lee, J. Basinski, L. Y. Juravel, and J. C. Woolley, *Can. J. Phys.* **58,** 923 (1980).
51. D. L. Rode, *Phys. Rev. B* **3,** 3287 (1971).
52. D. K. Hamilton, *Solid-State Electron.* **24,** 317 (1981).
53. B. R. Nag and G. M. Dutta, *J. Phys. C* **11,** 119 (1978).
54. V. V. Galavanov and N. V. Siukaev, *Phys. Status Solidi* **38,** 523 (1970).
55. F. Haga and H. Kimura, *J. Phys. Soc. Jpn.* **19,** 471 (1964).
56. S. Perkowitz, *J. Appl. Phys.* **40,** 3751 (1969).
57. K. Osamune and Y. Murakami, *Jpn. J. Appl. Phys.* **11,** 365 (1972).
58. H. J. Lee, J. Basinski, L. Y. Juravel, and J. C. Woolley, *Can. J. Phys.* **57,** 233 (1979).
59. P. Pfeffer, I. Gorczyca, and W. Zawadzki, *Solid State Commun.* **51,** 179 (1984).
60. E. E. Mendez, P. J. Price, and M. Heiblum, *Appl. Phys. Lett.* **45,** 294 (1984).
61. P. J. Price, *Phys. Rev. B* **32,** 2643 (1985).
62. K. Lee, M. S. Shur, T. J. Drummond, and H. Morkoç, *J. Appl. Phys.* **54,** 6432 (1983).
63. W. Walukiewicz, H. E. Ruda, J. Lagowski, and H. C. Gatos, *Phys. Rev. B* **30,** 4571 (1984); **32,** 2645 (1985).
64. B. Vinter, *Phys. Rev. B* **33,** 5904 (1986).
65. K. Hirakawa and H. Sakaki, *Appl. Phys. Lett.* **49,** 889 (1986).
66. T. Kawamura and S. Das Sarma, *Phys. Rev. B* **42,** 3725 (1990).
67. D. L. Rode, *Phys. Status Solidi B* **53,** 245 (1972).
68. S. Kocsis, *Phys. Status Solidi A* **28,** 133 (1975).
69. J. D. Wiley and M. DiDomenico, Jr., *Phys. Rev. B* **1,** 1655 (1970).
70. Y. Takeda, M. A. Littlejohn, J. A. Hutchby, and R. J. Trew, *Electron. Lett.* **17,** 686 (1981).
71. Y. Takeda, M. A. Littlejohn, and J. R. Hauser, *Appl. Phys. Lett.* **39,** 620 (1981).
72. J. H. Marsh, *Appl. Phys. Lett.* **41,** 732 (1982).
73. See, for instance, E. M. Conwell, *High Field Transport in Semiconductors*, Academic, New York, 1967.
74. W. Fawcett and D. C. Herbert, *Electron. Lett.* **9,** 308 (1973).

75. S. Zollner, S. Gopalan, and M. Cardona, *Solid State Commun.* **76,** 877 (1990).
76. R. Mickevičius and A. Reklaitis, *Semicond. Sci. Technol.* **5,** 805 (1990).
77. S. Zollner, S. Gopalan, and M. Cardona, *Appl. Phys. Lett.* **54,** 614 (1989).
78. S. Zollner, S. Gopalan, and M. Cardona, *J. Appl. Phys.* **68,** 1682 (1990).
79. S. N. Grinyaev, G. F. Karavaev, and V. G. Tyuterev, *Sov. Phys. Semicond.* **23,** 905 (1989).
80. S. Zollner, J. Kircher, M. Cardona, and S. Gopalan, *Solid-State Electron.* **32,** 1585 (1989).

8

OPTICAL PROPERTIES

8.1 OPTICAL DISPERSION RELATIONS

$In_{1-x}Ga_xAs_yP_{1-y}/InP$ heterostructures become increasingly interesting as light sources and detectors for optical fiber communication systems in the 1.4–1.7-μm wavelength region. Knowledge of optical properties is very important in the design and analysis of such optoelectronic devices.

The complex dielectric function

$$\epsilon(\omega) = \epsilon_1(\omega) + i\epsilon_2(\omega) \tag{8.1}$$

is known to describe the optical properties of the medium at all photon energies $E = \hbar\omega$. Real (ϵ_1) and imaginary (ϵ_2) parts of the dielectric function are connected by the well-known Kramers–Krönig (KK) relations:[1]

$$\epsilon_1(\omega) = 1 + \frac{2}{\pi}\int_0^\infty \frac{\omega'\epsilon_2(\omega')}{(\omega')^2 - \omega^2}\, d\omega' \tag{8.2a}$$

$$\epsilon_2(\omega) = -\frac{2}{\pi}\int_0^\infty \frac{\epsilon_1(\omega')}{(\omega')^2 - \omega^2}\, d\omega' \tag{8.2b}$$

The complex refractive index $n^*(\omega) = n(\omega) + ik(\omega)$ is now given by

$$n^* = n(\omega) + ik(\omega) = [\epsilon_1(\omega) + i\epsilon_2(\omega)]^{1/2} \tag{8.3}$$

where $n(\omega)$ is the real refractive index and $k(\omega)$ is the extinction coefficient, also called the *attenuation index*. The constants $n(\omega)$ and $k(\omega)$ are real and positive numbers and can be determined by optical measurements. From Eqs. (8.1) and (8.3), it follows that

$$\epsilon_1 = n^2 - k^2, \qquad \epsilon_2 = 2nk \tag{8.4}$$

and also that

$$n(\omega) = \frac{[\epsilon_1(\omega)^2 + \epsilon_2(\omega)^2]^{1/2} + \epsilon_1(\omega)}{2} \tag{8.5}$$

$$k(\omega) = \frac{[\epsilon_1(\omega)^2 + \epsilon_2(\omega)^2]^{1/2} - \epsilon_1(\omega)}{2} \tag{8.6}$$

The absorption coefficient $\alpha(\omega)$ depends not only on $\epsilon_2(\omega)$ but also on both parts of the dielectric function

$$\alpha(\omega) = \frac{4\pi}{\lambda} k(\omega) \tag{8.7}$$

where λ is the wavelength of light in the vacuum.

The normal-incidence reflectivity R is

$$R = \left| \frac{n^* - 1}{n^* + 1} \right|^2 = \frac{(n - 1)^2 + k^2}{(n + 1)^2 + k^2} \tag{8.8}$$

8.2 THE RESTSTRAHLEN REGION

The reststrahlen region is the region where the radiation field interacts with the fundamental lattice vibrations. Two effects resulted: (1) absorption or emission of the electromagnetic wave due to the creation or annihilation of lattice vibrations and (2) scattering of the electromagnetic wave by the lattice vibrational modes. Optical absorption due to lattice vibrations appears as absorption bands in the infrared region (with wavelength range of a few tens of micrometers) attributed either to a first- or second-order electronic dipole moment or to anharmonic terms in the potential energy. In polar crystals, as in InP, the first-order dipole moment gives rise to a very strong absorption band associated with optical modes having an of essentially zero **k** vector (i.e., long-wavelength optical phonons). A useful description of the theoretical and practical aspects of this subject is given by Balkanski.[2]

The complex dielectric permittivity $\epsilon(\omega)$ in the long-wavelength limit can be explained by a single harmonic oscillator model:

$$\epsilon(\omega) = \epsilon_\infty \left[1 + \frac{\omega_{\rm LO}^2 - \omega_{\rm TO}^2}{\omega_{\rm TO}^2 - \omega^2 - i\omega\gamma} \right] \tag{8.9}$$

where $\omega_{\rm LO}$ and $\omega_{\rm TO}$ are the long-wavelength LO and TO phonon frequencies, respectively and γ is the phonon damping constant. This expression can be modified using the Lyddane–Sachs–Teller relation of Eq. (5.3) as

$$\epsilon(\omega) = \epsilon_\infty + \frac{\omega_{\rm TO}^2(\epsilon_s - \epsilon_\infty)}{\omega_{\rm TO}^2 - \omega^2 - i\omega\gamma} \tag{8.10}$$

The reststrahlen parameters ϵ_s, ϵ_∞, $\omega_{\rm LO}$, $\omega_{\rm TO}$, and γ can be obtained from measurements of absorption, refractive index dispersion, reflection at normal and oblique incidence, and also from Raman scattering and other more complicated measurements.[3] Bairamov et al.[4] have recently performed room-temperature Raman and infrared–reflectivity measurements on semiinsulating InP doped with Fe and found the reststrahlen parameters of this material: $\omega_{\rm TO} = 303.7$ cm^{-1}, $\omega_{\rm LO} = 344.5$ cm^{-1}, and $\gamma = 1.2$ cm^{-1}. From the Lyddane–Sachs–Teller relationship and $\epsilon_s = 12.5$ (Table 5.1), we obtain $\epsilon_\infty = 9.71$, which is in acceptable agreement with the value of 9.66 obtained in Section 5.1 (see Table 5.1).

Figure 8.1 shows spectral variations of $\epsilon_1(\omega)$ and $\epsilon_2(\omega)$ from a numeric modeling of Eq. (8.9) [Eq. (8.10)] with the reststrahlen parameters of Bairamov et al.[4] The imaginary part of Eq. (8.9) is now modified as

$$\epsilon_2(\omega) = \frac{S\omega_{\rm TO}^2\omega\gamma}{(\omega_{\rm TO}^2 - \omega^2)^2 + \omega^2\gamma^2} \tag{8.11}$$

with $S \equiv \epsilon_\infty(\omega_{\rm LO}^2 - \omega_{\rm TO}^2)/\omega_{\rm TO}^2$. The frequency $\omega_{\rm TO}(=303.7$ cm$^{-1})$ in the present sample is much larger than $\gamma(=1.2$ cm$^{-1})$, and $\epsilon_2(\omega)$ then has a significant value only when ω is close to $\omega_{\rm TO}$. For such values of ω it is a good approximation to replace ω by $\omega_{\rm TO}$ everywhere in Eq. (8.11), except in the factor $\omega_{\rm TO} - \omega$, which gives

$$\epsilon_2(\omega) = \frac{S\omega_{\rm TO}\gamma/4}{(\omega_{\rm TO} - \omega)^2 + (\gamma/2)^2} \qquad (\gamma \ll \omega_{\rm TO}) \tag{8.12}$$

The $\epsilon_2(\omega)$ thus reaches a maximum value $S\omega_{\rm TO}/\gamma(=705)$ for frequency $\omega_{\rm TO}$ and drops to half that peak value for frequencies $\omega_{\rm TO} \pm \frac{1}{2}\gamma$. Expression (8.12) is known as the *Lorentzian lineshape*.

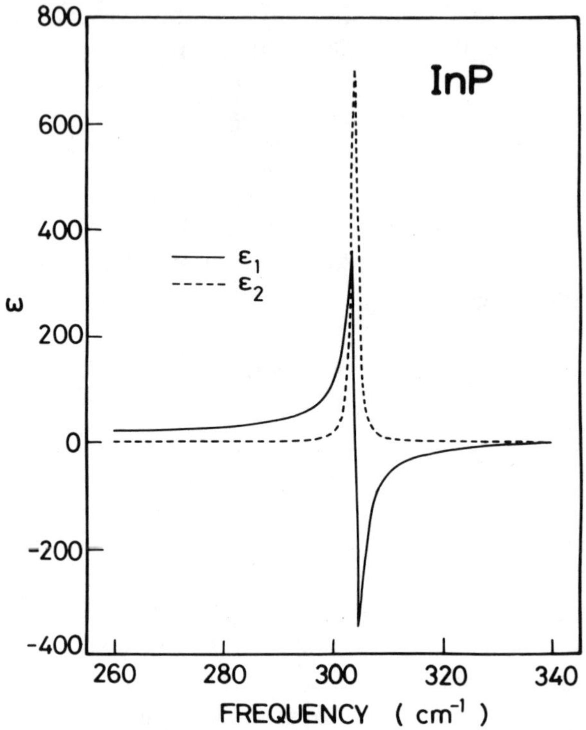

Figure 8.1 Spectral variation of $\epsilon(\omega)$ obtained from a numeric modeling of Eq. (8.9) [Eq. (8.10)] with the reststrahlen parameters of semiinsulating InP determined at room temperature by Bairamov et al.[4]

The corresponding variations of $n(\omega)$ and $k(\omega)$ obtained from Eqs. (8.5) and (8.6) are shown in Fig. 8.2. The absorption coefficient α and normal-incidence reflectivity R as calculated from Eqs. (8.7) and (8.8) are also shown in Figs. 8.3 and 8.4, respectively. Both n and $k(\alpha)$ dispersion curves show a strong peak at the resonance frequency $\omega = \omega_{TO}$. The reflectivity R at ω close to ω_{LO} is nearly zero.

The optical properties at the reststrahlen region in semiinsulating crystals are adequately well interpreted by the above-mentioned single harmonic oscillator model. If free carriers are present in the medium, we must take into account plasma contributions to the dielectric permittivity. When the frequency of free-carrier plasma excitations (plasmons) is close to the frequency of the LO phonons (ω_{LO}), the two excitations interact via their macroscopic electric fields. The dielectric permittivity in such case is given by

$$\epsilon(\omega) = \epsilon_\infty \left[1 + \frac{\omega_{LO}^2 - \omega_{TO}^2}{\omega_{TO}^2 - \omega^2 - i\omega\gamma} - \frac{\omega_p^2}{\omega(\omega + i\Gamma)} \right] \tag{8.13}$$

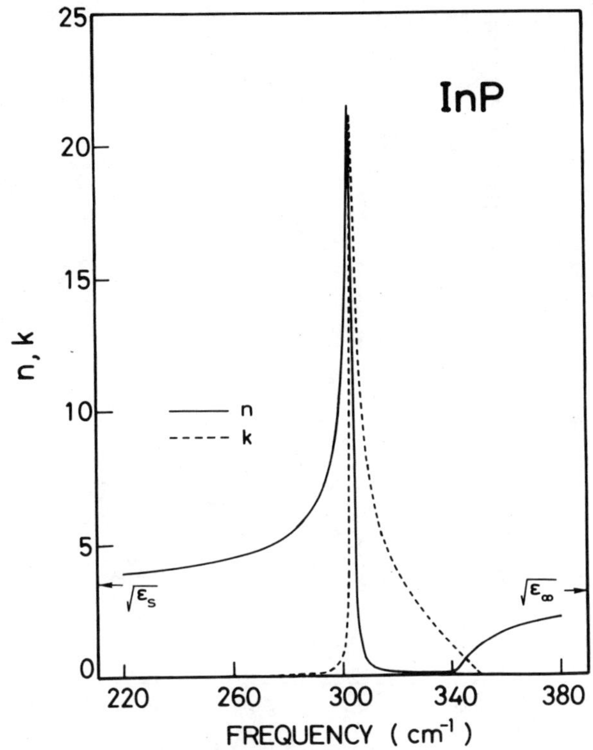

Figure 8.2 Numerically calculated spectral dependence of the real refractive index (n) and the extinction coefficient (k) at the reststrahlen region of InP using the same set of reststrahlen parameters as in Fig. 8.1.

where

$$\omega_p = \left(\frac{ne^2}{m^* \epsilon_\infty \epsilon_0} \right)^{1/2} \tag{8.14}$$

is the plasma frequency, Γ is the plasmon damping constant, and n and m^* are, respectively, the concentration and effective mass of the free carriers. Expression (8.13) has been used as the basis for sorting out plasmon–phonon effects in a number of investigations (see review by Blakemore[3]). Kukharskii,[5] however, showed that this expression oversimplifies the coupling of plasmons and phonons and can lead, principally, to errors in deducing the damping constants γ and Γ. In Kukharskii's model, $\epsilon(\omega)$ is given by

$$\epsilon(\omega) = \epsilon_\infty \frac{(\omega^2 + i\Gamma_-\omega - \omega_-^2)(\omega^2 + i\Gamma_+\omega - \omega_+^2)}{\omega(\omega + i\Gamma)(\omega^2 + i\gamma\omega - \omega_{TO}^2)} \tag{8.15}$$

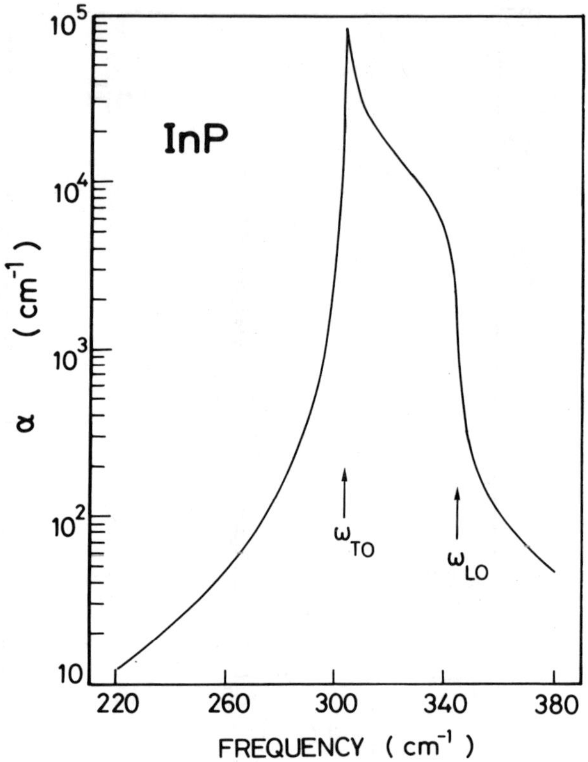

Figure 8.3 Numerically calculated spectral dependence of the absorption coefficient α at the reststrahlen region of InP using the same set of reststrahlen parameters as in Fig. 8.1.

where

$$\omega_{\pm}^2 = \tfrac{1}{2}\{\omega_{LO}^2 + \omega_p^2[(\omega_{LO}^2 + \omega_p^2)^2 - 4\omega_p^2\omega_{TO}^2]^{1/2}\} \tag{8.16}$$

In his model two additional parameters are the damping constants $\Gamma_{\pm}$ of the coupled plasmon–phonon modes. The plasmon and single-oscillator models are related to each other through the equation

$$\omega_+\omega_- = \omega_p\omega_{TO} \tag{8.17}$$

Bairamov et al.[4] have also measured room-temperature infrared–reflectivity spectra of *n*-type InP doped with Sn and S in the concentration range from 6.8×10^{17} to 1×10^{19} cm^{-1}. We reproduce their results in Fig. 8.5. The reflectance minimum at about 300 cm^{-1} is due to the LO phonon. The position of the second minimum, called the *plasma minimum*, depends on

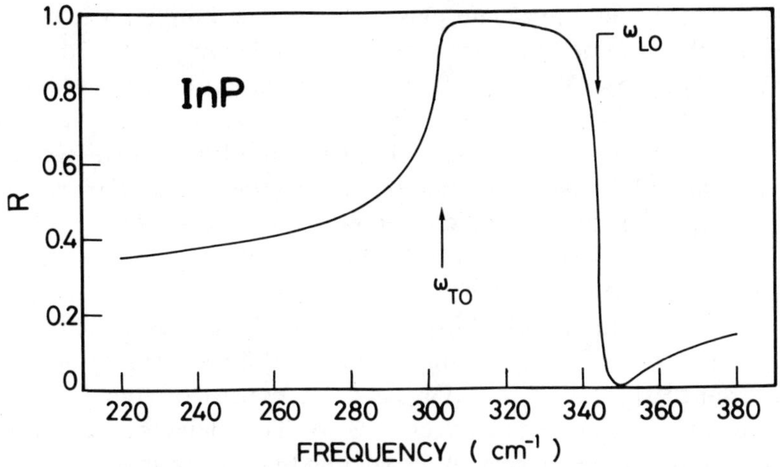

Figure 8.4 Numerically calculated spectral dependence of the normal-incidence reflectivity R at the reststrahlen region of InP using the same set of reststrahlen parameters as in Fig. 8.1.

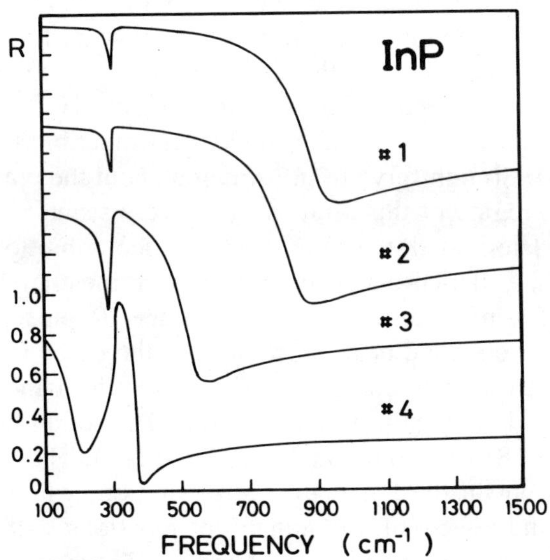

Figure 8.5 Room-temperature infrared reflectivity data for n-type InP. The carrier concentrations are 8.8×10^{18} cm^{-3} (#1), 7.1×10^{18} cm^{-3} (#2), 2.3×10^{18} cm^{-3} (#3), and 4.9×10^{17} cm^{-3} (#4). The minimum at about 300 cm^{-1} is due to the LO phonons. The position of the second minimum, called the *plasma minimum*, depends on the electron concentration. (From Bairamov et al.[4])

the electron concentration. The plasma resonance minimum is very close to the TO–phonon line in the spectrum of the relatively lightly doped sample (#1 in Fig. 8.5). When the carrier concentration is larger (#2–#4), the plasma minimum is shifted toward higher photon energies as seen in Fig. 8.5. The carrier concentrations are obtained from Eq. (8.14). Because InP has a nonparabolic conduction band, the effective mass $m^*(=m_e^\Gamma)$ in Eq. (8.14) depends on the carrier concentration (see Chapter 6). This nonparabolic-band correction is given by Eq. (6.10) and has been taken into consideration in the calculations.

Reflection spectra of bulk alloy samples usually consist of lattice bands, the number of which is equal to the number of different cells. For example, the reflectance spectra of $In_{1-x}Ga_xAs_yP_{1-y}$ consist of four lattice bands (i.e., four-mode behavior; see Section 3.3). The dielectric permittivity in such materials can be successfully explained by the following multioscillator model:

$$\epsilon(\omega) = \epsilon_\infty + \sum_j \frac{S_j \omega_{TOj}^2}{\omega_{TOj}^2 - \omega^2 - i\omega\gamma_j} - \frac{\epsilon_\infty \omega_p^2}{\omega(\omega + i\Gamma)} \qquad (8.18)$$

where $S_j[=\epsilon_\infty(\omega_{LOj}^2 - \omega_{TOj}^2)/\omega_{TOj}^2]$ is the oscillator strength, ω_{TOj} is the TO phonon frequency, and γ_j the damping constant of the jth lattice oscillator.

As mentioned in Section 3.3, many authors have performed optical studies in the reststrahlen region of the $In_{1-x}Ga_xAs_yP_{1-y}$ quaternary using the Raman and/or far-infrared–reflectivity techniques. However, such works were focused mainly on the phonon-mode determination (the four- or three-mode behavior). Straightforward information about the optical constants in the reststrahlen region of this alloy is, thus, very scant.[6–8]

Pickering[8] determined the four-mode lattice vibrational behavior in $In_{1-x}Ga_xAs_yP_{1-y}$ lattice-matched to InP over most of the composition range using a far-infrared–reflectivity technique. Representative examples of the spectra he obtained near the middle of the composition range (y = 0.35 and 0.50) are shown in Fig. 8.6. The four lattice bands and the plasma edge can be clearly recognized in the spectra. The solid lines are the results of fits to Eq. (8.18) with four oscillators (j = 1–4). The fits are seen to be fairly good except in the higher-frequency region, where the penetration depth is high and effects of any nonabrupt junctions will be most noticeable. The composition dependence of the oscillator strengths (S_j values) was also found to be proportional to the pair concentration [i.e., $(1-x)(1-y)$ for the InP-mode oscillator, $(1-x)y$ for the InAs-mode one, xy for the GaAs-mode one, and $x(1-y)$ for the GaP-mode one] for a random alloy.[8]

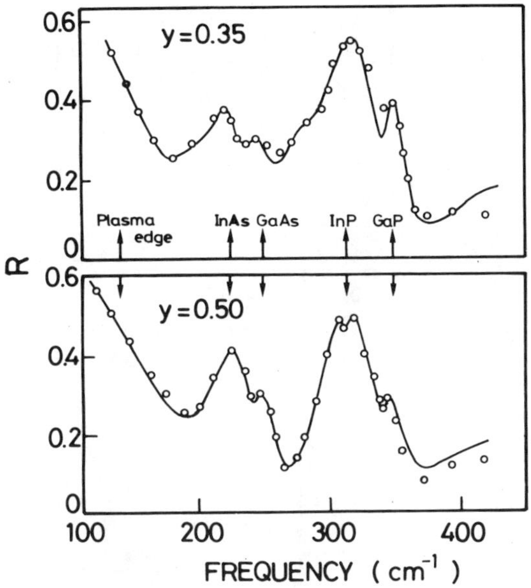

Figure 8.6 Room-temperature infrared reflectivity data for $In_{1-x}Ga_xAs_yP_{1-y}$ quaternary (y = 0.35 and 0.50). The solid lines represent the results of fits to Eq. (8.18) with four oscillators. (From Pickering.[8])

8.3 THE INTERBAND TRANSITION REGION

8.3.1 Model Dielectric Function

Let us first present various approximations[9] of the dielectric constants in the interband transition region of semiconductors. Figures 8.7*a–f* show the approximations used in the case of tetragonal semiconductors to describe the refractive-index dispersion below the fundamental absorption edge [$n(\omega) \simeq \epsilon_1(\omega)^{0.5}$]. The dashed curves are the spectral dependence $\epsilon_2(\omega)$ typical of the diamond- and zinc-blende-type structures. The continuous curves represent the various model representations of $\epsilon_2(\omega)$ used in Refs. 10–15.

The nature of the spectrum of $\epsilon_2(\omega)$ in the fundamental absorption region of direct-gap semiconductors is well characterized by a square-root dependence, $\epsilon_2(\omega) \propto (\hbar\omega - E_0)^{0.5}$ (Fig. 8.7*a*[10,11]). Wemple and DiDomenico have proposed a semiempirical single-effective-oscillator model to analyze refractive index dispersion in more than 100 widely different solids and liquids (Fig. 8.7*b*[12]). Their model requires two parameters, E_p and E_d, where the imaginary part of the dielectric constant ϵ_2 of the material was assumed to be a delta function at energy E_p and the strength of the effective

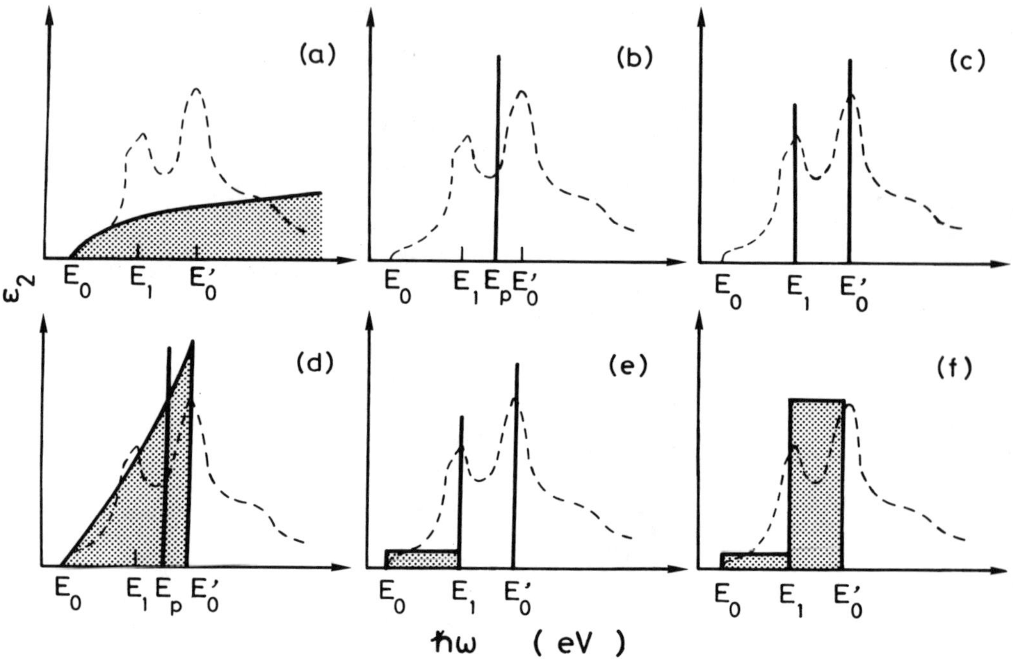

Figure 8.7 Imaginary part of the dielectric constant $\epsilon(\omega)$ (dashed lines) in the interband transition region and its model representations (solid lines) used in Refs. 10 and 11 (*a*), 12 (*b*), 13 (*c*), 14 (*d*), and 15 (*e*, *f*) for semiconductors with diamond and zinc-blend structures.

oscillator was defined to be $\pi E_d/2$. The KK transformation of this function gives

$$\epsilon_1(\omega) - 1 = \frac{E_p E_d}{E_p^2 - E^2} \tag{8.19}$$

where $E = \hbar\omega$. On the other hand, Stern considered a two-oscillator model that assumes effective oscillators at energies E_1 and E_2 (Fig. 8.7*c*[13]). Matters are, however, complicated by the lack of agreement of the data with the Wemple–DiDomenico or Stern model at the direct-band edge, which is the energy range of most interest for semiconductor optoelectronic devices. Afromowitz has therefore proposed a modified single-effective-oscillator model (Fig. 8.7*d*[14]), which takes into account the lowest-direct-gap contribution, and applied it to $Al_xGa_{1-x}As$, $GaAs_yP_{1-y}$, and $In_{1-x}Ga_xP$ ternaries. In the Afromowitz model the contribution of the lowest-direct gap E_0 to $\epsilon_2(\omega)$ was assumed to have the form of $\epsilon_2(\omega) \propto (\hbar\omega)^4$. His assumed $\epsilon_2(\omega)$ agrees closely with the actual spectrum in the region below E_1 (see Fig. 8.7*d*) but is fully empirically derived and therefore nonphysical. Pikhtin and coworkers' model representations of $\epsilon_2(\omega)$ also allow for the electronic density of states, which is assumed to be uniformly distributed in the interval between $\hbar\omega = E_0 - E_1$ (Fig. 8.7e[15]), $\hbar\omega = E_0 - E_1$ and $\hbar\omega = E_1 - E_2$ (Fig. 8.7f[15]). It seems that all these approximations oversimplify the actual $\epsilon_2(\omega)$ spectra.

Let us now present a new model of the dielectric functions $\epsilon_1(\omega)$ and $\epsilon_2(\omega)$, which covers the optical response of semiconductors over a wide range of photon energies.[16–24] The model is based on the KK transformation [Eq. (8.2)] and is strongly connected with the electronic energy-band structure of the medium (Chapter 6). This model predicts distinct structures at energies of E_0, $E_0 + \Delta_0$, E_1, $E_1 + \Delta_1$, and E_0' critical points (CP). The indirect-band-gap transitions also play an important part in the spectral dependence of both ϵ_1 and ϵ_2.

In Figs. 8.8 and 8.9 we show our calculated $\epsilon_2(\omega)$ and $\epsilon_1(\omega)$ spectra of

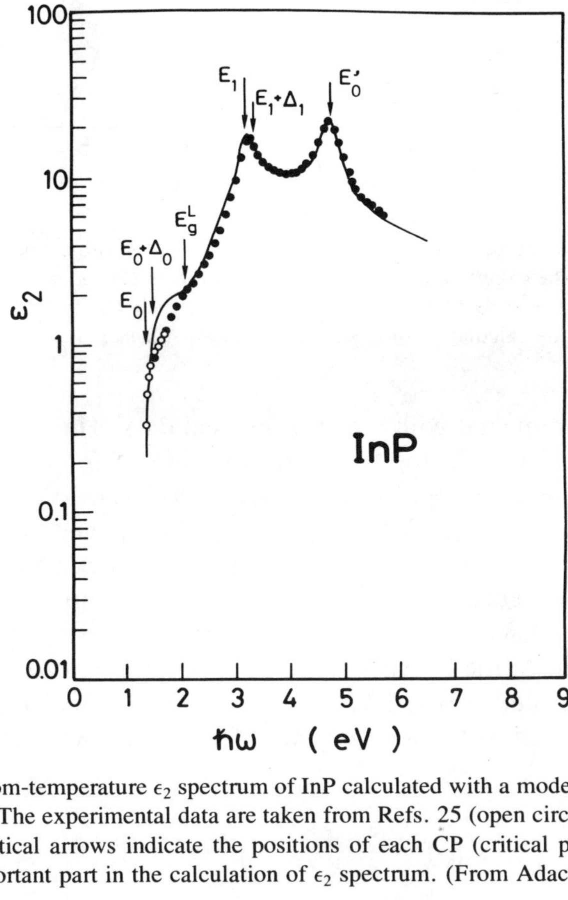

Figure 8.8 Room-temperature ϵ_2 spectrum of InP calculated with a model discussed in the text (solid line). The experimental data are taken from Refs. 25 (open circles) and 26 (solid circles). The vertical arrows indicate the positions of each CP (critical point) and E_g^L gap that play an important part in the calculation of ϵ_2 spectrum. (From Adachi.[16])

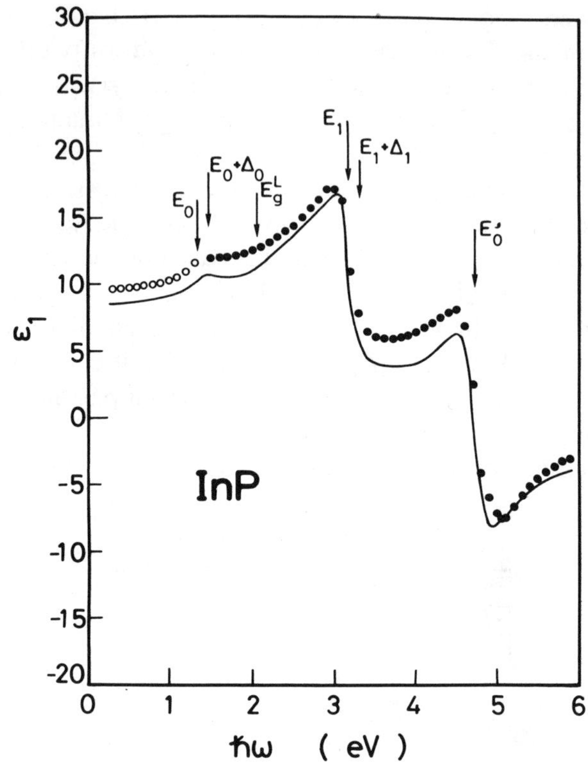

Figure 8.9 Room-temperature ϵ_1 spectrum of InP calculated with a model discussed in the text (solid line). The experimental data are taken from Refs. 27 (open circles) and 26 (solid circles). The vertical arrows indicate the positions of each CP and E_g^L gap that play an important part in the calculation of ϵ_1 spectrum. (From Adachi.[16])

InP at 300 K compared with the experimental data. The experimental data are taken from Refs. 25 (open circles, Fig. 8.8), 26 (solid circles, Figs. 8.8 and 8.9), and 27 (open circles, Fig. 8.9). The locations of several interband transitions, E_0, $E_0 + \Delta_0$, E_1, $E_1 + \Delta_1$, E_0', and E_g^L (indirect gap), which play an important part in the calculations of ϵ_1 and ϵ_2 spectra, are indicated in the figures by the arrows. Our model representation of $\epsilon_2(\omega)$ allows for joint density of states of the interband transitions from states in the valence band to the conduction band in the Brillouin zone (BZ). Agreement with the experimental data is good (Fig. 8.8). The constant $\epsilon_2(\omega)$ can be strongly related to the joint density-of-states (DOS) function $J_{cv}(\omega)$ as follows:

$$\epsilon_2(\omega) = \frac{4\hbar^2 e^2}{\pi \mu_0^2 \omega^2} |\langle c|p|v\rangle|^2 J_{cv}(\omega) \tag{8.20}$$

where μ_0 is the combined DOS mass and $\langle c|p|v\rangle$ is the momentum matrix element for v (valence) → c (conduction) transitions. Thus $J_{cv}(\omega)$ of paramount importance for the explanation of optical spectra, and its analytic behavior at various types of CP in the BZ can be calculated.[28] In the following we summarize the model dielectric functions for CP of each energy gap (E_0, $E_0 + \Delta_0$, E_1, $E_1 + \Delta_1$, E_0', and indirect-band gap), in III–V, zinc-blende-type materials. Combining all these contributions, we can obtain the spectral dependence of $\epsilon(\omega)$ of the materials in the entire range of photon energies.[16–24]

(a) E_0 and $E_0 + \Delta_0$ Transitions The E_0 and $E_0 + \Delta_0$ transitions are of the three-dimensional (3D) M_0 CPs. Assuming that the bands are parabolic and using the KK relation, we obtain the contribution of these gaps to $\epsilon(\omega)$:

$$\epsilon(\omega) = AE_0^{-1.5}\left\{f(\chi_0) + \frac{1}{2}\left(\frac{E_0}{E_0 + \Delta_0}\right)^{1.5} f(\chi_{so})\right\} \tag{8.21}$$

with

$$A = \tfrac{4}{3}(\tfrac{3}{2}\mu_0)^{1.5}P^2 \tag{8.22}$$

$$f(\chi_0) = \chi_0^{-2}[2 - (1 + \chi_0)^{0.5} - (1 + \chi_0)^{0.5}] \tag{8.23a}$$

$$f(\chi_{so}) = \chi_{so}^{-2}[2 - (1 + \chi_{so})^{0.5} - (1 + \chi_{so})^{0.5}] \tag{8.23b}$$

$$\chi_0 = \frac{\hbar\omega + i\Gamma}{E_0} \tag{8.24a}$$

$$\chi_{so} = \frac{\hbar\omega + i\Gamma}{E_0 + \Delta_0} \tag{8.24b}$$

where P^2 is the squared momentum matrix element and Γ is the broadening constant of the $E_0/(E_0 + \Delta_0)$ transitions. Here, for simplicity, the reduced mass μ_0 for the three valence bands is assumed to be equal.

In Fig. 8.10 we show our calculated $\epsilon(\omega)$ spectra of Eq. (8.21) [$\epsilon_2(\omega) = \text{Im}\,\epsilon(\omega)$; $\epsilon_1(\omega) = \text{Re}\,\epsilon(\omega)$]. The numeric parameters used in the calculations correspond to those for InP and are as follows: $E_0 = 1.35$ eV, $E_0 + \Delta_0 = 1.47$ eV, $A = 5.4$ eV$^{1.5}$, and $\Gamma = 0$ (solid lines) and 15 meV (dashed lines). This expression gives the spectral dependence $\epsilon_2(\omega)$ [i.e., Im $\epsilon(\omega)$] $\propto (\hbar\omega - E_0)^{0.5}$ [$\propto (\hbar\omega - E_0 - \Delta_0)^{0.5}$], which is well known for the majority of direct-gap semiconductors. The model is thus essentially the same as that used in Refs. 10 and 11 (Fig. 8.7*a*).

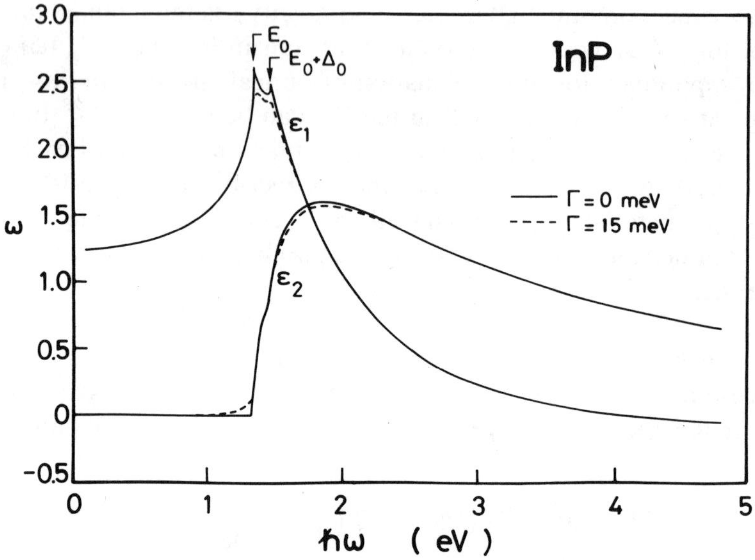

Figure 8.10 Lineshape of the $E_0/(E_0 + \Delta_0)$ gap contribution to $\epsilon(\omega)$ for InP [Eq. (8.21)] with two different broadening parameters [$\Gamma = 0$ meV (solid line) and $\Gamma = 15$ meV (dashed line)].

(b) E_1 and $E_1 + \Delta_1$ Transitions The E_1 and $E_1 + \Delta_1$ transitions are of the 3D M_1 CPs. The contributions to ϵ_2 of this type are

$$\epsilon_2(\omega) = \begin{cases} \pi\chi_1^{-2}[B_1 - B_{11}(E_1 - \hbar\omega)^{0.5}] & (\hbar\omega \le E_1) \\ \pi B_1 \chi_1^{-2} & (\hbar\omega \ge E_1) \end{cases} \tag{8.25a}$$

for the E_1 transitions, and

$$\epsilon_2(\omega) = \begin{cases} \pi\chi_{1s}^{-2}[B_2 - B_{21}(E_1 + \Delta_1 - \hbar\omega)^{0.5}] & (\hbar\omega < E_1 + \Delta_1) \\ \pi B_2 \chi_{1s}^{-2} & (\hbar\omega \ge E_1 + \Delta_1) \end{cases} \tag{8.25b}$$

for the $E_1 + \Delta_1$ transitions, where

$$\chi_1 = \frac{\hbar\omega}{E_1} \tag{8.26a}$$

$$\chi_{1s} = \frac{\hbar\omega}{E_1 + \Delta_1} \tag{8.26b}$$

The B values in Eq. (8.25) are the strength paramers of the E_1 and $E_1 + \Delta_1$ transitions.

Since the M_1-CP longitudinal effective mass m_L is much larger than its transverse counterparts, the m_T s, one can treat these 3D M_1 CPs as two-dimensional (2D) minima M_0. The contribution to ϵ_2 of this type of 2D minima is given by

$$\epsilon_2(\omega) = \pi[B_1\chi_1^{-2}H(\chi_1 - 1) + B_2\chi_{1s}^{-2}H(\chi_{1s}^{-1} - 1)] \tag{8.27}$$

where

$$H(z) = \begin{cases} 1 & \text{for } z \geq 0 \\ 0 & \text{for } z < 0 \end{cases} \tag{8.28}$$

The KK transformation of Eq. (8.27) gives

$$\epsilon_1(\omega) = -B_1\chi_1^{-2} \ln (1 - \chi_1^2) - B_2\chi_{1s}^{-2} \ln (1 - \chi_{1s}^2) \tag{8.29}$$

Calculation of Eqs. (8.27) and (8.29) satisfactorily interprets experimental ϵ spectra of various semiconductors in these transition regions. However, it is worth noting that these expressions do not account for any damping effect, that is, for a lifetime broadening. We thus introduce in these expressions such an effect in a phenomenologic manner by replacing $\hbar\omega$ by $\hbar\omega + i\Gamma$. The contribution to $\epsilon(\omega)$ of the E_1 and $E_1 + \Delta_1$ transitions is then written as

$$\epsilon(\omega) = -B_1\chi_{1d}^{-2} \ln (1 - \chi_{1d}^2) - B_2\chi_{1sd}^{-2} \ln (1 - \chi_{1sd}^2) \tag{8.30}$$

where

$$\chi_{1d} = \frac{\hbar\omega + i\Gamma}{E_1} \tag{8.31a}$$

$$\chi_{1sd} = \frac{\hbar\omega + i\Gamma}{E_1 + \Delta_1} \tag{8.31b}$$

We show in Fig. 8.11 our calculated $\epsilon(\omega)$ spectra of Eq. (8.30) [$\epsilon_2(\omega) = \text{Im}\,\epsilon(\omega)$; $\epsilon_1(\omega) = \text{Re}\,\epsilon(\omega)$]. The numeric values used in the calculations correspond to those for InP and are as follows: $E_1 = 3.16$ eV; $E_1 + \Delta_1 = 3.30$ eV, $B_1 = 3.6$, $B_2 = 0.9$, and $\Gamma = 0$ (solid lines) and 0.12 eV (dashed lines). As seen in Fig. 8.11, Eq. (8.30) with $\Gamma = 0$ eV exhibits a divergence in ϵ_1 at the E_1 CP. The damping effect can successfully decrease the

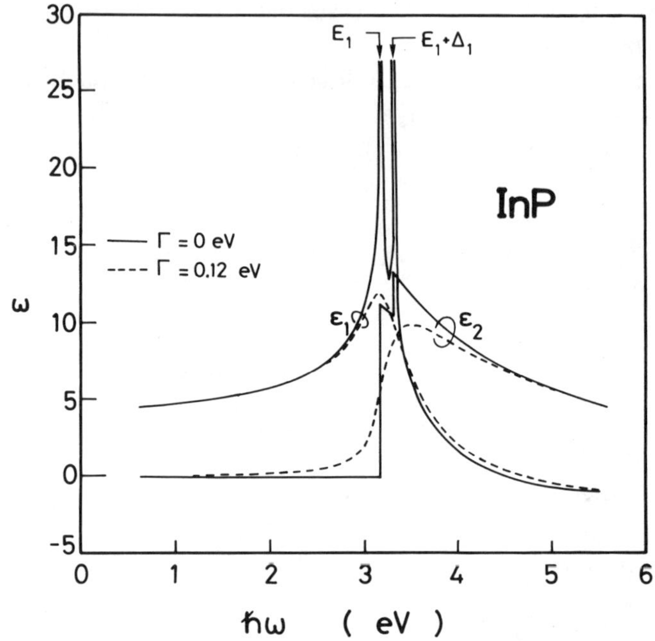

Figure 8.11 Lineshape of the $E_1/(E_1 + \Delta_1)$ gap contribution to $\epsilon(\omega)$ for InP [Eq. (8.30)] with two different broadening parameters [$\Gamma = 0$ eV (solid line) and $\Gamma = 0.12$ eV (dashed line)].

ϵ_1 strength of the E_1 structure and lead to agreement with experiment. In the limit $\Gamma \rightarrow 0$ eV, Eq. (8.30) [Im $\epsilon(\omega)$] also exactly agrees with Eq. (8.27).

(c) $E_0'(E_2)$ Transitions As discussed in Chapter 6, the more pronounced structure found in the $In_{1-x}Ga_xAs_yP_{1-y}$ system in the region higher in energy than E_1 is labeled E_0' (E_0' triplet). The E_0' triplet transitions in the $In_{1-x}Ga_xAs_yP_{1-y}$ system cannot usually be resolved in the rather broad CPs observed in optical measurements, such as reflectance spectroscopy and spectroscopic ellipsometry. We, therefore, consider that only the E_0' CP is the main dispersion source in this spectral region.

The nature of the E_0' transitions is more complicated. They do not correspond to a single, well-defined CP. If the transitions occur at $\Gamma(\Delta)$, then the CP should be a 3D M_0 (3D M_1) type. However, these fits show that neither the 3D M_0 nor the 3D M_1 model represents the peculiar lineshapes of ϵ_1 and ϵ_2 in the E_0' spectral region. The best fits are obtained with a damped harmonic oscillator (DHO) model.

The DHO gives

$$\epsilon(\omega) = \frac{C}{(1 - \chi_2^2) - i\chi_2\gamma} \tag{8.32}$$

with

$$\chi_2 = \frac{\hbar\omega}{E_0'} \tag{8.33}$$

where C is the nondimensional strength parameter and γ is the nondimensional broadening parameter. The calculated lineshapes of the DHO for InP are shown in Fig. 8.12. The numeric values are as follows: $E_0' = 4.72$ eV, $C = 1.30$, and $\gamma = 0$ (solid lines) and 0.093 (dashed lines). Note that in

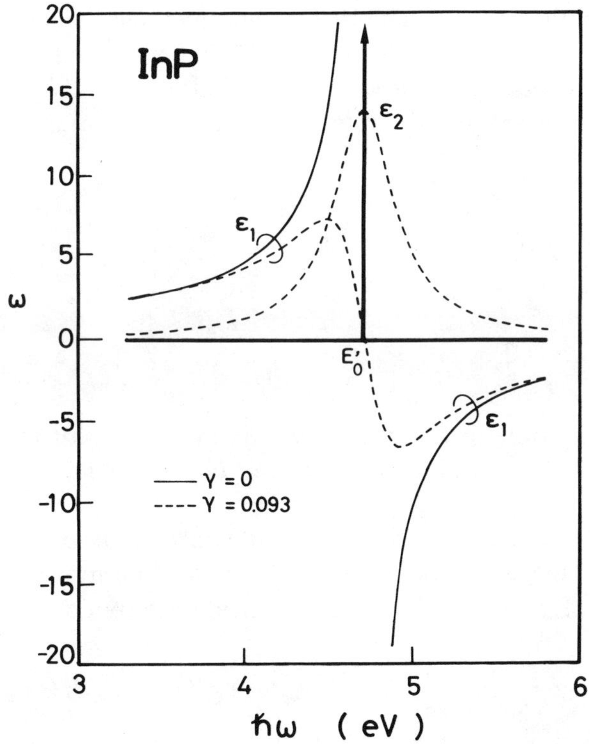

Figure 8.12 Lineshape of the E_0' gap contribution to $\epsilon(\omega)$ for InP [Eq. (8.32)] with two different broadening parameters [$\gamma = 0$ (solid line) and $\gamma = 0.093$ (dashed line)].

the limit $\gamma \rightarrow 0$ the ϵ_1 spectrum of the DHO is equivalent to the classic Drude–Lorentz formula: namely,

$$\epsilon_1(\omega) = \frac{E_0'^2 C}{E_0'^2 - (\hbar\omega)^2} \tag{8.34}$$

(d) Indirect-Band-Gap Transitions It is well known that not only the direct transitions at the CPs but also the indirect transitions at the band gap influence the optical dispersion relations of semiconductors. The indirect transitions in indirect-band-gap materials (e.g., GaP) take part at energies below the onset of the direct transitions, and vice versa above the onset of the direct transitions in direct-band-gap materials (e.g., InP). Because indirect transitions are higher order in the perturbation than direct ones, their strength is usually very weak, and one can expect to observe them only in spectra below the direct-threshold as a tail of the direct absorption edge in indirect-band-gap materials. However, previous work[21] required a considerable strength of the indirect-band-gap contribution for analyses of $\epsilon(\omega)$ spectra in both the indirect- and direct-band-gap materials.

The optical transition mechanism at the indirect band gap, E_g^{ID}, is expressed by a second-order process in the perturbation. This perturbation gives the contribution of the indirect transitions to $\epsilon_2(\omega)$ as

$$\epsilon_2(\omega) = \frac{D}{(\hbar\omega)^2}(\hbar\omega - E_g^{ID} \pm \hbar\omega_q)^2 H(1 - \chi_g) \tag{8.35}$$

with

$$\chi_g = \frac{E_g^{ID} \pm \hbar\omega_q}{\hbar\omega} \tag{8.36}$$

where D is the strength parameter, $\hbar\omega_q$ is the energy of the phonons involved in the indirect transitions, and the H is a function defined by Eq. (8.28).

The parabolic bands extending to infinite energies implied by Eq. (8.35) should be nonphysical. One must, thus, modify the model by taking into account a cutoff at energy E_c. This modification provides

$$\epsilon_2(\omega) = \frac{D}{(\hbar\omega)^2}(\hbar\omega - E_g^{ID} \pm \hbar\omega_q)^2 H(1 - \chi_g) H(1 - \chi_c) \tag{8.37}$$

with

$$\chi_c = \frac{\hbar\omega}{E_c} \tag{8.38}$$

Assuming that the strength parameter D is independent of the photon energy, the KK transformation of Eq. (8.37) gives

$$\epsilon_1(\omega) = \frac{2D}{\pi}\left\{ -\frac{(E_g^{ID})^2}{(\hbar\omega)^2} \ln\left(\frac{E_c}{E_g^{ID}}\right) + \frac{1}{2}\left(1 + \frac{E_g^{ID}}{\hbar\omega}\right)^2 \ln \frac{\hbar\omega + E_c}{\hbar\omega + E_g^{ID}} + \frac{1}{2}\left(1 - \frac{E_g^{ID}}{\hbar\omega}\right)^2 \ln \frac{\hbar\omega - E_c}{\hbar\omega - E_g^{ID}} \right\} \quad (8.39)$$

Damping effects are not taken into account in both Eqs. (8.37) and (8.39). The $\epsilon_1(\omega)$ spectrum of Eq. (8.39) therefore exhibits a divergence at E_c (see Fig. 8.13). We thus introduce in these expressions a phenomenologic damping by replacing $\hbar\omega$ by $\hbar\omega + i\Gamma$. The contribution of the indirect transitions to $\epsilon(\omega)$ is finally given by[21]

$$\epsilon(\omega) = \frac{2D}{\pi}\left\{ -\frac{(E_g^{ID})^2}{(\hbar\omega + i\Gamma)^2} \ln\left(\frac{E_c}{E_g^{ID}}\right) + \frac{1}{2}\left(1 + \frac{E_g^{ID}}{\hbar\omega + i\Gamma}\right)^2 \cdot \ln \frac{\hbar\omega + i\Gamma + E_c}{\hbar\omega + i\Gamma + E_g^{ID}} + \frac{1}{2}\left(1 - \frac{E_g^{ID}}{\hbar\omega + i\Gamma}\right)^2 \ln \frac{\hbar\omega + i\Gamma - E_c}{\hbar\omega + i\Gamma - E_g^{ID}} \right\} \quad (8.40)$$

In Fig. 8.13 we show $\epsilon(\omega)$ spectra calculated from Eq. (8.40) for InP. The numeric parameters used are $E_g^L = 2.05$ eV, $E_c = 3.16$ eV ($= E_1$), $D = 60.4$, and $\Gamma = 0.06$ eV. It is evident from the figure that the ϵ_1 spectrum exhibits no clear structure at the indirect edge E_g^{ID} (E_g^L). It is also noteworthy that in the limit $\Gamma \to 0$ eV the calculated ϵ_2 spectrum agrees exactly with that predicted by Eq. (8.37).

(e) Excitons and Exciton Parameters At low temperatures, optical spectra may not be explained within the framework of the one-electron approximation used in the above subsections [(a)–(d)], since the excitonic effect may profoundly modify the CP singularity structure at low temperatures. Excitonic states should, in principle, exist at each type of CP, because a Coulomb-like interaction is always present between electrons and holes. There are many theoretical papers dealing with excitonic effects on optical spectra of semiconductors.[29–41]

Let us present the contribution of exciton transitions to $\epsilon(\omega)$ at the lowest direct band edge. It is well known[28,30] that the discrete lines and continuum excitons in the neighborhood of the lowest direct band edge (3D M_0 CP)

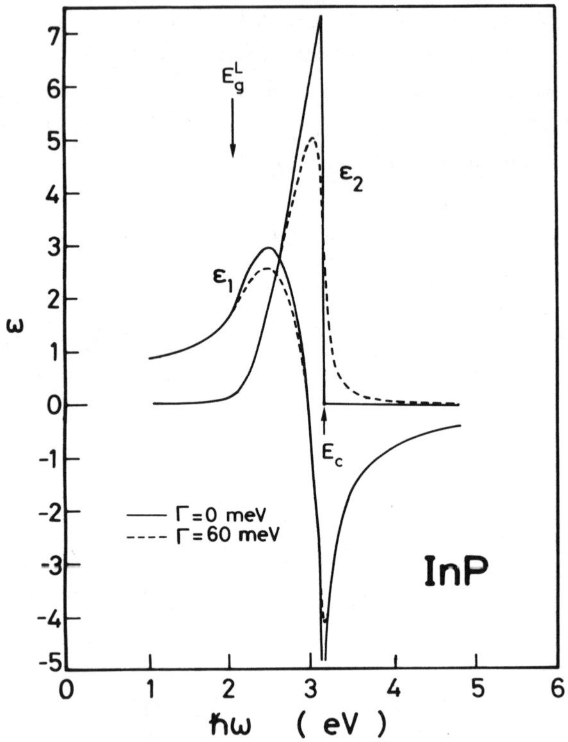

Figure 8.13 Lineshape of the indirect-band-gap (E_g^L) contribution to $\epsilon(\omega)$ for InP [Eq. (8.40)] with two different broadening parameters [$\Gamma = 0$ meV (solid line) and $\Gamma = 60$ meV (dashed line)]; E_c is a high-frequency cutoff energy (assumed to be $E_c = E_1$).

drastically change the optical spectrum. The discrete series of exciton lines at the E_0 gap can be simply given with a Lorentzian lineshape by

$$\epsilon(\omega) = \sum_{n=1}^{\infty} A_{0x}^n \left(E_0 - \frac{G^{3D}}{n^2} - \hbar\omega - i\Gamma \right)^{-1} \tag{8.41}$$

where A_{0x}^n is the exciton strength parameter and G^{3D} is the exciton Rydberg energy. The strength parameter A_{0x}^n is proportional to the envelope function of the nth exciton:

$$|\phi_n^{3D}(0)|^2 = \frac{V_0}{\pi (a_B^{3D})^3 n^3} \tag{8.42}$$

where a_B^{3D} is the exciton Bohr radius and V_0 is the volume of the unit cell.

The continuum–exciton transitions at the 3D M_0 CP exhibit behaviors like noninteracting electron–hole pair characteristics. The contribution of these transitions to $\epsilon(\omega)$ can, therefore, be expressed by an expression similar to Eq. (8.21).

The 3D exciton Rydberg energy G^{3D} ("isotropic" part) is now written as[41]

$$G^{3D} = \frac{\mu m_0 e^4}{2\hbar^2(\epsilon_0\epsilon_s)^2} = 13.6\,\frac{\mu}{\epsilon_s^2} \qquad \text{(in eV)} \tag{8.43}$$

where μm_0 is the exciton reduced mass (m_0 is the mass of a free electron). The corresponding exciton Bohr radius a_B^{3D} can then be given by

$$a_B^{3D} = \frac{\hbar^2\epsilon_0\epsilon_s}{\mu m_0 e^2} = 0.53\,\frac{\epsilon_s}{\mu} \qquad \text{(in Å)} \tag{8.44}$$

In Table 8.1 we list experimentally determined G values for InP, InAs, GaAs, and GaP. The free-exciton structure in InP has been studied by sev-

TABLE 8.1 Exciton Rydberg Energy *G* for InP, InAs, GaAs, and GaP (in meV)

InP	InAs	GaAs	GaP
4.0 (6 K)[a]	1.7 ± 0.1 (4.2 K)[h]	3.4 ± 0.2 (10 K)[i]	10.0 ± 1.0 (77 K)[l]
3.6 (298 K)[a]		2.5 ± 0.5 (294 K)[i]	5 ± 1[m]
4.8 (1.8 K)[b]		4.7 ± 0.4 (1.4 K)[j]	13 ± 1[n]
4.9 (2 K)[c]		4.2 ± 0.2 (2 K)[k]	21 ± 2[o]
5.2 ± 0.1 (4.2 K)[d]			22 ± 2 (2 K)[p]
4.8 ± 0.2 (1.2–2.1 K)[e]			20.5 (<4.2 K)[q]
5.1 ± 0.1 (1.6 K)[f]			24 ± 3 (1.5 K)[r]
5.0 ± 0.1 (2 K)[g]			

[a]W. J. Turner et al., *Phys. Rev.* **136,** A1467 (1964).
[b]A. M. White et al., *J. Phys. C* **5,** 1727 (1972).
[c]F. Evangelisti et al., *Phys. Rev. B* **9,** 1516 (1974).
[d]P. Rochon and E. Fortin, *Phys. Rev. B* **12,** 5803 (1975).
[e]S. B. Nam et al., *Phys. Rev. B* **13,** 1643 (1976).
[f]H. Mathieu et al., *Phys. Rev. B* **32,** 4042 (1985).
[g]M. A. Abdullaev et al., *Sov. Phys. Semicond.* **23,** 724 (1989).
[h]A. V. Varfolomeev et al., *Sov. Phys. Semicond.* **9,** 530 (1975).
[i]M. D. Sturge, *Phys. Rev.* **127,** 768 (1962).
[j]M. A. Gilleo et al., *Phys. Rev.* **174,** 898 (1968).
[k]D. D. Sell, *Phys. Rev. B* **6,** 3750 (1972).
[l]Indirect excitons [P. J. Dean and D. G. Thomas, *Phys. Rev.* **150,** 690 (1966)].
[m]Direct excitons [P. J. Dean et al., *J. Appl. Phys.* **38,** 3551 (1967)].
[n]Indirect excitons [D. Auvergne et al., *Phys. Rev. B* **12,** 1371 (1975)].
[o]Indirect excitons [A. A. Kopylov and A. N. Pikhtin, *Sov. Phys. Semicond.* **11,** 510 (1977)].
[p]Indirect excitons [M. D. Sturge et al., *Appl. Phys. Lett.* **32,** 49 (1978)].
[q]Indirect excitons [R. G. Humphreys et al., *Phys. Rev. B* **18,** 5590 (1978)].
[r]Indirect excitons [B. Monemar and L. Samuelson, *Solid State Commun.* **26,** 165 (1978)].

eral investigators.[42–48] Turner et al.[43] observed sharp peaks in intrinsic absorption coefficient data for undpoded InP at 6 and 77 K. From these data they determined that the exciton Rydberg energies are 4.0 ± 0.2 meV at 6 K and 3.6 ± 0.4 meV at 298 K. White et al.,[44] using photoluminescence measurements on high-quality epilayers, determined the Rydberg energy in InP to be 4.8 meV at 1.8 K. Reflectance spectra for pure InP have been reported by Evangelisti et al.[45] The G value they obtained is 4.9 meV. Photovoltaic spectra measured by Rochon and Fortin[42] for n-type InP also give $G = 5.2 \pm 0.1$ meV. Nam et al.[46] have determined the exciton Rydberg energy to be $G = 4.8 \pm 0.2$ meV and the reduced mass of the exciton to be $\mu = 0.052$. Mathieu et al.[47] have recently investigated the reflectivity spectra of InP at normal incidence, at 1.6 K, and under [100] uniaxial stress. They found three transverse exciton energies associated with the 1s ($n = 1$) ground state and the 2s ($n = 2$) and 3s ($n = 3$) excited states. They determined the exciton Rydberg energy $G = 5.1 \pm 0.1$ meV, exchange energy $\Delta = 0.04 \pm 0.02$ meV, and longitudinal–transverse splitting $E_{LT} = 0.17 \pm 0.02$ meV by resolving the fine structure of the exciton states. The surface dead layer corresponding with the best experimental fit was found to be twice the exciton Bohr radius, as expected for an intrinsic exciton-free surface layer. More recently, Abdullaev et al.[48] studied an exciton structure at the absorption edge of high-purity epitaxial InP films at 2 K. They found the Rydberg energy $G^* = G + \Delta G(=5.0 \pm 0.1$ meV) where $G(=4.82$ meV) is the "isotropic" part given by Eq. (8.43) and $\Delta G(=0.17$ meV) is the "anisotropic" correction found from perturbation theory.[41]

Since ϵ_s in Eq. (8.43) is a relatively strong function of the temperature T, the Rydberg energy G is also a function of T. As mentioned in Section 5.1, the values of ϵ_s for semiconductors usually decrease with decreasing T. For InP a value of $\epsilon_s = 12.56 \pm 0.2$ was obtained at $T = 300$ K and an extrapolated value of 11.93 ± 0.2 was obtained at $T = 77$ K.[49] Assuming $\mu^{-1} = (m_e^\Gamma/m_0)^{-1} + (m_{hh}/m_0)^{-1}$ and $m_e^\Gamma/m_0 = 0.079$ and $m_{hh}/m_0 = 0.46$ for InP (see Section 6.2), we calculate $G = 5.8$ meV at $T = 300$ K and 6.4 meV at $T = 77$ K. These values are considerably larger than those determined experimentally (3.6 ± 0.2 meV at 298 K and 4.0 ± 0.4 meV at 77 K[43]). If we assume $\mu^{-1} = (m_e^\Gamma/m_0)^{-1} + (m_{lh}/m_0)^{-1}$ and $m_{lh}/m_0 = 0.120$ for InP (also see Section 6.20), agreement is better ($G = 4.1$ meV at 300 K and 4.6 meV at 77 K).

In Fig. 8.14 we show the interpolated G as a function of y for $In_{1-x}Ga_xAs_yP_{1-y}$ lattice-matched to InP. The end-binary data used are 4.8 meV for InP,[44,46] 1.7 meV for InAs,[50] 4.2 meV for GaAs,[51] and 5 meV for GaP.[52] The calculated curve, as a function y, is given by the quadratic

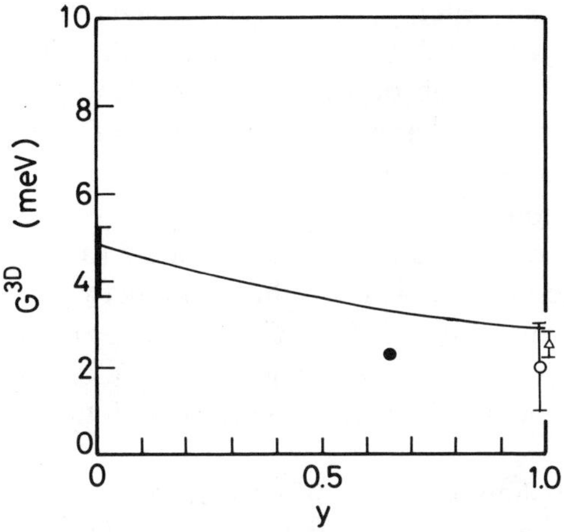

Figure 8.14 Three-dimensional exciton Rydberg energy G^{3D} as a function of y for $In_{1-x}Ga_xAs_yP_{1-y}$ lattice-matched to InP. The experimental data are taken from Table 8.1 (InP), from Refs. 53 (solid circle), 54 (open circle), and 63 (open triangle). The solid line shows the interpolated result of Eq. (2.3) using the end-binary data (InP,[44,46] InAs,[50] GaAs,[51] and GaP[52]).

form (in millielectron volts):

$$G^{3D}(y) = 4.8 - 2.9y + 1.0y^2 \tag{8.45}$$

Satzke et al.[53] have studied optical absorption of an $In_{1-x}Ga_xAs_yP_{1-y}$/InP double heterostructure ($y \simeq 0.65$). Although no discrete exciton levels were observed even at low temperatue (10 K), an exciton Rydberg energy of 2.3 meV is obtained from the continuum states by fitting the spectrum with a Lorentzian function. The steepness of the absorption at the band edge provided a value of 4 meV for the continuum–exciton linewidth. This value is larger than the Rydberg energy and explains why no discrete exciton levels are observed. The solid circle in Fig. 8.14 is the plot of their determined G value. The corresponding exciton reduced mass μ/m_0 is 0.029. They also considered that this reduced mass is in good agreement with the value of 0.030 derived from an electron mass of 0.051 and a light-hole mass of 0.07 reported for material of similar composition.

Chen and Kim[54] have observed a free exciton absorption peak in the transmission spectra of $In_{0.53}Ga_{0.47}As$ and determined an exciton Rydberg energy of 2 ± 1 meV at 5 K for this alloy. Their result is plotted in Fig. 8.14 by the open circle. Adding the measured Rydberg energy to the exciton absorption peak energy, they determined the E_0 gap energy of $In_{0.53}Ga_{0.47}As$ at 5 K to be 0.819 ± 0.0015 eV. At 77 K, the exciton peak was still resolved in the absorption spectrum. From the position of the exciton absorption peaks a band-gap energy shift of 4 meV was determined for $In_{0.53}Ga_{0.47}As$ between 5 and 77 K.

Detailed absorption measurements on $In_{0.53}Ga_{0.47}As$ have also been carried out in the temperature range between 5 and 340 K by Zielinski et al.[55] They observed free exciton peaks in the range between 5 and 100 K. The exciton parameters derived at 5 K are $E_0 = 0.8215 \pm 0.0002$ eV, $G = 2.5 \pm 0.3$ meV, $\Gamma = 5.1 \pm 0.3$ meV, and $P^2 = 20.7 \pm 1$ eV (squared momentum matrix element). Their G value is also plotted in Fig. 8.14 by the open triangle.

In Fig. 8.15 we show the spectral dependence of the absorption coefficients $\alpha(\omega)$ in the neighborhood of the E_0 edge of InP. The open circles are the measured spectra taken from Ref. 43 (T = 6, 77, 298 K). The solid lines represent the calculated results taking into account the discrete exciton states [i.e., Eq. (8.41)]. The numeric parameters used are $E_0 - 1.421$ eV, $G = 4.8$ meV, $\Gamma = 5.3$ meV, and $A_{0x}^1 = 3 \times 10^{-3}$ eV ($A_{0x}^n = A_{0x}^1/n^3$) at 6 K, and $E_0 = 1.414$ eV, $G = 4.6$ meV, $\Gamma = 5.4$ meV, and $A_{0x}^1 = 3 \times 10^{-3}$ eV at 77 K. The broadening parameters Γ determined here are found to agree well with those presented in Fig. 6.13.

There may be only two analytic expressions that enable us to treat the excitonic effects at the $E_1/(E_1 + \Delta_1)$ spectral region: (1) the effective-mass (EM) approximation[32,33] and (2) the Koster–Slater (KS) method.[34,36]

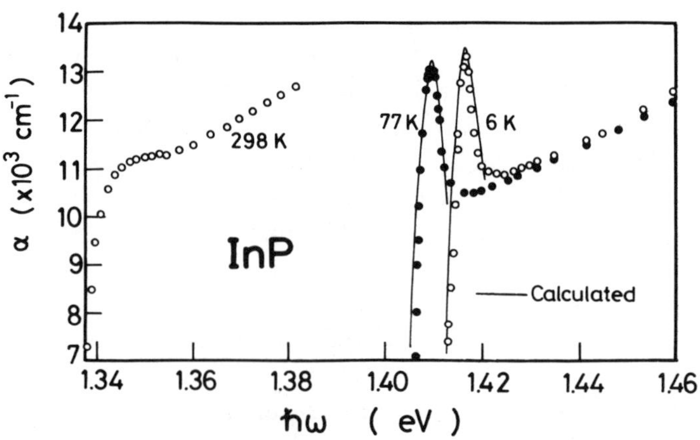

Figure 8.15 Spectral dependence of $\alpha(\omega)$ in the neighborhood of the E_0 edge of InP as reported by Turner et al.[43] The solid lines show the numerically calculated results taking into account the discrete exciton series [i.e., Eq. (8.41)].

In the case of the 3D M_1 CPs (i.e., the saddlepoint excitons or hyperbolic excitons), the EM equation is much more difficult to solve. However, in the limit $m_L^{-1} \sim 0$ the equation predicts a series of the 2D Wannier-type excitons[32]

$$E_{x1}^{2D} = E_1 - \frac{G_1^{2D}}{(n - \frac{1}{2})^2} \tag{8.46a}$$

$$E_{x\Delta}^{2D} = E_1 + \Delta_1 - \frac{G_\Delta^{2D}}{(n - \frac{1}{2})^2} \tag{8.46b}$$

where $G_1^{2D}(G_\Delta^{2D})$ is the 2D-exciton Rydberg energy. The contribution of the 2D-exciton transitions to $\epsilon(\omega)$ is written with a Lorentzian lineshape as

$$\epsilon(\omega) = \sum_{n=1}^{\infty} [B_{1x}^n(E_{x1}^{2D} - \hbar\omega - i\Gamma)^{-1} + B_{2x}^n(E_{x\Delta}^{2D} - \hbar\omega - i\Gamma)^{-1}] \tag{8.47}$$

where $B_{1x}^n(B_{2x}^n)$ is the exciton strength parameter proportional to the envelope function of the nth exciton. The envelope function can now be given by

$$|\phi_n^{2D}(0)|^2 = \frac{16V_0}{\pi(a_B^{2D})^3(2n - 1)^3} \tag{8.48}$$

In Eq. (8.48) a_B^{2D} is the 2D-exciton Bohr radius. The 2D EM approximation also gives the continuum part of the excitonic states.[32] One can, however, consider that the contribution of this part is similar to that of the one-electron approximation [i.e., Eq. (8.30)].

In the KS method, the electron–hole interaction (Coulomb potential) is evaluated in the tight-binding approximation (TBA) and its attraction is approximated by an on-site contact interaction. An approximate expression of $\epsilon(\omega)$ based on this model is written as[34,36]

$$\epsilon(\omega) - 1 = \frac{\tilde{\epsilon}(\omega) - 1}{1 - g[\tilde{\epsilon}(\omega) - 1]}, \qquad g > 0 \tag{8.49}$$

where $\tilde{\epsilon}$ is the one-electron dielectric function and the parameter g is proportional to the depth of the assumed square-well potential. This expression was successfully used to show the presence of the excitonic effects in the E_1 structure of low-temperature electroreflectance data of Ge.[34]

As we can see in Refs. 56 and 57, optical spectra in the $E_1/(E_1 + \Delta_1)$ structure region of InP and GaAs become sharp when the temperature is lowered. Such a spectral change cannot be explained within the framework of the one-electron approximation with lifetime broadening corrections. This fact clearly suggests evidence for the contribution of the excitonic effects to the $E_1/(E_1 + \Delta_1)$ transitions. Not only the EM approximation but also the KS method dramatically modifies and sharpens the $E_1/(E_1 + \Delta_1)$ CP structure. The degree of sharpness is, however, larger for the EM approximation than for the KS method.[57] We find better agreement with the experiment using the EM approximation.

Multiparticle effects on CPs in the interband continuum of semiconductors with their detailed electronic-energy band structures have been treated.[38,39] Results have shown that by introducing the excitonic interaction the absorption at $E_0'(E_2)$ CP is markedly reduced with no drastic change in its shape. Unfortunately, however, no analytic lineshape suitable for fitting the excitonic-effect-influenced E_0' lineshape has been reported to date. In Refs. 56–59 the DHO model of Eq. (8.32) is found to be a good representation for the $E_0'(E_2)$ CP both with and without the presence of excitonic interaction. The strength parameter C in this expression may, in principle, be independent of the temperature. Thus, a change in C values can account for the excitonic interaction strength at the $E_0'(E_2)$ CP in semiconductors.[56–59]

8.3.2 Experimental ϵ_1 and ϵ_2 Spectra

Spectroscopic ellipsometry is known to be highly suitable for investigating the optical response of semiconductors over a wide range of photon energies.[60] This technique is very surface-sensitive, but the surface-contamination effects can, in principle, be mathematically removed in the analysis. Aspnes and Studna[61] have measured room-temperature optical constants of undoped InP at energies between 1.5 and 6.0 eV by spectroscopic ellipsometry. They have presented numeric information on the optical constants, ϵ_1, ϵ_2, n, k, R, and α, in tabular form. Their results illustrate the importance of proper sample preparation and the capability of separately determining both ϵ_1 and ϵ_2 in optical measurements.

Lautenschlager et al.[62] have measured the temperature dependence of the dielectric function, $\epsilon = \epsilon_1 + i\epsilon_2$, of compensated, high-resistivity InP in a temperature range between 30 and 750 K and a photon-energy range between 1.3 and 5.5 eV by spectroscopic ellipsometry. The structures observed in the ϵ spectra were attributed to interband CPs that can be analyzed in terms of the following standard analytic lineshapes:

$$\epsilon(\omega) = C_0 - A_0 e^{i\phi}(\hbar\omega - E_{CP} + i\Gamma)^n \tag{8.50}$$

They determined the CP parameter amplitude A_0, energy threshold E_{CP}, broadening Γ, and excitonic phase angle ϕ for each CP [E_0, $E_0 + \Delta_0$, E_1, $E_1 + \Delta_1$, and $E_0'/(E_0' + \Delta')$] by fitting the numerically obtained second-derivative spectra $d^2\epsilon(\omega)/d\omega^2$ of the experimental $\epsilon(\omega)$. Unfortunately, however, these expressions are not suitable for describing the dielectric behavior $\epsilon(\omega)$ over the entire range of photon energies.[59]

The dielectric functions of high-purity InP and $In_{0.53}Ga_{0.47}As$ layers grown by metal–organic vapor-phase epitaxy have been measured by Erman et al.[63] in the 2.0–3.5-eV range using spectroscopic ellipsometry. They demonstrated that the ellipsometry technique can provide a useful, nondestructive, and rapid assessment of the $In_{0.53}Ga_{0.47}As$/InP heterostructure qualities. The analysis of the data using the multilayer modeling gave information on the chemical nature and the thickness of the interface region. The sharpest interface as measured by this modeling was 30 Å of InGaAsP at the interface of InP grown on $In_{0.53}Ga_{0.47}As$ and 9 Å of InAs for $In_{0.53}Ga_{0.47}As$ grown on InP.

Burkhard et al.[64] have measured n and α of $In_{1-x}Ga_xAs_yP_{1-y}$ lattice-matched to InP ($0 \leqq y \leqq 1.0$) and of GaAs and GaP by ellipsometry in the wavelength range between 0.365 and 1.1 μm. High-quality layers were grown by liquid-phase epitaxy (LPE). They also determined the thickness d_{ox} of the temporal growth of the natural oxide layer on various InP and GaAs samples by the ellipsometry. The increase of d_{ox} for InP was found to be 3.9 Å for every factor of 10 increase in time when measured in minutes.

Kelso et al.[26] have ellipsometrically measured high-precision pseudodielectric function spectra from 1.5 to 6.0 eV of $In_{1-x}Ga_xAs_yP_{1-y}$ lattice-matched to InP. A series of $In_{1-x}Ga_xAs_yP_{1-y}$ samples that covered the range $y = 0$ to 1.0 were grown by LPE. Each layer had a thickness larger than or equal to 1 μm. The compositions were determined from X-ray measurements and room-temperature photoluminescence spectra and should be accurate to ± 0.02 for y. As mentioned in Section 6.1, they assigned the feature previously attributed E_2 in InP to $E_0'(E_0' + \Delta_0')$.

Knowledge of the structure, on the atomic scale, of the interface region between various layers is important for understanding the electronic properties of heterostructures. High sensitivity makes spectroscopic ellipsometry a useful tool for addressing heterostructure problems. This has been illustrated by an investigation of the structural nature of

$Al_xGa_{1-x}As/GaAs$[65] and $In_{0.53}Ga_{0.47}As/InP$ heterojunctions.[63] Similar work on $In_{1-x}Ga_xAs_yP_{1-y}/InP$ heterojunctions has been made by Drevillon et al.[66] using polarization–modulation spectroscopic ellipsometry. The measured spectral range was 1.5–5.3 eV, and the optical-constant data obtained agreed with the results of Kelso et al.[26] They compared the steepness of the heterojunctions between two types of structures: those produced by the growth of $In_{1-x}Ga_xAs_yP_{1-y}$ (or $In_{0.53}Ga_{0.47}As$) on InP and those obtained by the reverse growth sequence. The sharpest interface was achieved for InP growth on quaternary where the interface region was estimated to be 10 ± 10 Å thick.

The data of Kelso et al.[26] and Drevillon et al.[66] are very useful for understanding fundamental optical properties of the $In_{1-x}Ga_xAs_yP_{1-y}$ alloy. These data seem to have one disadvantage, however; they are not expressed as continuous analytic functions of photon energies.

The fits with the dielectric-function model proposed here to the experimental $\epsilon_2(\omega)$ and $\epsilon_1(\omega)$ spectra of InP at 300 K are shown in Figs. 8.8 and 8.9, respectively. The solid circles represent the spectroscopic–ellipsometry data of Kelso et al.[26] The solid lines were obtained from the sum of Eq. (8.21) (Fig. 8.10), Eq. (8.30) (Fig. 8.11), Eq. (8.32) (Fig. 8.12), Eq. (8.40) (Fig. 8.13), and Eq. (8.47). The 2D exciton strength and broadening parameters determined by these fits are $B_{1x}^1 = 1.00$ eV, $B_{2x}^1 = 0.25$ eV, and $\Gamma[E_1/(E_1 + \Delta_1)] = 0.12$ eV. Since the 2D ground-state exciton term ($n = 1$) in Eq. (8.47) contains 95% of the total oscillator strength, one can neglect the excited-state terms ($n \geq 2$) in the calculation. The 3D discrete-exciton states may also be present only in the limited region close to the E_0 edge, and their strength is considerably weaker than those of the ensuring E_1, $E_1 + \Delta_1$, and E_0'. The contribution of these states is thus not so important, and can be neglected in the analysis.

As seen in Fig. 8.8, the transitions at the M_0 edges $[E_0/(E_0 + \Delta_0)]$ yield a continuous absorption obeying the well-known $\frac{1}{2}$-power law [i.e., $\propto (\hbar\omega - E_0)^{0.5}$]. The transitions at the E_g^L gap provide a gradually increasing absorption spectrum characterized by a power law of $(\hbar\omega - E_g^L)^2$. Because of the low probability for indirect transitions, one can realize them only in the spectral region between E_g^L and E_1. The E_1 and $E_1 + \Delta_1$ gaps are of the 3D M_1 (2D M_0) type. Hence, the lineshape of corresponding ϵ_2 spectrum [Eq. (8.30)] should be characterized by a steep low-energy side and a less steep high-energy side, in good agreement with the experimental data. Introduction of the damping effect into Eq. (8.30) can successfully decrease the strength of the E_1 and $E_1 + \Delta_1$ peaks (see Fig. 8.11) and produces a result that agrees with experiment ($\Gamma = 0.12$ eV). The experimental ϵ_2 spectrum shows a relatively strong, symmetric peak in the E_0' CP region. This peak can be interpreted well by the DHO model of Eq. (8.32).

It is also evident that the strong negative peak of ϵ_1 observed in the E_0' structure region can be interpreted well by the DHO model (see Fig. 8.9).

We show in Figs. 8.16 and 8.17 our calculated $\epsilon_2(\omega)$ and $\epsilon_1(\omega)$ spectra of $In_{0.53}Ga_{0.47}As$ at 300 K compared with experimental data. The experimental data are taken from Refs. 25 (open circles, Fig. 8.16), from 26 (solid circles, Figs. 8.16 and 8.17), and from 27 (open circles, Fig. 8.17). The solid lines are the results of the sum of Eqs. (8.21), (8.30), (8.32), (8.40), and (8.47). The numeric parameters used are as follows: $E_0 = 0.75$ eV, $E_0 + \Delta_0 = 1.04$ eV, $A = 1.20$ eV$^{1.5}$, $\Gamma[E_0/(E_0 + \Delta_0)] = 15$ meV, $E_1 = 2.57$ eV, $E_1 + \Delta_1 = 2.83$ eV, $B_1 = 3.84$ eV, $B_2 = 1.48$ eV, $\Gamma[E_0/(E_0 + \Delta_0)] = 0.12$ eV, $B_{1x}^1 = 0.90$ eV, $B_{2x}^1 = 0.35$ eV, Γ(2D exciton) $= 0.12$ eV, $E_0' = 4.41$ eV, $C = 2.90$, $\gamma = 0.225$, $E_g^L = 1.20$ eV, $D = 20.7$, and $\Gamma(E_g^L) = 0.06$ eV.

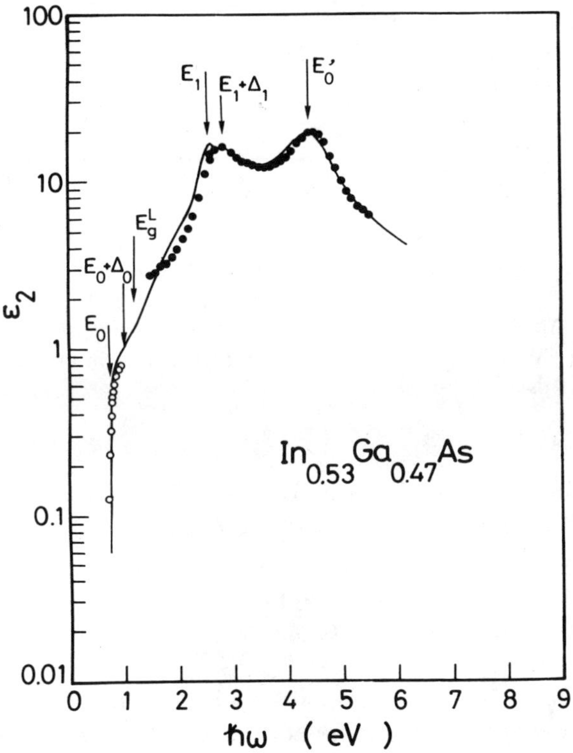

Figure 8.16 The ϵ_2 spectrum of $In_{0.53}Ga_{0.47}As$ (300 K). The solid line shows the result of the sum of Eqs. (8.21), (8.30), (8.32), (8.40), and (8.47). The experimental data are taken from Refs. 25 (open circles) and 26 (solid circles). The vertical arrows indicate the positions of each CP and E_g^L gap that play an important part in the calculation of the ϵ_2 spectrum. (From Adachi.[16])

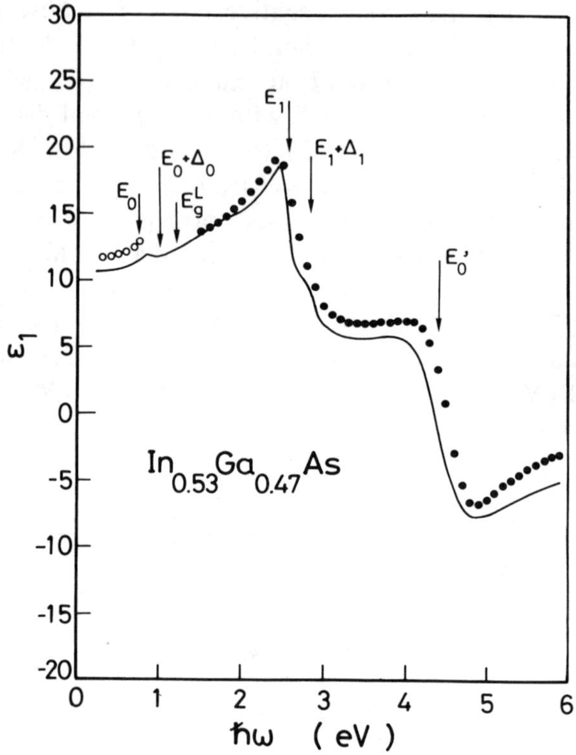

Figure 8.17 The ϵ_1 spectrum of $In_{0.53}Ga_{0.47}As$ (300 K). The solid line represents the result of the sum of Eqs. (8.21), (8.30), (8.32), (8.40), and (8.47). The experimental data are taken from Refs. 27 (open circles) and 26 (solid circles). The vertical arrows indicate the positions of each CP and E_g^L gap that play an important part in the calculation of the ϵ_1 spectrum. (From Adachi.[16])

It is clear that the 3D M_0 term [Eq. (8.21)] and the indirect-band-gap term [Eq. (8.40)] interpret the 0.75–3.0-eV region of the ϵ_2 spectrum well. The calculated value of ϵ_2 at the E_2 peak is 19.1, in good agreement with the experimental value (19.3). As in the case of InP, we see a better fit of our ϵ_1 model to the experimental data over the entire range of photon energies.

As mentioned in Section 8.3.1, the dielectric function is expressed by the electronic energy-band parameters (band-gap energies) and the corresponding strength and broadening parameters. In Ref. 16 such parameters for $In_{1-x}Ga_xAs_yP_{1-y}$ lattice-matched to InP are specified in terms of y alone. Using these results, we can easily calculate the spectral dependence of ϵ_1 and ϵ_2 for an optional composition of the $In_{1-x}Ga_xAs_yP_{1-y}$ quaternary. The temperature dependence of $\epsilon(\omega)$ for InP was also analyzed by the same method.[56] The excitonic effects accounted for the discrepancies in the ϵ_2 strengths of the E_1 and E_0' peaks in InP; thus, the inclusion of the excitonic effects in the one-electron band model changed the strength of the peaks in the correct direction to agree with the experimental result (it led to strengthening of the E_1 peak and weakening of the E_0' peak).

8.3.3 Dispersion of the Refractive-Index, Extinction, and Absorption Coefficients

The analysis presented in Section 8.3.2 is based on a simplified model of the energy-band structure of the materials. The model covers the $\epsilon(\omega)$ spectra of the materials in the entire range of photon energies. These results prompt us to apply the model to other important optical constants for practical-device engineering that are related to the dielectric function [such as the refractive index (n), extinction coefficient (k), absorption coefficient (α), and normal-incidence reflectivity (R)].[20,22,24,67,68]

Figure 8.18 shows the experimentally determined n and k for InP at 300 K. The data are taken from Refs. 69 (open-circles) and 61 (solid circles).

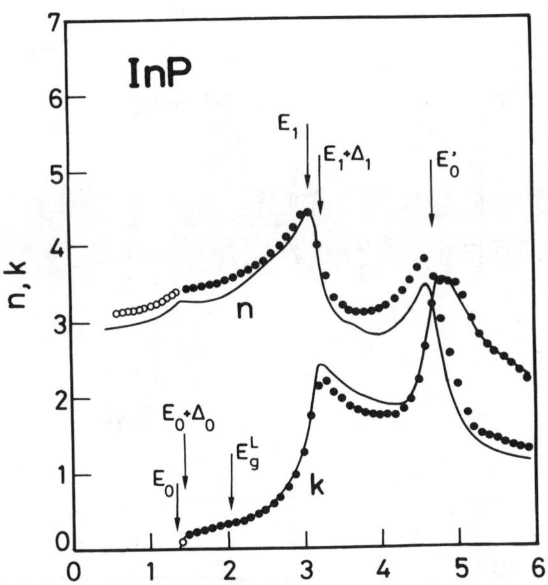

Figure 8.18 Numerically calculated spectral dependence of n and k for InP (300 K). The solid and open circles show the experimental data taken from Refs. 61 and 69, respectively. (From Adachi.[68])

The solid lines show the numerically calculated spectral dependence of n and k for InP. They are obtained from Eq. (8.5) for n and from Eq. (8.6) for k.

The strongest peak in n at ~3.1 eV is related mainly to the E_1 transitions. The experimental value of this peak is about 4.4. The calculation of Eq. (8.5) gives $n = 4.38$, in good agreement with this value. In k, the spectrum above ~1.4 eV is associated with the onset of the $E_0/(E_0 + \Delta_0)$ transitions. The strongest peak in k occurs at ~4.8 eV. This peak is related to the $E_0'(E_2)$ transitions.

A comparison of the experimental data of α[61] with our calculated values for InP at 300 K is shown in Fig. 8.19. The solid line is calculated from Eq. (8.7). Since InP is a direct-band-gap semiconductor, the indirect transitions (E_g^L) take part above the onset of the direct transitions that occur at 1.35 eV (E_0). If we do not take into account the E_g^L gap contribution, the fit with the experiment becomes very poor in the 2–3-eV region. The the-

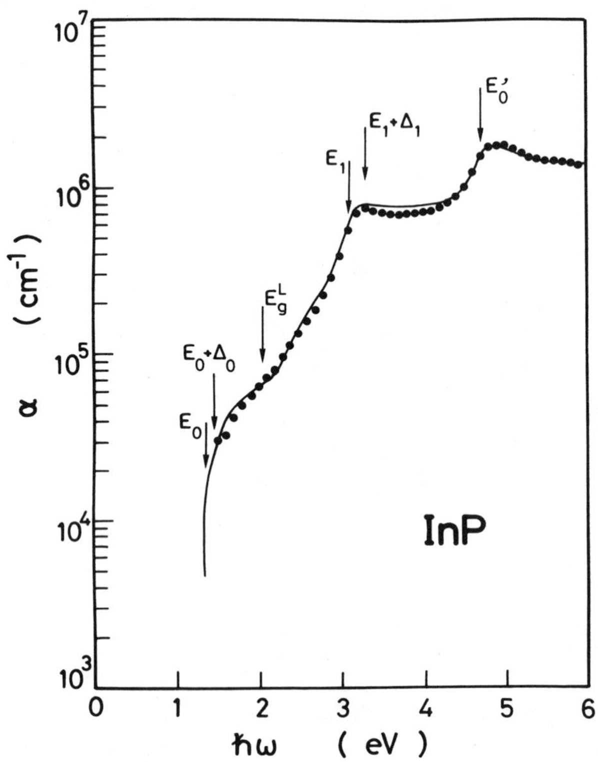

Figure 8.19 Numerically calculated spectral dependence of α for InP (300 K). The experimental data (solid circles) are taken from Ref. 61. (From Adachi.[68])

oretical E_1 contribution provides a plateau-like lineshape in α, in good agreement with experimental results. The strongest peak in α occurs at ~4.8 eV. This peak can be successfully interpreted by the E_0'-gap contribution.

In Fig. 8.20 we show a comparison of our calculated normal-incidence reflectivity R to the experimental data for InP at 300 K. The data are taken from Ref. 61. The solid line is obtained from Eq. (8.8). The overall features of our R is in reasonably good agreement with experimental data.

Figure 8.21 shows the experimentally determined n and k for $In_{0.53}Ga_{0.47}As$ at 300 K. The data are taken from Refs. 27 (open circles) and 26 (solid circles). The solid lines are the calculated spectral dependence of n and k for $In_{0.53}Ga_{0.47}As$. As clearly seen in the figure, our model reasonably interprets the experimental data of both n and k over a wide range of photon energies.

Figures 8.22 and 8.23 show a comparison of our calculated α and R to the experimental data of $In_{0.53}Ga_{0.47}As$ at 300 K, respectively. The data are taken from Ref. 26. As in the case of InP, our model interprets the characteristic experimental dispersion of α and R for $In_{0.53}Ga_{0.47}As$ well.

In order to design $In_{1-x}Ga_xAs_yP_{1-y}$/InP waveguiding devices, it is necessary to specify the refractive indices or refractive-index steps of the

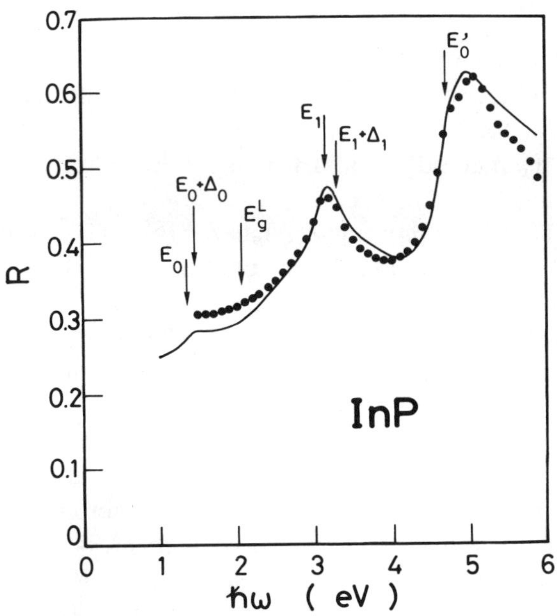

Figure 8.20 Numerically calculated spectral dependence of the normal-incidence reflectivity R for InP (300 K). The experimental data (solid circles) are taken from Ref. 61.

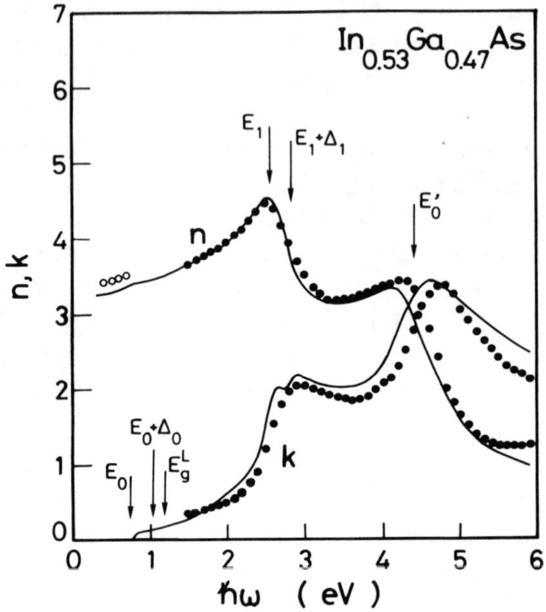

Figure 8.21 Numerically calculated spectral dependence of n and k for $In_{0.53}Ga_{0.47}As$ (300 K). The solid and open circles are the experimental data taken from Refs. 26 and 27, respectively. (From Adachi.[68])

heterojunction structures. In the energy region below the fundamental absorption edge, where the material is nearly transparent, the n measurements were carried out for InP by a prism method[70,71] and by a near-Brewster-angle reflectivity method[72,73] and for $In_{1-x}Ga_xAs$ ternary ($x > 0.53$) by surface reflection.[74]

Chandra et al.[75] have presented n data for $In_{1-x}Ga_xAs_yP_{1-y}$ layers with y between 0.276 and 1.0 taken as a function of the wavelength from 2.5 μm to about 0.1 μm below their respective absorption edges. Interference fringes were the basis of the measurements. The layers were grown by HCl-transport vapor-phase epitaxy on (001) InP and were unintentionally doped. Broberg and Lindgren[73] have also measured the n values of $In_{1-x}Ga_xAs_yP_{1-y}$ layers on (001) InP using a Brewster-angle method. The layers used were all undoped. The compositions of the layers were in the range 0–0.742 and were determined by electron microprobe X-ray analysis. The measurements were made at the wavelengths $\lambda = 1.15$, 1.30, and 1.55 μm. These wavelengths are of particular interest for optical fiber communication systems. They have also found that a lattice mismatch variation of 10^{-3} corresponds to a change in n of approximately 0.02 for a $y = 0.474$ layer at $\lambda = 1.3$ μm. The n dispersion in the near-band-gap region of

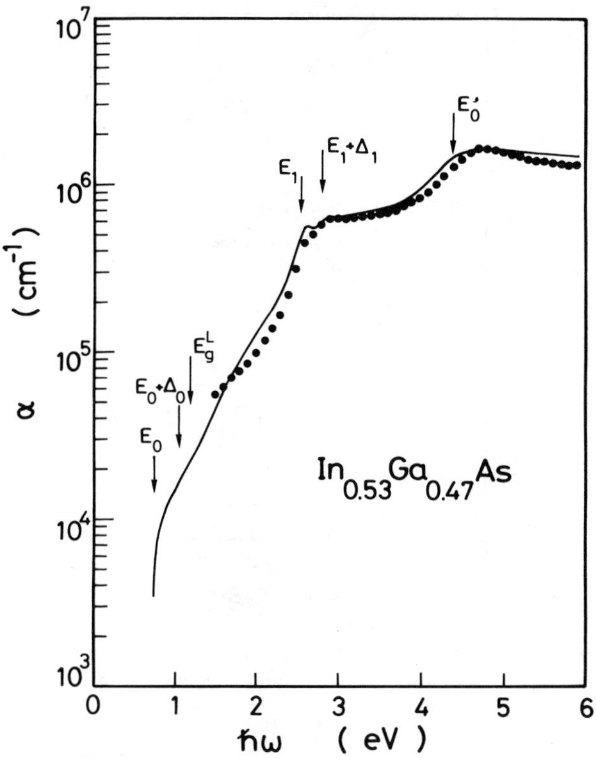

Figure 8.22 Numerically calculated spectral dependence of α for $In_{0.53}Ga_{0.47}As$ (300 K). The experimental data (solid circles) are taken from Ref. 26. (From Adachi.[68])

$In_{1-x}Ga_xAs_yP_{1-y}$ quaternary has also been determined by Kowalsky et al.[25,76] from transmission and reflection measurements.

A new method of detecting optical waveguide modes of semiconductor epitaxial layers was developed by Henry et al.,[77] who applied it to $In_{1-x}Ga_xAs_yP_{1-y}$ layers lattice-matched to InP. The modes were excited by photoluminescence and coupled out of the semiconductor wafer by a grating etched on the surface of the guiding layer. The quaternary layers studied were grown by two-phase LPE, and were of eight different compositions. This method was used to measure the n dispersion in the transparent region and also the change in the mode index of a laser in going from low current up to threshold.

A diffraction grating formed on an InP substrate has also been used by Fronts et al.[78] for measuring the refractive index of $In_{1-x}Ga_xAs_yP_{1-y}$ at the wavelength for stimulated emission. It is possible to determine the refractive index of a guiding layer by measuring the cutoff wavelengths of the supported modes in a waveguide. This method has been applied by

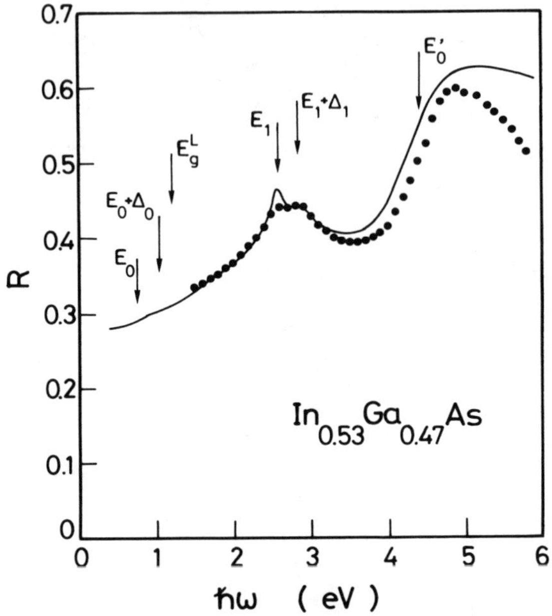

Figure 8.23 Numerically calculated spectral dependence of the normal-incidence reflectivity R for $In_{0.53}Ga_{0.47}As$ (300 K). The experimental data (solid circles) are taken from Ref. 26.

Morasca et al.[79] to $In_{1-x}Ga_xAs_yP_{1-y}$/InP waveguides. The n values they obtained were accurate to the third decimal plane.

The refractive index n of the $In_{1-x}Ga_xAs_yP_{1-y}$ quaternary varies both with photon energy and mole composition y. Accurate experimental n values for optional composition of this alloy at energies below the fundamental band gap are not yet available. This necessitates the use of some sort of calculation (interpolation). Different approaches have been made to calculate the n dispersion for the $In_{1-x}Ga_xAs_yP_{1-y}$ quaternary at energies below the direct band edge. They are based on a simple interpolation scheme,[80–82] a modified, semiempirical single-effective-oscillator model,[83] and a quantum-density-matrix approach.[84] These models interpreted the published experimental data well.

Let us now present a method for calculating the refractive index of the $In_{1-x}Ga_xAs_yP_{1-y}$ quaternary in the region below the lowest direct band edge. As can be seen in Fig. 8.10, the $E_0/(E_0 + \Delta_0)$ transitions strongly contribute to the optical dispersion, but not to their values. On the contrary, the higher-band-gap contributions (E_1, $E_1 + \Delta_1$, and E_0') in the transparency region do not contribute to the optical dispersion, but to their absolute values. Since $\epsilon_2(\omega)$ may be taken as zero in the transparency region, we can successfully assume that

$$n(\omega) \simeq [\epsilon_1(\omega)]^{0.5} \tag{8.51}$$

The $n(\omega)$ in the transparency region can then be simply expressed, using Eq. (8.21), as[27,85,86]

$$n(\omega) = \left\{ A^* \left[f(\chi_0) + \frac{1}{2}\left(\frac{E_0}{E_0 + \Delta_0}\right)^{1.5} f(\chi_{so}) \right] + B^* \right\}^{0.5} \tag{8.52}$$

where A^* is the renewed strength parameter of the E_0 and $(E_0 + \Delta_0)$ transitions and B^* corresponds to the nondispersive contribution (i.e., constant) arising from the higher-lying band gaps, such as E_1, $E_1 + \Delta_1$, E_0', E_2, $E_2 + \Delta_2$, E_1', and $E_1' + \Delta_1'$. As seen in Figs. 8.9 and 8.17, our calculated values of ϵ_1 are usually somewhat smaller than the experimental values over the entire range of photon energies. This is because we cannot take into account other, higher-lying gaps, such as E_2, $E_2 + \Delta_2$, E_1', and $E_1' + \Delta_1'$, in the calculations.

Chandra et al.[75] have measured room-temperature n values for the $In_{1-x}Ga_xAs_yP_{1-y}$ quaternary of target compositions $y = 0$–1.0 in increments of 0.3 for wavelengths from 2.5 μm to about 0.1 μm below their respective absorption edges. An attempt is made first to fit $\epsilon(\omega)\,[=n(\omega)^2]$ of Eq. (8.52) to the experimental data of Pettit and Turner[70] and Chandra et al.[75] Figures 8.24 and 8.25 show the calculated dispersion of ϵ_1 [Eq. (8.52)] compared with the experimental data of Refs. 70 and 75, respectively. These comparisons show quite good agreement between calculation and experiment. The numeric values of A^* and B^*, required to fit Eq. (8.52) with the experimental data, are shown as a function of y in Fig. 8.26. These results provide

$$A^*(y) = 8.40 - 3.40y \tag{8.53a}$$

$$B^*(y) = 6.60 + 3.40y \tag{8.53b}$$

The dielectric constant $\epsilon_1(\omega)$, in other words $n(\omega)$, of $In_{1-x}Ga_xAs_yP_{1-y}$ can then be specified in terms of y alone.

Introducing Eq. (8.53) into Eq. (8.52) and using Eq. (8.51), we can present the refractive indices for an optional composition of the $In_{1-x}Ga_xAs_yP_{1-y}$ quaternary. The calculated refractive-index dispersion of this system with y-composition increments of 0.1 is shown in Fig. 8.27. The impurities or, more generally, the lattice defects in semiconductors, reduce the lifetime of the excited electronic states, resulting in a change in

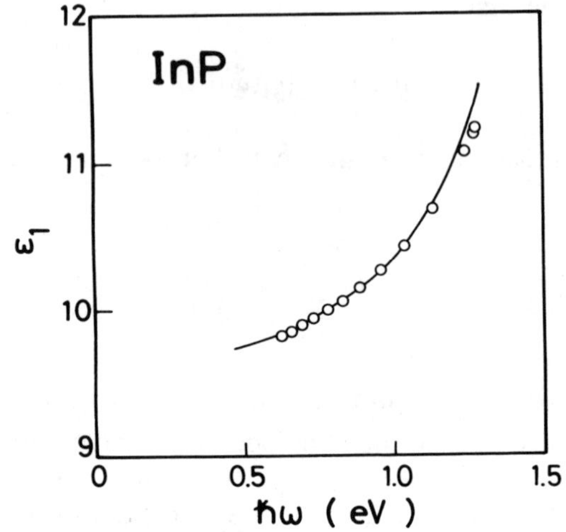

Figure 8.24 Calculated spectral dependence of ϵ_1 [Eq. (8.52)] for InP compared with the experimental data of Ref. 70.

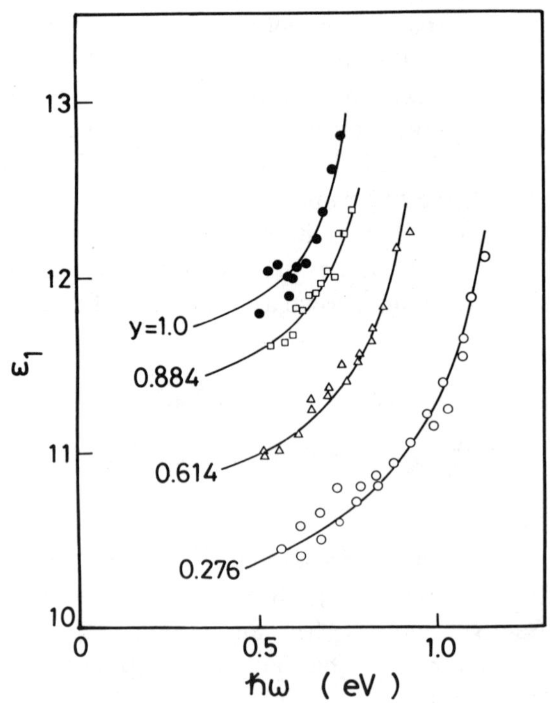

Figure 8.25 Calculated spectral dependence of ϵ_1 [Eq. (8.52)] for $In_{1-x}Ga_xAs_yP_{1-y}$ lattice-matched to InP compared with the experimental data of Ref. 75. (From Adachi.[27])

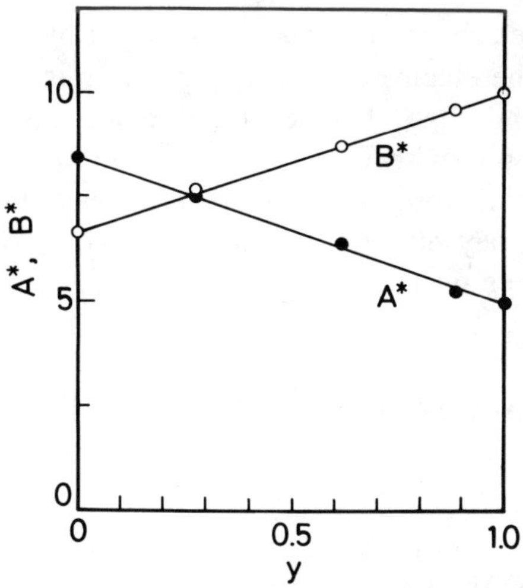

Figure 8.26 Variation of A^* and B^*, required to fit Eq. (8.52) with the experimental data, versus y for $In_{1-x}Ga_xAs_yP_{1-y}$ lattice-matched to InP. (From Adachi.[27])

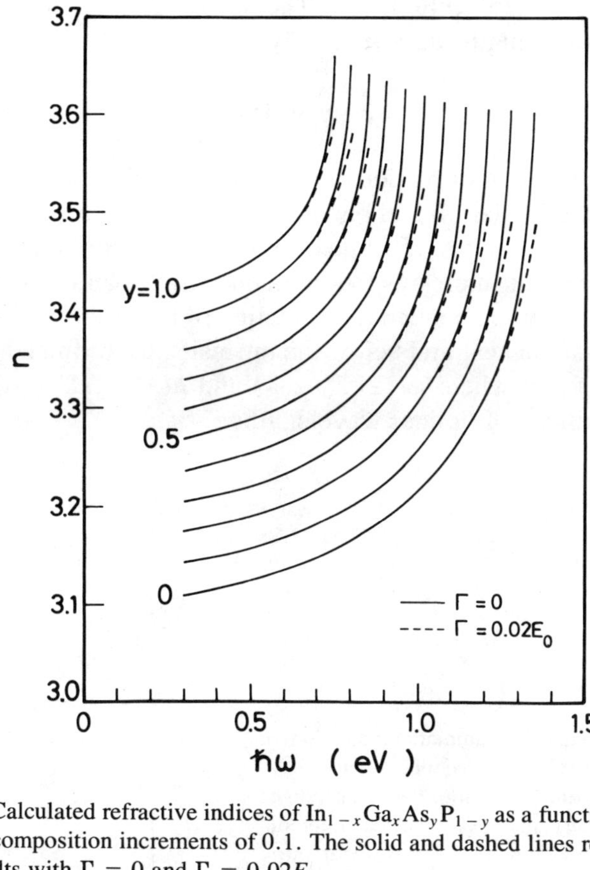

Figure 8.27 Calculated refractive indices of $In_{1-x}Ga_xAs_yP_{1-y}$ as a function of the photon energy with y-composition increments of 0.1. The solid and dashed lines represent, respectively, the results with $\Gamma = 0$ and $\Gamma = 0.02E_0$.

optical properties. The refractive-index dispersion may be affected by the lifetime broadening (damping) Γ of the electronic states, especially in the photon-energy region close to the band edge. The solid and dashed lines in the figure are, respectively, the results with $\Gamma = 0$ and $\Gamma = 0.02E_0$. It is obvious that the damping effect considerably reduces the n values at the E_0 gap region. One must, therefore, pay attention to the analysis of the refractive-index dispersion in such a region.

8.4 FREE-CARRIER EFFECTS ON OPTICAL PROPERTIES

8.4.1 Optical Absorption

The optical absorption of interest here can be divided into three separate processes, which are shown in Fig. 8.28. The direct (indirect) transitions from the valence band to the conduction band give rise to the absorption contribution α_{VC} (see Section 8.3). The normal free-carrier absorption α_{FC} depends on wavelengths approximately as

$$\alpha_{FC}(\omega) \propto (\hbar\omega)^{-m} \tag{8.54}$$

where m is dependent on various intraband scattering processes, that is, $m \simeq 2.5$ limited by polar optical phonons (α_{op}), $m \simeq 1.5$ limited by acoustic phonons (α_{ac}), and $m \simeq 3.5$ limited by ionized-impurity scattering (α_{im}).[1] The final process of interest is the interconduction-band absorption α_{CC}. In Fig. 8.29 we show schematic representation of these absorption coefficients as a function of $\hbar\omega$ near and below the intrinsic absorption edge.[87] The α_{VC} and α_{CC} takes part when $\hbar\omega \geq E_0(\alpha_{VC})$ and $\hbar\omega \geq \Delta E_{CC}(\alpha_{CC})$. The α_{FC} becomes significant at longer wavelengths.

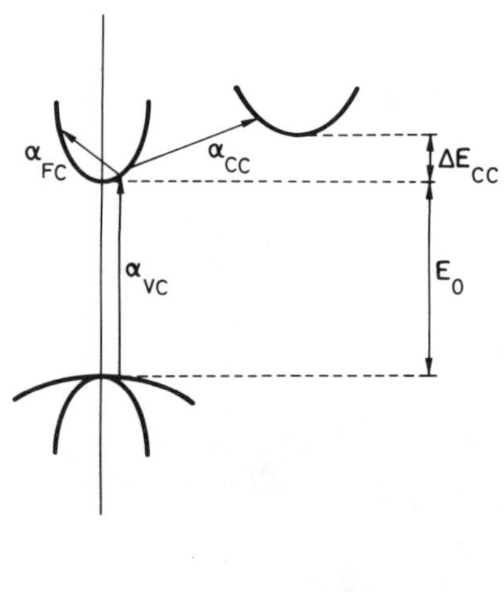

Figure 8.28 Band-diagrammatic indication of the optical absorption in a semiconductor (α_{VC}, direct interband transition from the valence band to the conduction band; α_{CC}, indirect intercon-duction-band transition; α_{FC}, free-carrier intra-band transition).

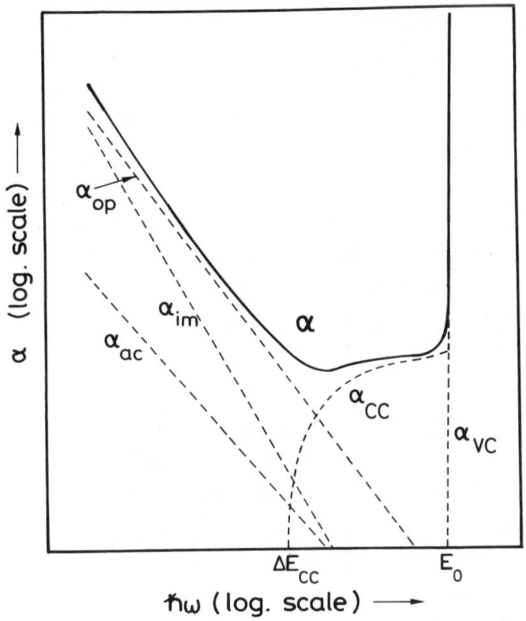

Figure 8.29 Schematic representation of the absorption coefficient α in dependence on photon energy $\hbar\omega$ near and below the intrinsic absorption edge (α_{VC}, direct interband transition from the valence band to the conduction band; α_{CC}, indirect interconduction-band transition; α_{op}, free-carrier intraband transition involving optical phonons; α_{ac}, free-carrier intraband transition involving acoustic phonons; α_{im}, free-carrier intraband transition involving ionized impurities).

Near-band-gap absorption in InP-based semiconductors has been measured by many authors (InP,[43,48,88,89] $In_{1-x}Ga_xAs_yP_{1-y}$,[25,53,76,90–94] and $In_{0.53}Ga_{0.47}As$[54,55,95–98]). Optical absorption in the free-carrier absorption range has also been measured for InP[99–103] and $In_{1-x}Ga_xAs$.[104] We will not attempt to review the considerable theoretical work on optical absorption in the free-carrier absorption range,[105] but will present the relevant experimental results.

As in the case of the refractive index (see Section 8.3.3), the absorption spectrum in the vicinity of the fundamental edge is known to depend strongly on crystalline quality. There may appear two prominent features in the fundamental-absorption-edge spectra of doped samples: (1) the whole absorption spectrum shifts to higher energies as the free-electron concentration increases (Moss–Burnstein shift) and (2) a pronounced absorption tail appears (Urbach tail[106]) such that the Urbach parameter defined as $U_0 = (d\ \ln(\alpha)/dE)^{-1}$ increases with the doping concentration. Fornari and Kumar[107] have found that the absorption spectra in bulk Fe-doped InP differ considerably from those recorded on undoped samples. Transition metals, in particular Fe, are often used as dopants in order to obtain semiinsulating properties of InP[108] and $In_{0.53}Ga_{0.47}As$.[109] To understand whether the effect was caused by the overall impurity concentration in the crystals rather than by the nature of the impurity, they repeated the measurements on S- and Sn-doped wafers.[107] We reproduce their results in Fig. 8.30. All three spectra are practically featureless at low energies (≤ 7500 cm^{-1}), but those for the Fe-doped and undoped (or S-doped) samples differ considerably in the tail regions. The authors concluded that the tail in the Fe-doped samples is due to absorption by Fe-related deep centers. They also determined that the absorption coefficient at 1×10^4 cm^{-1} can be taken as a sensitive measure of the Fe concentration and plotted the absorption coefficient against the Fe concentration (as determined independently with chemical analysis) to obtain a graph that is useful for routine characterization of Fe-doped InP crystals. Furthermore, the doping with Fe was found to give rise to considerable band-gap narrowing in this material.

The infrared absorption in InP has been studied as a function of free-carrier concentration in the spectral range from a few tens of micrometers to the fundamental absorption edge for both n-type[99–102] and p-type samples.[101,103] In Figure 8.31 shows a typical example for n-InP.[102] The data indicate that an absorption minimum exists at about 1.6 μm, in agreement with the results of Lorenz et al.[99] and Dumke et al.[100] At wavelengths $\lambda > 1.6$ μm free-carrier absorption dominates in the spectra. The dashed line in the figure gives a slope that is limited by ionized-impurity scattering

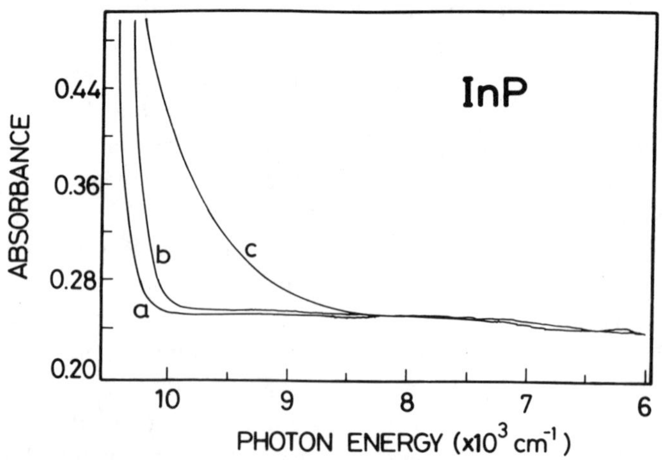

Figure 8.30 Absorption spectra of (a) undoped, (b) S-doped, and (c) Fe-doped InP. The absorption tail in the Fe-doped sample is due to absorption by Fe-related deep centers. (From Fornari and Kumar.[107])

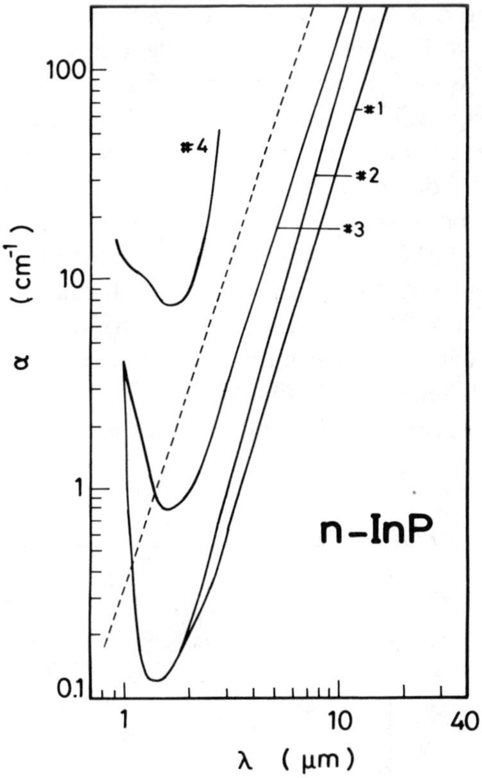

Figure 8.31 Absorption coefficient α versus wavelength λ for n-type InP. The carrier concentrations are 3.3×10^{16} cm^{-3} (#1), 4.0×10^{17} cm^{-3} (#2), 8.8×10^{17} cm^{-3} (#3), and 5.0×10^{18} cm^{-3} (#4). The dashed line gives a slope that is limited by ionized-impurity scattering. (From Ballman et al.[102])

[i.e., $\lambda \propto (\hbar\omega)^{3.5}$]. The power index m taken with a heavily doped sample (#3) is about 3.5 (i.e., the dominance of ionized-impurity scattering). Yamaguchi et al.[101] also found that for n-type InP m increases with increasing free-electron concentration from about 2.0 ($n \simeq 1 \times 10^{16}$ cm^{-3}) to 3.5 ($n \simeq 1 \times 10^{18}$ cm^{-3}). For wavelengths near 1.3 μm, the absorption coefficients shown in Fig. 8.31 are still proportional to the free-electron concentration. This fact suggests the dominance of interconduction-band absorption in this spectral region. Detailed studies of interconduction-band absorption in n-InP has been carried out by Lorenz et al.[99] and Dumke et al.[100]

The infrared absorption in p-type InP has been studied by several authors.[101,103] Casey and Carter[103] have measured the absorption spectra in p-InP between wavelengths of 1.0 and 3.4 μm for hole concentrations from

$p = 3.1 \times 10^{17}$ to 6.6×10^{18} cm^{-3} at 297 K. The spectra they obtained are shown in Fig. 8.32. Ballman et al.[102] have also measured infrared absorption in lightly doped, p-type InP ($p = 6.0 \times 10^{16}$ cm^{-3}). Their result is also shown in Fig. 8.32 by the dashed line. The figure clearly shows that the absorption coefficient increases as the hole concentration increase. Casey and Carter[103] determined the concentration variation of the absorption coefficient to be $\alpha = 14(p/10^{18}\ \text{cm}^{-3})\ \text{cm}^{-1}$ at a wavelength of 1.3 μm and $\alpha = 20(p/10^{18}\ \text{cm}^{-3})\ \text{cm}^{-1}$ at a wavelength of 1.5 μm. Heating to 100°C increased α by a measurable, although small, amount. Peaks that appeared at about 0.4 eV (#1 and #2) are due to intervalence-band absorption caused by transitions between the heavy-hole and the spin–orbit splitoff valence bands [see Section 8.5 (Fig. 8.34)].

Aliev et al.[104] have studied infrared absorption as a function of free-

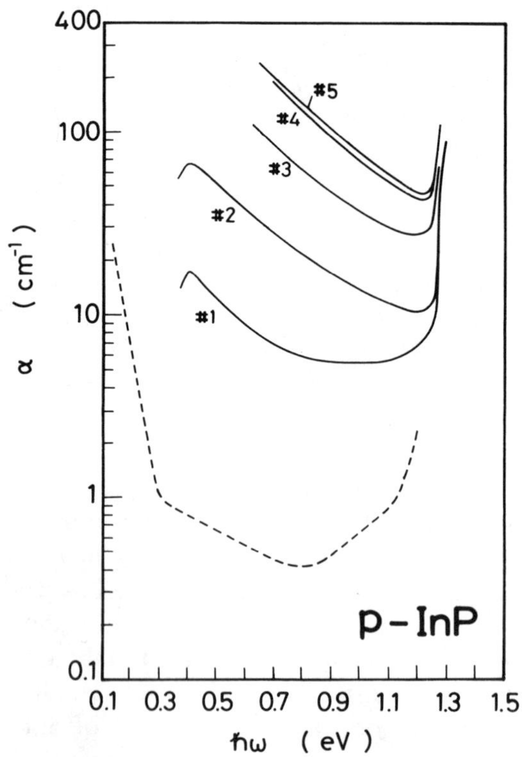

Figure 8.32 Absorption coefficient α versus wavelength λ for p-type InP. The solid and dashed lines represent the experimental data taken from Refs. 103 and 102, respectively. The carrier concentrations are 3.1×10^{17} cm^{-3} (#1), 8.6×10^{17} cm^{-3} (#2), 1.9×10^{18} cm^{-3} (#3), 5.3×10^{18} cm^{-3} (#4), 6.6×10^{18} cm^{-3} (#5), and 6.0×10^{16} cm^{-3} (dashed line).

carrier concentration for n-type $In_{1-x}Ga_xAs$ ($x \leq 0.1$) at 92 and 300 K and found that the absorption curves rise smoothly with wavelength and carrier concentration. This is a typical characteristic of free-carrier absorption. The power indices obtained varied with alloy composition x, carrier concentration, and temperature ($2.4 \leq m \leq 4.0$). The experimental findings for the alloys were found to be in good agreement with the results obtained from the formula $\alpha = \alpha_{dis} + \alpha_{InAs}$, where α_{dis} is an extra contribution of the alloy disorder to α as pointed out theoretically by Sieránski and Szatkowski[110] and α_{InAs} is the experimental value of the absorption coefficient of the InAs binary.

Rajalakshmi and Arora[111] have analyzed the absorption spectra of Ge-doped $In_{1-x}Ga_xAs_yP_{1-y}$ lattice-matched to InP ($y = 0.28$) by including an Urbach tail due to transitions from disorder-induced localized states at the valence-band edge to the conduction band. Different types of disorder have been taken into consideration in their analyses. The dependence of the absorption coefficients on electron concentration for a variety of compositions in the $In_{1-x}Ga_xAs_yP_{1-y}$/InP system at a wavelength of 1.3 μm have been calculated by Fiedler and Schlachetzki,[87] who used values for the end-point binaries and took into account the Moss–Burstein band-gap shift. Brosson et al.[112] have also shown that measurements of threshold current density and external quantum efficiency for a broad-area laser structure lead to a determination of the carrier-induced optical loss and differential loss. The differential loss they determined was $d\alpha/dN \simeq 1.1$–2.3×10^{-17} cm^2 at $\lambda = 1.53$ μm in a $\lambda_g = 1.30$-μm n-$In_{1-x}Ga_xAs_yP_{1-y}$ layer.

8.4.2 Carrier-Induced Change in Refractive Index

There has been an increasing interest in intensity and phase modulators lately because of their importance in the development of integrated optoelectronics. Several authors have demonstrated light switches and modulators based on refractive-index changes in InGaAsP/InP heterojunction p–n devices.[113–115] The operation of these devices has been explained in terms of the linear electrooptic effect and higher-order field effects[115] or by the effects of a free-carrier-induced change in the refractive index.[113,114] The linear and nonlinear electrooptic effects change the refractive index in regions of the p–n device where high electric fields are present. Details of these effects will be presented in the next section. The free-carrier effects can be classified into two categories: (1) a refractive index change that is basically caused by free-carrier plasma dispersion (intraband free-carrier absorption) and (2) a refractive-index change that originates from the change in the fundamental absorption gap (Moss–Burstein shift and/or band-gap shrinkage).

Intraband free-carrier absorption can be described, in the simplest way, in terms of the classic dispersion theory, which starts with an equation of motion for an electron of charge e and mass $m_e(=m_e^\Gamma$; see Section 6.2.1)

$$m_e(\ddot{x} + \gamma\dot{x} + \omega_0^2 x) = eE_f \exp(i\omega t) \tag{8.55}$$

with the damping constant γ, the restoring force $m_e\omega_0^2 x$, and the field of a electromagnetic wave $E = E_f \exp(i\omega t)$. This leads to the following result for the dielectric constant[1]

$$\epsilon(\omega) = 1 + \left(\frac{Ne^2}{m_e \epsilon_0}\right) \frac{1}{\omega_0^2 - \omega^2 + i\gamma\omega} \tag{8.56}$$

where N is the number of the electrons per unit volume. Here we can put $\omega_0 \to 0$, because there is no restoring force for free electrons. One can, then, obtain

$$\epsilon_1(\omega) = 1 - \left(\frac{Ne^2}{m_e \epsilon_0}\right) \frac{1}{\omega^2} \tag{8.57a}$$

$$\epsilon_2(\omega) = -\left(\frac{Ne^2}{m_e \epsilon_0}\right) \frac{\gamma}{\omega^3} \tag{8.57b}$$

If we replace ω by ω_p and ϵ_1 by ϵ_∞ (i.e., the long-wavelength limit), Eq. (8.57a) provides a relation between the plasma frequency ω_p and the carrier concentration given by Eq. (8.14). In the limit $\gamma \to 0$ and using $2n\,\Delta n \simeq \Delta\epsilon_1$ [see Eq. (8.51)], we can obtain the change in the refractive index

$$\Delta n(\lambda) = -\frac{e^2\lambda^2 N}{8\pi^2 n(\lambda)\epsilon_0 c^2 m_e} \tag{8.58}$$

where c is the light velocity in the vacuum and λ is the wavelength of light. Equation (8.58) suggests that the change Δn is directly proportional to λ^2, N, and $(1/n)$. When both electrons and holes are present in a substance, as is usual for injection lasers, the change in n can be written, by modifying Eq. (8.58), as

$$\Delta n(\omega) = -\frac{e^2\lambda^2}{8\pi^2 n(\lambda)\epsilon_0 c^2}\left(\frac{N}{m_e} + P\,\frac{m_{lh}^{1/2} + m_{hh}^{1/2}}{m_{lh}^{3/2} + m_{hh}^{3/2}}\right) \tag{8.59}$$

where P is the number of holes per unit volume.

A decrease in absorption for photon energies slightly above the fundamental band gap can be observed for several semiconductors when they are doped. This phenomenon, the so-called Moss–Burstein shift, is explained by band filling and is partly compensated by band-gap shrinkage due to carrier–carrier interactions at high concentrations. The relative contributions of band filling, band-gap shrinkage, and free-carrier absorption to the refractive-index change Δn in InGaAsP/InP have been studied by Botteldooren and Baets[116] and Bennett et al.[117]

In Fig. 8.33 we show a typical result for InP taken from Bennett et al.[117] Figure 8.33*a* illustrates the individual contributions to Δn of the band filling

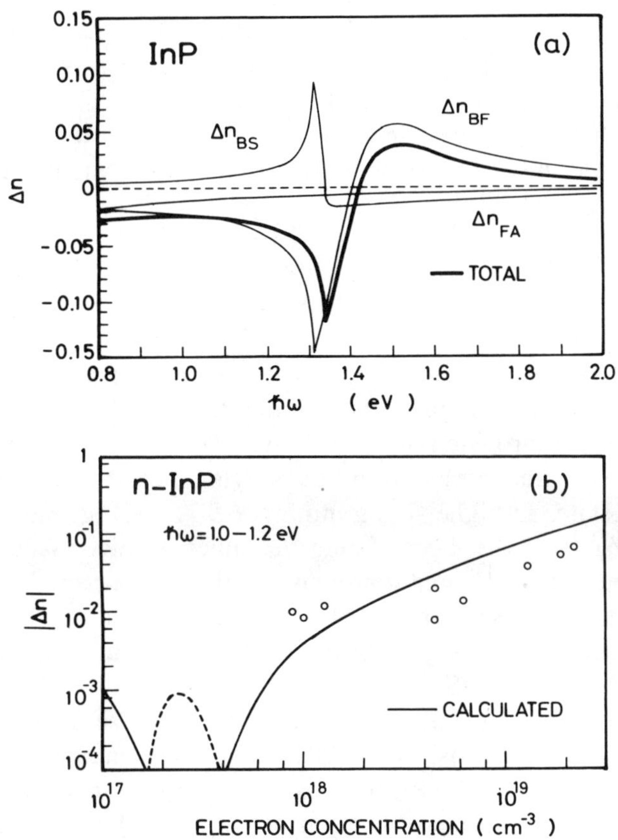

Figure 8.33 Carrier-induced change in the refractive index: (*a*) predicted change in the refractive index Δn for InP ($N = P = 3 \times 10^{18}$ cm^{-3}) from the band-filling (Δn_{BF}), band-gap shrinkage (Δn_{BS}), and free-carrier absorption (Δn_{FA}); (*b*) comparison of the calculated and experimental Δn for *n*-type InP. The solid and dashed lines show the calculated results [$\Delta n < 0$ (solid lines) and $\Delta n > 0$ (dashed line)]. The solid circles show the experimental data. (From Bennett et al.[117])

(Δn_{BF}), band-gap shrinkage (Δn_{BS}), and free-carrier absorption (Δn_{FA}) for InP with carrier concentrations $N = P = 3 \times 10^{18}$ cm^{-3}. The bold line represents the simple sum of these contributions. The curves for the Δn_{BS} and Δ_{BS} pass through zero at photon energies just above (Δn_{BF}) or below (Δn_{BS}) the direct band edge E_0 undergoing a reversal in sign. The value of $|\Delta n_{FA}|$ increases gradually with decreasing photon energy because of its λ^2 dependence [Eq. (8.59)]. Figure 8.33*b* shows a comparison of the estimated and available experimental Δn for *n*-InP at photon energies between 1.0 and 1.2 eV. The prediction of Δn is seen to be in fairly good agreement with the experimental data.

Whalen and Stone[72] have measured the refractive index of *n*-type InP samples with carrier concentrations ranging from 10^{15} to 10^{19} cm^{-3} at the two wavelengths, $\lambda = 0.633$ and 1.15 μm, using a near-Brewster-angle reflectivity technique. Their data reveal that the index n is relatively insensitive to changes in carrier concentration below 10^{18} cm^{-3}, but for impurity levels above 5×10^{18} cm^{-3} there is an increase in the index change $|\Delta n|$ with an increase in carrier concentration that can be explained in terms of the conduction-band filling.

8.5 INTERVALENCE-BAND ABSORPTION

As indicated in Fig. 8.31 (also see Fig. 8.29), the infrared absorption due to free electrons in *n*-type InP usually exhibits absorption band on the low-energy side of the intrinsic absorption edge. This band is attributed to free-electron indirect transitions from states lying near the conduction-band minimum to states in higher-lying minima (α_{CC}; see Fig. 8.28). The same absorption bands have been found in other semiconductors such as *n*-GaAs[118] and *n*-GaP,[119] and consequently the evidence indicates that the bands are not due to the presence of unknown impurities or lattice defects. A similar structure was also found in *p*-type semiconductors, such as Ge,[120] InAs,[121] GaAs,[122] and InSb,[123] and explained in terms of hole transitions between various branches of the valence band in the semiconductors (i.e., the intervalence-band transitions). The absorption bands peaking at $\hbar\omega = 0.4$ eV seen in Fig. 8.32 (*p*-InP) originate from these intervalence-band transitions. In Fig. 8.34 possible types of the intervalence-band transitions in the zinc-blende cyrstals are illustrated schematically. We can expect from the figure three types of the hole transitions: (1) the heavy-hole (HH) band $\rightarrow$ spin splitoff (SO) hole band, (2) the light-hole (LH) band $\rightarrow$ SO hole band, and (3) the HH band $\rightarrow$ LH band.

The zinc-blende-type crystal has a T_d point-group symmetry. The HH and LH bands have Γ_8 symmetry, while the SO band has Γ_7 symmetry.

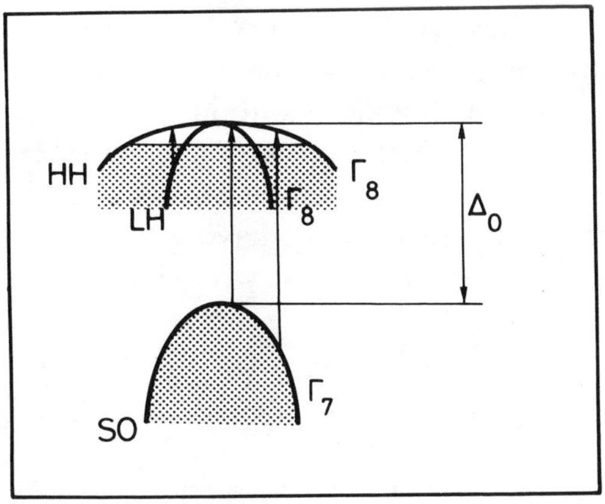

Figure 8.34 Schematic representation of the intervalence-band transitions in *p*-type, zinc-blend semiconductors (HH, LH, SO = (respectively), heavy-hole, light-hole, spin–orbit splitoff valence bands).

The light polarization vector $E \perp (x, y, z)$ of the point group T_d belongs to T_5 symmetry. The selection rules of the optical transitions between the valence bands can be given by the following direct products:[124,125]

$$\text{HH} \rightarrow \text{SO:} \quad \Gamma_8 \times \Gamma_7 = \Gamma_3 + \Gamma_4 + \Gamma_5 \tag{8.60a}$$

$$\text{LH} \rightarrow \text{SO:} \quad \Gamma_8 \times \Gamma_7 = \Gamma_3 + \Gamma_4 + \Gamma_5 \tag{8.60b}$$

$$\text{HH} \rightarrow \text{LH:} \quad \Gamma_8 \times \Gamma_8 = \Gamma_1 + \Gamma_2 + \Gamma_3 + 2\Gamma_4 + 2\Gamma_5 \tag{8.60c}$$

These three products contain the representation of Γ_5 symmetry. This means that all the transitions are possible.

The joint-density-of-states (DOS) function $J_{\alpha\beta}(\omega)$ measures the product density of full and empty states of equal energy difference. This quantity mainly determines the interband contribution to $\epsilon_2(\omega)$, and thus to the optical constants of solids (see Section 8.3). According to different combinations of the sign of

$$\frac{1}{\mu_i} = \left(\frac{1}{m_\alpha} + \frac{1}{m_\beta}\right)_{k=k_0} \qquad (i = x, y, z) \tag{8.61}$$

one has to distinguish between four types of CPs, M_0, M_1, M_2, and M_3 (the Van Hove singularities). The states α and β mean empty and full bands, respectively. The effective mass in the α band has the relation

$$m_\alpha \text{ (electrons)} = -m_\alpha \text{ (holes)} \tag{8.62}$$

The type of singularities for $In_{1-x}Ga_xAs_yP_{1-y}$ quaternary can be determined by inserting numeric values of the effective hole masses (see Section 6.2.2; assumed simple, parabolic, isotropic bands) into Eq. (8.61). The results suggest that $J_{HH\text{-}SO}$ (HH → SO), $J_{LH\text{-}SO}$ (LH → SO), $J_{HH\text{-}LH}$ (HH → LH) are, respectively, represented by M_0-, M_3-, and M_0-type CPs. The M_0-type (M_3-type) CP is realized by having all $\mu_i > 0$ ($\mu_i < 0$). The ω-dependent states functions are, thus, written as[124]

$$J_{HH\text{-}SO}(\omega) = \begin{cases} C_0(\hbar\omega - \Delta_0)^{0.5} & (\hbar\omega > \Delta_0) \\ 0 & (\hbar\omega < \Delta_0) \end{cases} \tag{8.63a}$$

$$J_{LH\text{-}SO}(\omega) = \begin{cases} 0 & (\hbar\omega > \Delta_0) \\ C_1(\Delta_0 - \hbar\omega)^{0.5} & (\hbar\omega < \Delta_0) \end{cases} \tag{8.63b}$$

$$J_{HH\text{-}LH}(\omega) = C_2(\hbar\omega)^{0.5} \tag{8.63c}$$

The LH → SO transition is not so important because of a low density of states in the LH band and a very low hole occupancy of final states for transitions at the band-gap energy. The larger density of states and hole occupancy in the HH band may produce much greater absorption for the HH → SO transition than for the LH → SO transition.

Intervalence-band absorption has been observed in various p-type semiconductors (e.g., Ge, InP, InAs, GaAs, InSb). Near room temperature, the threshold current of $In_{1-x}Ga_xAs_yP_{1-y}$/InP lasers usually increases as exp (T/T_0), where $T_0 \sim 65$ K, whereas theoretical descriptions of ideal lasers give $T_0 \sim 200$ K.[126] Several loss mechanisms have been proposed to explain this temperature sensitivity:[127] (1) carrier leakage over the heterobarrier into the confining layers, (2) nonradiative losses due to Auger recombination, (3) nonradiative recombination at defects and interface states, and (4) intervalence-band absorption causing optical losses. Yamanishi et al.[128] have found spectral burning in spontaneous emission spectra from $In_{1-x}Ga_xAs_yP_{1-y}$/InP lasers ($\lambda_g = 1.3$ μm). The results are successfully understood on the basis of population-burning of holes associated with the saturation of the intervalence-band absorption. Henry et al.[126] have made measurements of the intervalence-band absorption spectra in p-type InP, GaAs, and $In_{0.53}Ga_{0.47}As$. The measured spectra were broader and had a smaller temperature dependence. In Fig. 8.35 we show their measured results for p-$In_{0.53}Ga_{0.47}As$. The solid line is the estimated fit based on an elemental $\boldsymbol{k} \cdot \boldsymbol{p}$ band model. Note that HH → SO (LH → SO) absorption occurs when $\hbar\omega > \Delta_0$ ($\hbar\omega < \Delta_0$) [see also Eq. (8.63)]. The authors discussed the effects of intervalence-band absorption on the thermal behavior of $In_{1-x}Ga_xAs_yP_{1-y}$/InP lasers. They concluded that because of its low

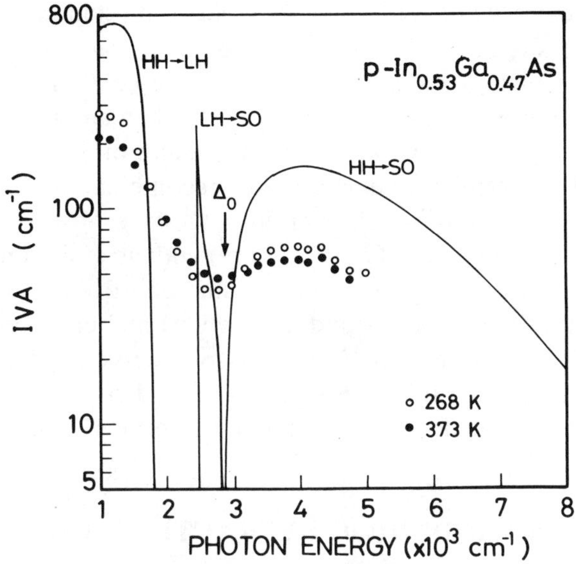

Figure 8.35 Measured intervalence-band absorption in *p*-type $In_{0.53}Ga_{0.47}As$ (1×10^{18} cm^{-3}) at 268 (open circles) and 373 K (solid circles). The solid lines are the theoretical curves predicted from an empirical $\boldsymbol{k} \cdot \boldsymbol{p}$ band model. (From Henry et al.[126])

strength and weak temperature dependence, intervalence-band absorption should have only a minor effect on the temperature dependence of laser threshold.

Childs et al.[127] concentrated on intervalence-band absorption calculating its magnitude using a pseudopotential band structure and matrix elements evaluated using 21 ***k***-space directions through the Brillouin zone. They calculated that at 1.6 μm the absorption coefficient for intervalence-band absorption is $\alpha = 39$ cm^{-1} for 1×10^{18} cm^{-3} holes at room temperature. This is in very good agreement with the magnitude of α determined experimentally from laser characteristics[129] that require an absorption coefficient in the active region of about 100 cm^{-1} at the threshold current density (N_{th}), which is typically $N_{th} = 2 \sim 3 \times 10^{18}$ cm^{-3}. However, they concluded that the temperature sensitivity of intervalence-band absorption $(1/\alpha)(d\alpha/dT) = 0.45\%$ K^{-1} at 300 K that they derived is insufficient to explain the observed temperature sensitivity of lasers. Adams and coworkers[130] have recently reevaluated the results of Childs et al.[127] and showed that when the rate of change of N_{th} with temperature is taken into account, the intervalence-band absorption can account almost entirely for the temperature sensitivity of 1.6-μm devices. Haug[131] has also discussed the intervalence-band absorption as a phenomenon that affects the threshold current of semiconductor lasers. However, his calculated values of the absorption coefficient were too small to yield a substantial effect. But he showed that the situation is changed if phonon-assisted intervalence-band absorption is taken into account. The absorption coefficient, in that case, was considerably greater, for instance by more than an order of magnitude in $In_{1-x}Ga_xAs_yP_{1-y}/InP$ with a laser wavelength of 1.3 μm.

In principle, emissions associated with recombinations in the intervalence bands are also possible. Mozer et al.[132] have observed emission in $In_{1-x}Ga_xAs_yP_{1-y}/InP$ lasers ($\lambda_g = 1.3$ μm) with an emission energy close to $E_0 + \Delta_0$, which is attributed to recombination of holes in the SO band (generated by the intervalence-band absorption) and electrons in the conduction band. Nonomura et al.[133] have also observed 4–5-μm emissions from $In_{1-x}Ga_xAs_yP_{1-y}/InP$ lasers ($\lambda_g = 1.3$, 1.5 μm) associated with recombinations of holes in the SO band and electrons in the LH band.

8.6 NONLINEAR INDEX OF REFRACTION AND ABSORPTION

Nonlinear optical properties of InP-based semiconductors and quantum-well structures are of considerable interest for various optical device applications. Much work has concentrated on the near-band-gap nonlinear index of refraction and absorption in various semiconductor materials. Reintjes et al.[134] have reported observation of dynamic saturation of optical absorption in $In_{1-x}Ga_xAs$ and $InAs_yP_{1-y}$ alloys at 300 K using mode-locked pulses from a Nd:glass laser. Increase in the transmission coefficient of epitaxial films by more than four orders of magnitude above the equilibrium value has been observed. Islam et al.[135,136] also reported picosecond near-band-gap nonlinear absorption at $\lambda \sim 1.5$ μm and degenerate four-wave mixing experiments in $In_{1-x}Ga_xAs_yP_{1-y}$ at 300 K using a mode-locked color center laser. Nonlinear absorption cross sections σ_{eh} as large as -5.7×10^{-15} cm^2 were obtained from the pump–probe results, while an effective nonlinear cross section σ_{eff} as large as 7.8×10^{-16} cm^2 (corresponding to a steady $|\chi^{(3)}| \sim 3.8 \times 10^{-3}$ esu for a 20-ns relaxation time) were measured in degenerate four-wave mixing experiments. Nonlinear optical absorption in an $In_{1-x}Ga_xAs_yP_{1-y}/InP$ heterostructure laser ($y = 0.257$) has also been observed by Jokerst and Garmire.[137] They found a factor of 2 change in absorption at peak power densities of 0.2 W/cm^2.

Values of the nonlinear refractive index of a $In_{0.53}Ga_{0.47}As/InP$ multiple quantum-well structure obtained by transforming intensity-dependent absorption data (the KK transformation) have recently been reported by Fisher.[138] He obtained the absorption data by a pump–probe technique. The results indicated that the variation of the refractive-index change with pump intensity shows a saturation that can be well described by a simple band-filling model. Two-photon absorption coefficients in InP has also been reported and discussed in Refs. 139 and 140.

At a high light intensity (>1 MW/cm^2), the optical absorption due to intervalence-band transitions at wavelengths near 10 μm has been shown to saturate in *p*-type Ge,[141–143] Si,[141,143] and GaAs.[141,144] This nonlinear absorption property has used to control CO_2 lasers to obtain large peak output powers.[145] However, to our knowledge, there has been no report on InP-related compounds.

REFERENCES

1. See, for instance, G. Harbake, in *Optical Properties of Solids*, F. Abelés, ed., North-Holland, Amsterdam, 1972, p. 21.
2. M. Balkanski, in Ref. 1 (above), p. 529.
3. J. S. Blakemore, *J. Appl. Phys.* **53,** R123 (1982).
4. B. H. Bairamov, I. P. Ipatova, V. A. Milorava, V. V. Toporov, K. Naukkarinen, T. Tuomi, G. Irmer, and J. Monecke, *Phys. Rev. B* **38,** 5722 (1988).
5. A. A. Kukharskii, *Sov. Phys. Solid State* **14,** 1501 (1972).
6. G. M. Zinger, I. P. Ipatova, and A. I. Ryskin, *Sov. Phys. Semicond.* **18,** 13 (1984).
7. P. M. Amirtharaj, G. D. Holah, and S. Perkowitz, *Phys. Rev. B* **21,** 5656 (1980).
8. C. Pickering, *J. Electron. Mater.* **10,** 901 (1981).
9. A. N. Pikhtin and A. D. Yas'kov, *Sov. Phys. Semicond.* **22,** 613 (1988).
10. L. I. Korovin, *Sov. Phys. Solid State* **1,** 1202 (1960).
11. M. Cardona, in *Solid State Physics, Nuclear Physics and Particle Physics*, I. Saavedra, ed., Benjamin, New York, 1968, p. 737.
12. S. H. Wemple and M. DiDomenico, Jr., *Phys. Rev. B* **3,** 1338 (1971).
13. F. Stern, *Phys. Rev.* **133,** A1653 (1964).
14. M. A. Afromowitz, *Solid State Commun.* **15,** 59 (1974).
15. A. N. Pikhtin and A. D. Yas'kov, *Sov. Phys. Semicond.* **12,** 622 (1978).
16. S. Adachi, *Phys. Rev. B* **39,** 12612 (1989).
17. S. Adachi, *Phys. Rev. B* **35,** 7454 (1987).
18. S. Adachi, *Phys. Rev. B* **38,** 12345 (1988).
19. S. Adachi, *Phys. Rev. B* **38,** 12966 (1988).
20. S. Adachi, *J. Appl. Phys.* **66,** 813 (1989).
21. S. Adachi, *Phys. Rev. B* **41,** 3504 (1990).
22. S. Adachi, *J. Appl. Phys.* **67,** 6427 (1990).
23. S. Adachi, *J. Appl. Phys.* **68,** 1192 (1990).

24. S. Adachi and T. Taguchi, *Phys. Rev. B* **43,** 9569 (1991).
25. W. Kowalsky, H.-H. Wehmann, F. Fiedler, and A. Schlachetzki, *Phys. Status Solidi A* **77,** K75 (1983).
26. S. M. Kelso, D. E. Aspnes, M. A. Pollack, and R. E. Nahory, *Phys. Rev. B* **26,** 6669 (1982).
27. S. Adachi, *J. Appl. Phys.* **53,** 5863 (1982).
28. G. Harbake, Ref. 1 (above), p. 21.
29. D. C. Reynolds and T. C. Collins, *Excitons: Their Properties and Uses*, Academic, New York, 1981.
30. R. J. Elliott, *Phys. Rev.* **108,** 1384 (1957).
31. Y. Toyozawa, *J. Phys. Chem. Solids* **25,** 59 (1964).
32. B. Velický and J. Sak, *Phys. Status Solidi* **16,** 147 (1966).
33. E. O. Kane, *Phys. Rev.* **180,** 852 (1969).
34. J. E. Rowe and D. E. Aspnes, *Phys. Rev. Lett.* **25,** 162 (1970).
35. Y. Pétroff and M. Balkanski, *Phys. Rev. B* **3,** 3299 (1971).
36. R. M. Martin, J. A. Van Vechten, J. E. Rowe, and D. E. Aspnes, *Phys. Rev. B* **6,** 2500 (1972).
37. M. Welkowski and R. Braunstein, *Phys. Rev. B* **5,** 497 (1972).
38. W. Hanke and L. J. Sham, *Phys. Rev. B* **21,** 4656 (1980).
39. M. del Castillo-Mussot and L. J. Sham, *Phys. Rev. B* **31,** 2092 (1985).
40. A. D'Andrea and R. D. Sole, *Phys. Rev. B* **38,** 1197 (1988).
41. A. Baldereschi and N. O. Lipari, *Phys. Rev. B* **3,** 439 (1971).
42. P. Rochon and E. Fortin, *Phys. Rev. B* **12,** 5803 (1975).
43. W. J. Turner, W. E. Reese, and G. D. Pettit, *Phys. Rev.* **136,** A1467 (1964).
44. A. M. White, P. J. Dean, L. L. Taylor, R. C. Clarke, D. J. Ashen, and J. B. Mullin, *J. Phys. C* **5,** 1727 (1972).
45. F. Evangelisti, J. U. Fischbach, and A. Frova, *Phys. Rev. B* **9,** 1516 (1974).
46. S. B. Nam, D. C. Reynolds, C. W. Litton, T. C. Collins, P. J. Dean, and R. C. Clarke, *Phys. Rev. B* **13,** 1643 (1976).
47. H. Mathieu, Y. Chen, J. Camassel, and J. Allegre, *Phys. Rev. B* **32,** 4042 (1985).
48. M. A. Abdullaev, S. I. Kokhanovskii, Y. M. Makushenko, and R. P. Seisyan, *Sov. Phys. Semicond.* **23,** 724 (1989).
49. L. G. Meiners, *J. Appl. Phys.* **59,** 1611 (1986).
50. A. V. Varfolomeev, R. P. Seisyan, and R. N. Yakimova, *Sov. Phys. Semicond.* **9,** 530 (1975).
51. D. D. Sell, *Phys. Rev. B* **6,** 3750 (1972).
52. P. J. Dean, G. Kaminsky, and R. B. Zetterstrom, *J. Appl. Phys.* **38,** 3551 (1967).
53. K. Satzke, G. Weiser, R. Höger, and W. Thulke, *J. Appl. Phys.* **63,** 5485 (1988).

54. Y.-S. Chen and O. K. Kim, *J. Appl. Phys.* **52,** 7392 (1981).
55. E. Zielinski, H. Schweizer, K. Streubel, H. Eisele, and G. Weimann, *J. Appl. Phys.* **59,** 2196 (1986).
56. S. Adachi, *Jpn. J. Appl. Phys.* **28,** 1536 (1989).
57. S. Adachi, *Phys. Rev. B* **41,** 1003 (1990).
58. T. Miyazaki and S. Adachi, *Phys. Status Solidi B* **163,** 299 (1991).
59. T. Aoki and S. Adachi, *J. Appl. Phys.* **69,** 1574 (1991).
60. D. E. Aspnes, in *Optical Properties of Solids: New Developments*, B. O. Seraphin, ed., North-Holland, Amsterdam, 1976, p. 799.
61. D. E. Aspnes and A. A. Studna, *Phys. Rev. B* **27,** 985 (1983).
62. P. Lautenschlager, M. Garriga, and M. Cardona, *Phys. Rev. B* **36,** 4813 (1987).
63. M. Erman, J. P. Andre, and J. LeBris, *J. Appl. Phys.* **59,** 2019 (1986).
64. H. Burkhard, H. W. Dinges, and E. Kuphal, *J. Appl. Phys.* **53,** 655 (1982).
65. M. Erman, J. B. Theeten, N. Vodjdani, and Y. Demay, *J. Vacuum Sci. Technol. B* **1,** 328 (1983).
66. B. Drevillon, E. Bertran, P. Alnot, J. Olivier, and M. Razeghi, *J. Appl. Phys.* **60,** 3512 (1986).
67. S. Adachi, *J. Appl. Phys.* **66,** 3224 (1989).
68. S. Adachi, *J. Appl. Phys.* **66,** 6030 (1989).
69. B. O. Seraphin and H. E. Bennett, in *Semiconductors and Semimetals*, Vol. 3, R. K. Willardson and A. C. Beer, eds., Academic, New York, 1967, p. 499.
70. G. D. Pettit and W. J. Turner, *J. Appl. Phys.* **36,** 2081 (1965).
71. V. B. Bogdanov, V. T. Prokopenko, and A. D. Yaskov, *Opt. Spectrosc.* (USSR) **62,** 551 (1987).
72. M. S. Whalen and J. Stone, *J. Appl. Phys.* **53,** 4340 (1982).
73. B. Broberg and S. Lindgren, *J. Appl. Phys.* **55,** 3376 (1984).
74. T. Takagi, *Jpn. J. Appl. Phys.* **17,** 1813 (1978).
75. P. Chandra, L. A. Coldren, and K. E. Strege, *Electron. Lett.* **17,** 6 (1981).
76. W. Kowalsky, A. Schlachetzki, and F. Fiedler, *Phys. Status Solidi A* **68,** 153 (1981).
77. C. H. Henry, L. F. Johnson, R. A. Logan, and D. P. Clarke, *IEEE J. Quantum Electron.* **QE-21,** 1887 (1985).
78. K. Fronts, N. I. Maiorova, V. A. Mishurnyi, V. I. Kuchinskii, E. L. Portnoi, and V. B. Smirnitskii, *Sov. Tech. Phys. Lett.* **12,** 342 (1986).
79. S. Morasca, B. Sordo, C. D. Bernardi, and M. Meliga, *Appl. Phys. Lett.* **52,** 1593 (1988).
80. G. H. Olsen, T. J. Zamerowski, R. T. Smith, and E. P. Bertin, *J. Electron. Mater.* **9,** 977 (1980).
81. R. E. Nahory and M. A. Pollack, *Electron. Lett.* **14,** 727 (1978).

82. H. Burkhard, *J. Appl. Phys.* **55,** 503 (1984).
83. J. Buus and M. J. Adams, *IEE Solid-State Electron Dev.* **3,** 189 (1979).
84. B. Jensen and A. Torabi, *J. Appl. Phys.* **54,** 3623 (1983).
85. S. Adachi, *J. Appl. Phys.* **58,** R1 (1985).
86. S. Adachi, *J. Appl. Phys.* **61,** 4869 (1987).
87. F. Fiedler and A. Schlachetzki, *Solid-State Electron.* **30,** 73 (1987).
88. R. Newman, *Phys. Rev.* **111,** 1518 (1958).
89. H. Müller, R. Trommer, M. Cardona, and P. Vogl, *Phys. Rev. B* **21,** 4879 (1980).
90. H. H. Wieder, A. R. Clawson, and G. E. McWilliams, *Appl. Phys. Lett.* **31,** 468 (1977).
91. H. H. Caspers and H. H. Wieder, *Solid State Commun.* **29,** 403 (1979).
92. G. D. Holah, E. L. Meeks, and F. L. Eisele, *J. Vacuum Sci. Technol. B* **1,** 182 (1983).
93. V. P. Evtikhiev, D. Z. Garbuzov, and A. T. Gorelenok, *Sov. Phys. Semicond.* **17,** 891 (1983).
94. Y. K. Su, M. C. Wu, K. Y. Cheng, and C. Y. Chang, *J. Cryst. Growth* **67,** 477 (1984).
95. Y. Takeda, A. Sasaki, Y. Imamura, and T. Takagi, *J. Appl. Phys.* **47,** 5405 (1976).
96. J. Y. Marzin, J. L. Benchimol, B. Sermage, B. Etienne, M. Voos, and M. Balkanski, *Solid State Commun.* **45,** 79 (1983).
97. D. A. Humphreys, R. J. King, D. Jenkins, and A. J. Moseley, *Electron. Lett.* **21,** 1187 (1985).
98. F. R. Bacher, J. S. Blakemore, J. T. Ebner, and J. R. Arthur, *Phys. Rev. B* **37,** 2551 (1988).
99. M. R. Lorenz, W. Reuter, W. P. Dumke, R. J. Chicotka, G. D. Pettit, and J. M. Woodall, *Appl. Phys. Lett.* **13,** 421 (1968).
100. W. P. Dumke, M. R. Lorenz, and G. D. Pettit, *Phys. Rev. B* **1,** 4668 (1970).
101. M. Yamaguchi, A. Yamamoto, S. Shinoyama, and H. Sugiura, *J. Appl. Phys.* **53,** 633 (1982).
102. A. A. Ballman, A. M. Glass, R. E. Nahory, and H. Brown, *J. Cryst. Growth* **62,** 198 (1983).
103. H. C. Casey, Jr. and P. L. Carter, *Appl. Phys. Lett.* **44,** 82 (1984).
104. M. I. Aliev, K. A. Khalilov, and G. B. Ibragimov, *Phys. Status Solidi B* **140,** K83 (1987).
105. E. Haga and H. Kimura, *J. Phys. Soc. Jpn.* **19,** 658 (1964); **19,** 1596 (1964).
106. F. Urbach, *Phys. Rev.* **92,** 1324 (1953).
107. R. Fornari and J. Kumar, *Appl. Phys. Lett.* **56,** 638 (1990).
108. F. Vidimari, M. Caldironi, A. Di Paola, R. Chen, and S. Pellegrino, *Electron. Lett.* **27,** 817 (1991) and references cited therein.

109. G. Guillot, G. Bremond, T. Benyattou, F. Ducroquet, B. Wirth, M. Colombet, A. Louati, and A. Bencherifa, *Semicond. Sci. Technol.* **5,** 391 (1990) and references cited therein.

110. K. Sieránski and J. Szatkowski, *Phys. Status Solidi B* **104,** 57 (1981).

111. R. Rajalakshmi and B. M. Arora, *J. Appl. Phys.* **67,** 3533 (1990).

112. P. Brosson, C. Labourie, L. L. Gouezigou, J. L. Lievin, J. Jacquet, F. Leblond, A. Olivier, and D. Leclerc, *Electron. Lett.* **25,** 1623 (1989).

113. K. Ishida, H. Nakamura, H. Matsumura, T. Kadoi, and H. Inoue, *Appl. Phys. Lett.* **50,** 141 (1987).

114. O. Mikami and H. Nakagome, *Opt. Quantum Electron.* **17,** 449 (1985).

115. U. Koren, T. L. Koch, H. Presting, and B. I. Miller, *Appl. Phys. Lett.* **50,** 368 (1987).

116. D. Botteldooren and R. Baets, *Appl. Phys. Lett.* **54,** 1989 (1989).

117. B. R. Bennett, R. A. Soref, and J. A. del Alamo, *IEEE J. Quantum Electron.* **26,** 113 (1990).

118. W. G. Spitzer and J. M. Whelan, *Phys. Rev.* **114,** 59 (1959).

119. W. G. Spitzer, M. Gershenzon, C. J. Frosch, and D. F. Gibbs, *J. Phys. Chem. Solids* **11,** 339 (1959).

120. H. B. Briggs and R. C. Fletcher, *Phys. Rev.* **91,** 1342 (1953).

121. F. Matossi and F. Stern, *Phys. Rev.* **111,** 472 (1958).

122. R. Braunstein, *J. Phys. Chem. Solids* **8,** 280 (1959).

123. G. W. Gobeli and H. Y. Fan, *Phys. Rev.* **119,** 613 (1960).

124. S. Adachi, *J. Appl. Phys.* **53,** 8775 (1982).

125. The group-theoretical symbols used here are taken from G. F. Koster, J. O. Dimmock, R. G. Wheeler, and H. Statz, *Properties of the Thirty-Two Point Groups*, MIT Press, Cambridge, MA, 1963.

126. C. H. Henry, R. A. Logan, F. R. Merritt, and J. P. Luongo, *IEEE J. Quantum Electron.* **QE-19,** 947 (1983).

127. G. N. Childs, S. Brand, and R. A. Abram, *Semicond. Sci. Technol.* **1,** 116 (1986) and references cited therein.

128. M. Yamanishi, I. Suemune, K. Nonomura, and N. Mikoshiba, *Jpn. J. Appl. Phys.* **22** (suppl. **22-1**), 303 (1983).

129. A. R. Adams, D. Patel, P. D. Greene, and G. D. Henshall, *Electron. Lett.* **18,** 919 (1982).

130. A. R. Adams, K. C. Heasman, and J. Hilton, *Semicond. Sci. Technol.* **2,** 761 (1987).

131. A. Haug, *Semicond. Sci. Technol.* **5,** 557 (1990).

132. A. Mozer, K. M. Romanek, W. Schmid, M. H. Pilkuhn, and E. Schlosser, *Appl. Phys. Lett.* **41,** 964 (1982).

133. K. Nonomura, I. Suemune, M. Yamanishi, and N. Mikoshiba, *Jpn. J. Appl. Phys.* **22,** L556 (1983).

134. J. F. Reintjes, J. C. McGroddy, and A. E. Blakeslee, *J. Appl. Phys.* **46,** 879 (1975).

135. M. N. Islam, E. P. Ippen, E. G. Burkhardt, and T. J. Bridges, *Appl. Phys. Lett.* **47,** 1042 (1985).

136. M. N. Islam, E. P. Ippen, E. G. Burkhardt, and T. J. Bridges, *J. Appl. Phys.* **59,** 2619 (1986).

137. N. M. Jokerst and E. Garmire, *Appl. Phys. Lett.* **53,** 897 (1988).

138. M. A. Fisher, *J. Appl. Phys.* **67,** 543 (1990).

139. S. S. Mitra, L. M. Narducci, R. A. Shatas, Y. F. Tsay, and A. Vaidyanathan, *Appl. Opt.* **14,** 3038 (1975).

140. A. Vaidyanathan, T. Walker, A. H. Guenther, S. S. Mitra, and L. M. Narducci, *Phys. Rev. B* **21,** 743 (1980).

141. A. F. Gibson, C. A. Rosito, C. A. Raffo, and M. F. Kimmitt, *Appl. Phys. Lett.* **21,** 356 (1972).

142. M. Kawai and T. Miyakawa, *Jpn. J. Appl. Phys.* **20,** 369 (1981).

143. R. B. James, E. Schweig, D. L. Smith, and T. C. McGill, *Appl. Phys. Lett.* **40,** 231 (1982).

144. R. B. James, W. H. Christie, R. E. Eby, B. E. Mills, and L. S. Darken, Jr., *J. Appl. Phys.* **59,** 1323 (1986).

145. R. B. James and Y.-C. Chang, *Phys. Rev. B* **38,** 9778 (1988) and references cited therein.

9

ELASTOOPTIC AND ELECTROOPTIC EFFECTS

9.1 ELASTOOPTIC EFFECT

9.1.1 Model Representation

Investigation of the elastooptic (photoelastic) behavior in solids is an old topic that arises in close connection with the fundamental optical properties of the solids. Knowledge of the photoelastic behavior of III–V compounds forms an important role not only in the design of elastooptic devices, such as light modulators, deflectors, and switches,[1] but also in the analysis of lattice-mismatched semiconductor heterostructures.

The distribution of stress in heteroepitaxial semiconductor structures is a subject of perennial interest since internal stresses arise normally in thin epitaxial films during preparation of the films by heteroepitaxial growth. The photoelastic phenomenon has been successfully applied by Reinhart and Logan[2] to analyze interface stress in GaAs–AlGaAs layer structures. They demonstrated that the elastic nature and magnitude of the stress is consistent with predictions based on the different thermal expansion coefficients of the layers. Similar analysis has also been carried out by several authors[3,4] in investigations of various strain fields and misfit-induced dislocations in III–V heteroepitaxial layers (see Chapter 11).

It is noted that photoelasticity is a function of light wavelength (i.e., it has a wavelength dependence). There are several theoretical works dealing with the spectral dependence of photoelasticity.[5–9] The application of an external uniaxial stress to a solid produces a change in its crystal symmetry that results in significant changes in its electronic and optical properties.[5,8] An optically isotropic semiconductor, like InP, becomes birefringent under the action of this stress. The uniaxial stress effect can be expressed as

$$\alpha_{\mathrm{pe}} = \frac{\Delta \epsilon_{ij}}{X} = -\sum_{mn} \epsilon_{ii}\epsilon_{jj} p_{ijkl} S_{klmn} \tag{9.1}$$

where α_{pe} is the linear photoelastic coefficient, $\epsilon_{ii} (=\epsilon_{jj})$ is the component of the dielectric constant tensor in the absence of the stress (see Chapter 8), p_{ijkl} is the component of the fourth-rank photoelastic tensor, and S_{klmn} is the component of the elastic compliance tensor (see Section 3.2.1); $\Delta\epsilon_{ij}$ is the change in the real part of the dielectric constants parallel $[\epsilon_1(\|)]$ and perpendicular $[\epsilon_1(\perp)]$ to the direction of the stress X:

$$\Delta\epsilon_{ij} = \epsilon_1(\|) - \epsilon_1(\perp) = -\frac{1}{\alpha_{\mathrm{r}}}[n(\|) - n(\perp)] \tag{9.2}$$

where n is the refractive index and α_{r} is the fractional coefficient defined by[10]

$$\alpha_{\mathrm{r}} = \frac{1}{4}\left(\frac{\epsilon_1 + (\epsilon_1^2 + \epsilon_2^2)^{0.5}}{2}\right)^{-0.5} [1 + (\epsilon_1^2 + \epsilon_2^2)^{-0.5}\epsilon_1] \tag{9.3}$$

In the region of transparency (i.e., $\epsilon_1 >> \epsilon_2$) we can rightly reduce the coefficient to $\alpha_{\mathrm{r}} \simeq \frac{1}{2}\epsilon_1^{-0.5}$.

The linear photoelastic tensor in a zinc-blende structure can now be written as

$$[p] = \begin{bmatrix} p_{11} & p_{12} & p_{12} & 0 & 0 & 0 \\ p_{12} & p_{11} & p_{12} & 0 & 0 & 0 \\ p_{12} & p_{12} & p_{11} & 0 & 0 & 0 \\ 0 & 0 & 0 & p_{44} & 0 & 0 \\ 0 & 0 & 0 & 0 & p_{44} & 0 \\ 0 & 0 & 0 & 0 & 0 & p_{44} \end{bmatrix} \tag{9.4}$$

The photoelastic component p_{ijkl} can also be defined by using the inverse dielectric constant $(1/\epsilon_1)$ as

$$\Delta\left(\frac{1}{\epsilon_1}\right)_{ij} = \frac{-\Delta\epsilon_{ij}}{\epsilon_{ii}\epsilon_{jj}} = \sum_{kl} p_{ijkl} e_{kl} \tag{9.5}$$

where e_{kl} is the strain component that is connected with the stress through Eq. (3.26).

The first-order change in the real part of the dielectric constant $\Delta\epsilon_1$ with applied stress can be given by

$$\Delta\epsilon_1(\omega) = \sum_i \left(\frac{\partial \epsilon_1}{\partial M_i} \Delta M_i + \frac{\partial \epsilon_1}{\partial E_{gi}} \Delta E_{gi} \right) \tag{9.6}$$

where M_i and E_{gi} are, respectively, the strength parameter and interband transition energy of the ith transition (E_0, $E_0 + \Delta_0$, E_1, $E_1 + \Delta_1$, E_0', etc.). Introducing Eq. (8.52) into Eq. (9.6), we finally obtain the expression for the photoelastic coefficient:[8]

$$\alpha_{pe} = C^* \left\{ -g(\chi) + \frac{4E_0}{\Delta_0} \left[f(\chi) - \left(\frac{E_0}{E_0 + \Delta_0} \right)^{3/2} f(\chi_{so}) \right] \right\} + D^* \tag{9.7}$$

with

$$C^* = \begin{cases} (\frac{3}{2}\mu_0)^{1.5} P^2 b(S_{11} - S_{12}) E_0^{-5/2} & \text{for } X \parallel [100] \\ (\frac{3}{2}\mu_0)^{1.5} P^2 d S_{44} E_0^{-5/2}/2\sqrt{3} & \text{for } X \perp [111] \end{cases} \tag{9.8}$$

$$g(\chi) = \chi^{-2}[2 - (1 + \chi)^{-0.5} - (1 - \chi)^{-0.5}] \tag{9.9}$$

where b and d are the shear deformation potentials of the valence bands (see Section 7.1) and $f(\chi)$ is a function defined by Eq. (8.23). In Eq. (9.7), the first term corresponds to the contributions arising from the E_0 and $E_0 + \Delta_0$ gaps and the second term D^* corresponds to the contributions arising from other, far-off critical points in the band structure (E_1, $E_1 + \Delta_1$, E_0', E, $E_2 + \Delta_2$, etc.). The term D^* is assumed to be nondispersive (i.e., constant). This assumption is essentially the same as for B^* used in the expression for the unperturbed ϵ_1 [Eq. (8.52)]. The first and second terms in the curly bracket of Eq. (9.7) come from the change in E_g (transition energy) and M (strength parameter), respectively.

Figure 9.1 shows the calculated results of the dispersive components, $f(\chi)$ and $g(\chi)$. It is worth noting as seen in this figure that $g(\chi)$ shows a very sharp dispersion near the band edge E_0 ($\chi = 1.0$) compared with that of $f(\chi)$. The parameters C^* and D^* can be treated as adjustable parameters to fit the experimental data.

9.1.2 Analyses

The photoelastic coefficient can be obtained from a stress-induced birefringence (piezobirefringence) measurement. Piezobirefringence data have

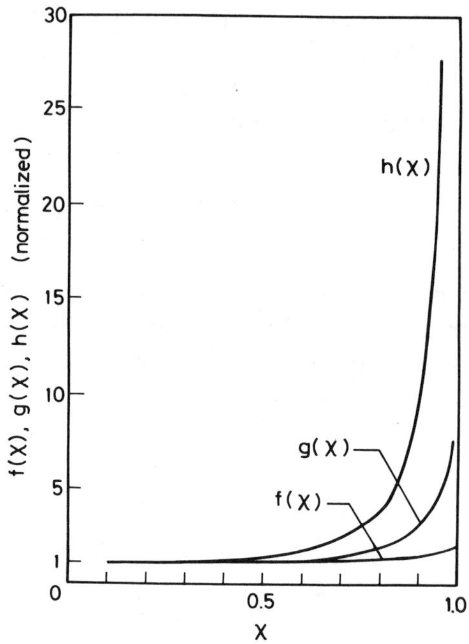

Figure 9.1 Calculated lineshapes of the dispersive components, $f(\chi)$, $g(\chi)$, and $h(\chi)$. Each curve was normalized to unity at $\chi \rightarrow 0$.

been reported for a variety of crystals, and for InP,[11] InAs,[12] GaAs,[5,13] and GaP,[11,14] These data are accurate only in the region below or near the fundamental edge. This is because the measurement employs transmission of light through the sample and is consequently limited to a frequency range where the material is transparent. Materials whose lowest gap is direct, or with a direct gap only slightly above the indirect one (e.g., Ge), have a strong dispersion of the photoelastic coefficients in the region near this gap.[5,15] On the other hand, materials whose lowest gap is indirect and far removed from a direct gap, such as Si and GaP, have a very weak photoelastic dispersion near the indirect-gap region.[5,14] Strong enhancement of the photoelastic coefficients in the neighborhood of the direct gap was first observed by Tell et al.[16] in ZnO and CdS.

Results of the experimental dispersion for InP, InAs, GaAs, and GaP are shown in Figs. 9.2 and 9.3. The experimental data are taken from Refs. 11 (InP), 12 (InAs), 5 (GaAs), and 11 (GaP). The solid lines are the calculated results of Eq. (9.7). Numeric values of the dispersion parameters C^* and D^* determined for these materials are given in Table 9.1.

It is clear from the figures that the theoretical and experimental dispersions agree quite accurately at photon energies close to E_0. In the present model, the photoelastic coefficient α_{pe} can be expressed by a sum of the dispersive contribution arising from the lowest direct gaps $E_0/(E_0 + \Delta_0)$

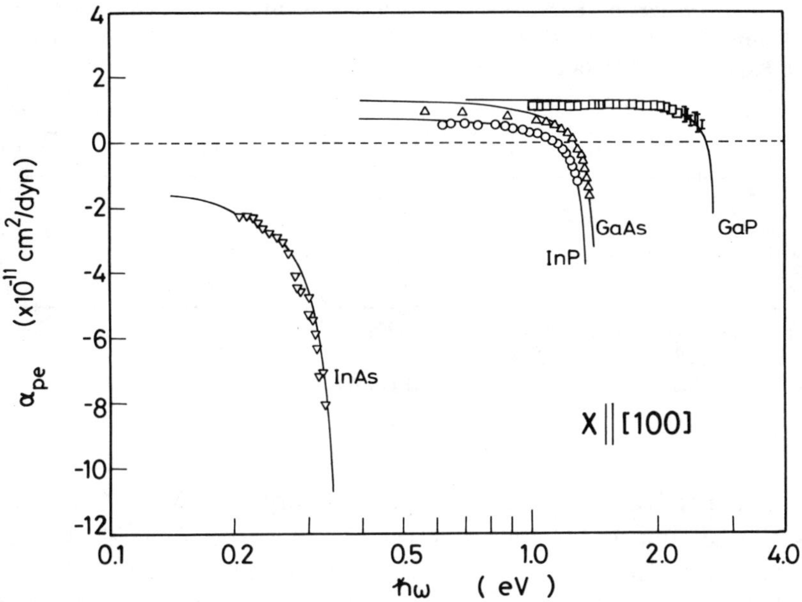

Figure 9.2 Dispersion of the photoelastic coefficients in InP, InAs, GaAs, and GaP with the [100] compressive stress. The experimental data are taken from Refs. 11 (InP), 12 (InAs), 5 (GaAs), and 11 (GaP). The solid lines are the calculated results of Eq. (9.7).

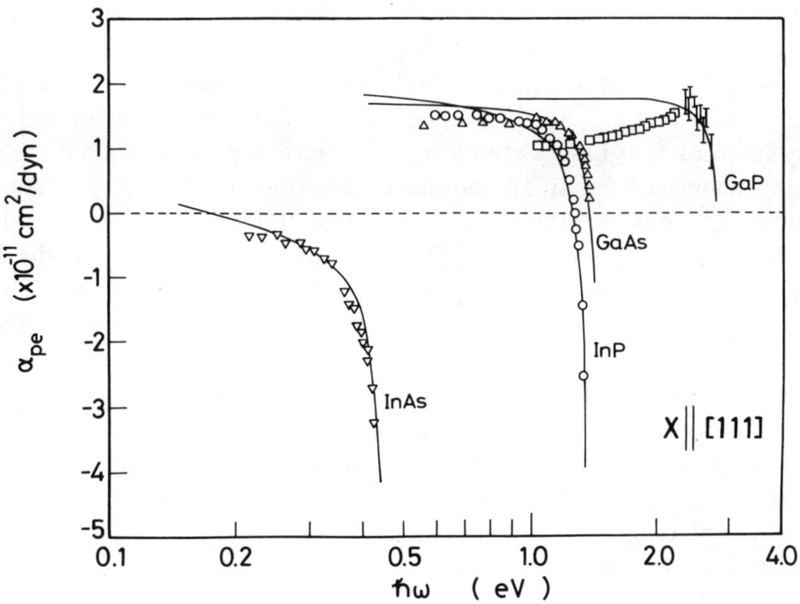

Figure 9.3 Dispersion of the photoelastic coefficients in InP, InAs, GaAs, and GaP with the [111] compressive stress. The experimental data are taken from Refs. 11 (InP), 12 (InAs), 5 (GaAs), and 11 (GaP). The solid lines show the calculated results of Eq. (9.7).

TABLE 9.1 Dispersion Parameters, C^* and D^* (in 10^{-11} cm^2/dyn), Used for Calculation of Photoelastic Coefficient of InP, InAs, GaAs, and GaP [Obtained by Fitting Experimental Data to Eq. (9.7)]

	C^*		D^*	
Material	$X \parallel [100]$	$X \parallel [111]$	$X \parallel [100]$	$X \parallel [111]$
InP[a]	−0.29	−0.36	1.39	2.60
InAs[b]	−2.58	−1.48	2.19	2.32
GaAs[c]	−0.46	−0.21	2.22	2.12
GaP[a]	−0.18	−0.06	1.73	1.92

[a]F. Canal et al., *Solid State Commun.* **29,** 523 (1979).
[b]P. Y. Yu et al., *Phys. Rev. B* **3,** 340 (1971).
[c]C. W. Higginbotham et al., *Phys. Rev.* **184,** 821 (1969).

and nondispersive background contribution arising from the higher-lying gaps. These contributions are opposite in sign (see Table 9.1). One can thus expect $\alpha_{pe}(\omega) = 0$ at which the lowest-gap contribution is exactly canceled by the higher-lying gap contribution. The parameter C^* is a strong function of E_0 [i.e., $C^* \propto E_0^{-5/2}$; see Eq. (9.8)]. The E_0 gap varies widely from material to material for a family of the III–V compounds. Expression (9.8) indicates that smaller E_0 gap material has a larger E_0 gap contribution ($-C^*$). This is confirmed in Table 9.1 (also see Ref. 8). If a material has a smaller ($-C^*$) value (e.g., InP, GaAs, GaP), then the sign of the photoelastic coefficient is positive for long wavelengths and negative when approaching the lowest direct gap. On the other hand, material having a larger ($-C^*$) value (e.g., InAs, InSb) does not exhibit this sign reversal.[12]

By interpolating binary strength parameters, we can calculate the spectral dependence of α_{pe} for an optional composition of the $In_{1-x}Ga_xAs_yP_{1-y}$ quaternary.[8] These results are shown in Fig. 9.4 with the stress parallel to the [100] direction and in Fig. 9.5 with the stress parallel to the [111] direction. It can be seen that α_{pe} for $In_{1-x}Ga_xAs_yP_{1-y}$ has a strong wavelength dependence. As the photon energy approaches the lowest direct gap E_0, the coefficient increases dramatically. As in the case of $\epsilon_1(\omega)$ (Fig. 8.27), the experimental α_{pe} values in this spectral region will be strongly influenced by the lifetime-broadening effect of the electronic states.[17] This effect cannot be taken into consideration in the calculations, although we can easily introduce it in the same manner as that mentioned in Chapter 8 (i.e., by replacing $\hbar\omega$ by $\hbar\omega + i\Gamma$).

From Eq. (9.1) we can obtain

$$\alpha_{pe}(\omega) = -\epsilon_1(\omega)^2(p_{11} - p_{12})(S_{11} - S_{12}) \tag{9.10a}$$

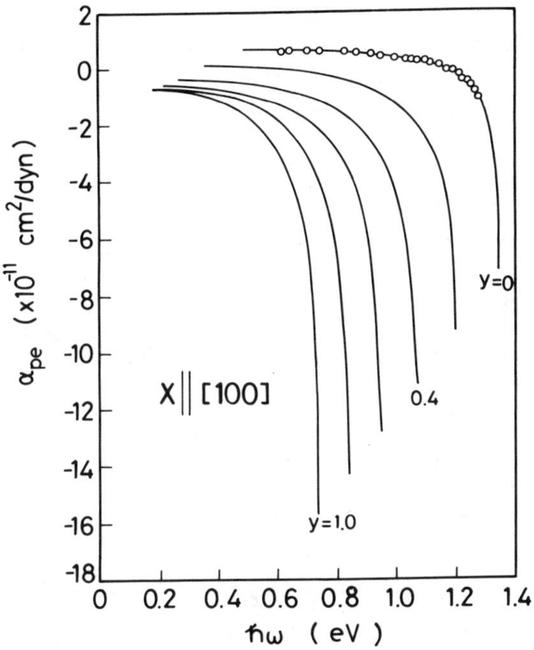

Figure 9.4 Calculated photoelastic coefficients in $In_{1-x}Ga_xAs_yP_{1-y}$ lattice-matched to InP for $X \| [100]$ as a function of the photon energy with y-composition increments of 0.2. The open circles are the experimental data for InP ($y = 0$) taken from Ref. 11. (From Adachi and Oe.[8])

for the [100] uniaxial stress and

$$\alpha_{pe}(\omega) = -\epsilon_1(\omega)^2 p_{44} S_{44} \tag{9.10b}$$

for the [111] uniaxial stress. By introducing numeric values of S_{ij} (Table 3.1) and $\epsilon_1(\omega)$ ($\epsilon_1 \simeq n^2$; Fig. 8.27) into Eq. (9.10), it is also easy to calculate the spectral dependence of $p_{11} - p_{12}$ and p_{44}. The results of this calculation are shown in Figs. 9.6 ($p_{11} - p_{12}$) and 9.7 (p_{44}). The photoelastic constants, p_{11}, p_{12}, and p_{44}, in InP have also been determined by Suzuki and Tada[18] at several wavelengths (1.06, 1.208, 1.306, and 1.5 μm) using the Dixon–Cohen method.

As discussed above, the linear photoelastic coefficient can be obtained by fitting $\Delta\epsilon_1 = \epsilon_1(\|) - \epsilon_1(\perp)$ to the expression $\alpha_{pe}X$. Increasing the stress X, however, usually increases the birefringence in the manner $\Delta\epsilon_1 = \alpha_{pe}X + \beta_{pe}X^2$ where β_{pe} is the so-called quadratic photoelastic coefficient. The quadratic coefficient β_{pe} has been measured in several semiconductors, such as Ge,[5] GaAs,[5] InAs,[12] and GaSb.[12]

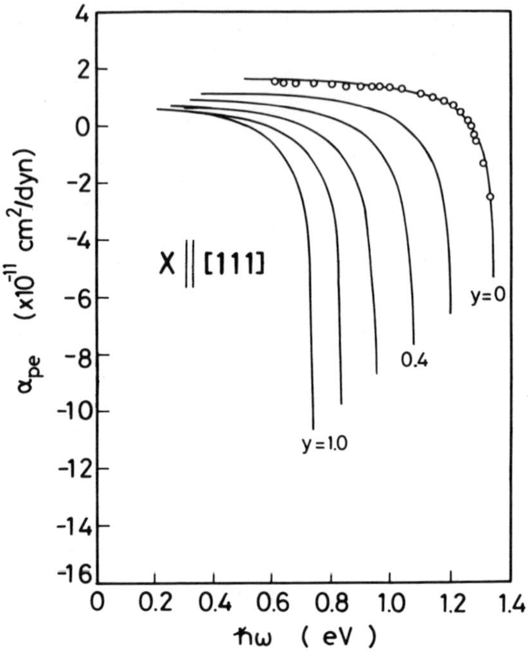

Figure 9.5 Calculated photoelastic coefficients of $In_{1-x}Ga_xAs_yP_{1-y}$ lattice-matched to InP for $X \| [111]$ as a function of the photon energy with y-composition increments of 0.2. The open circles represent the experimental data for InP ($y = 0$) taken from Ref. 11. (From Adachi and Oe.[8])

Chandrasekhar et al.[19,20] have developed a new method to measure stress-induced birefringence in an opaque region that employs the Raman-scattering technique as a probe. Using this new method, they measured the photoelastic coefficient α_{pe} in Si (0.5–3.38 eV)[20] and GaP (1.0–2.6 eV)[11] above the fundamental absorption edges. In their analyses, the contribution from the imaginary part of the dielectric constant was neglected, although they pointed out that its contribution should be taken into account in order to discuss the spectral dependence above 3.0 eV in Si. Their data have been successfully analyzed by Adachi and Hamaguchi[10] by introducing new coefficients that determine the fractional contributions of $\Delta\epsilon_1$ and $\Delta\epsilon_2$ to α_{pe}.

Large photoelastic coefficients of III–V compounds can cause significant changes in the refractive index for relatively small strains. Kirkby et al.[21] formed a photoelastic waveguide in GaAs/AlGaAs by making small variations in the refractive index of the active layer beneath the stripe contact of the laser. The refractive-index variations were caused by the effect of a strain field in the semiconductor surrounding the stripe window. Similar

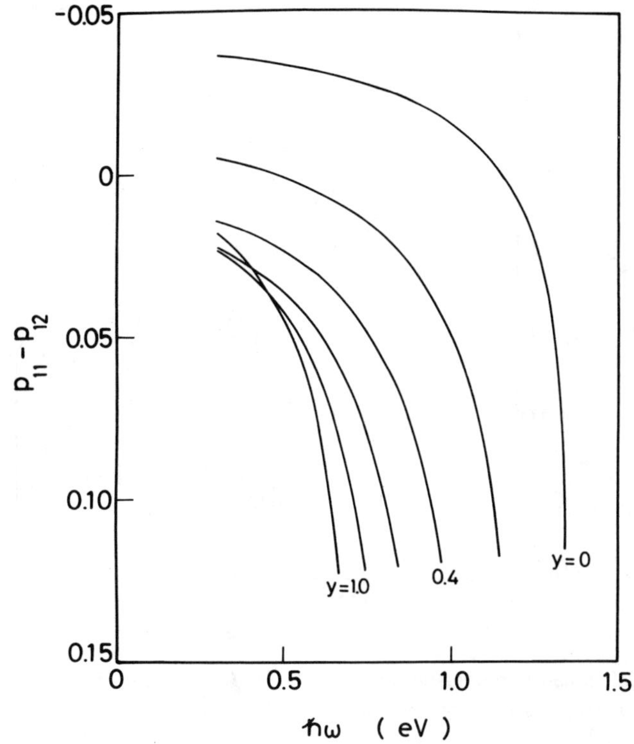

Figure 9.6 Calculated photoelastic constant p_{11}–p_{12} for $In_{1-x}Ga_xAs_yP_{1-y}$ lattice-matched to InP as a function of photon energy with *y*-composition increments of 0.2.

photoelastic waveguides have been fabricated on GaAs-based semiconductors by several authors.[22–26]

9.2 LINEAR ELECTROOPTIC EFFECT

9.2.1 Model Representation

It is of considerable interest to investigate the linear electrooptic effect in semiconductors. The effect affords a convenient and widely used means of controlling the intensity of the phase of optical radiation in crystals. This property has been extensively explored and used for a wide variety of optoelectronic devices.[27] Much progress has recently been made in developing the best potential devices for use in optical communication systems that take advantage of low-loss, low-dispersion optical fibers. Recent studies[28–30] have, therefore, required accurate values of the electrooptic con-

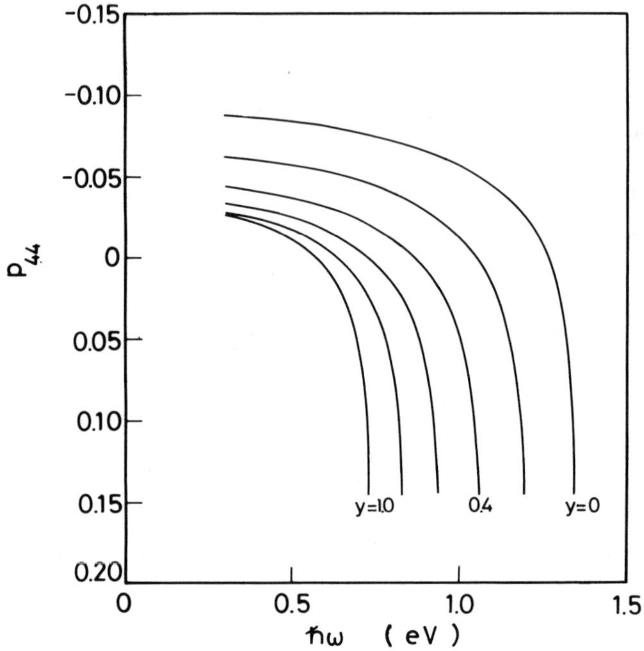

Figure 9.7 Calculated photoelastic constant p_{44} for $In_{1-x}Ga_xAs_yP_{1-y}$ lattice-matched to InP as a function of photon energy with y-composition increments of 0.2.

stant as a function of wavelength. Of course, fundamentals of the electrooptic effects are of interest from the aspect of the solid-state physics.

DiDomenico and Wemple[31] have proposed a microscopic tensor theory of the electrooptic and nonlinear optical effects in ferroelectrics in terms of energy-band pictures. Garrett[32] has also presented a one-dimensional anharmonic oscillator model to explain a number of nonlinear optical phenomena. Garrett's model was extended by Sugie and Tada[33] to a three-dimensional oscillator model so that it is applicable to materials with arbitrary crystal structures.

The electrooptic effect in the region of interest close to the band edge has also been analyzed by Adachi and Oe[34] in terms of electric-field-induced modulations of electronic energy-band structures. Recently, Hernández-Cabrera et al.[35] theoretically analyzed the linear electrooptic effect by means of a microscopic simple model (tight-binding scheme) without any fitting of parameters. These results have shown satisfactory agreement with available experimental information.

The linear electrooptic (Pockels) effect refers to a change in relative optical dielectric immpermeability proportional to an applied electric field.

The linear electrooptic effect can, thus, be defined by

$$\Delta\left(\frac{1}{\epsilon}\right)_{ij} = r_{ij,k}E_k \tag{9.11}$$

where ϵ is the relative dielectric constant, E_k is the applied electric field, and the indices i, j, and k refer to the rectangular coordinate axes 1 (x), 2 (y), and 3 (z). In order to describe the linear electrooptic effect rigorously, we must take into account a possible piezoelectrically induced elastooptic contribution to the impermeability change. In this case, the electrooptic constant can be given by

$$r^{\mathrm{T}}_{ij,k} = r^{\mathrm{S}}_{ij,k} + r^{\mathrm{P}}_{ij,k} \tag{9.12}$$

where $r^{\mathrm{T}}_{ij,k}$ is the free value determined at constant stress (e.g., by making a measurement at low frequencies well below acoustic resonances of the sample), $r^{\mathrm{S}}_{ij,k}$ is the clamped value determined at constant strain (e.g., at high frequencies well above the acoustic resonances of the sample), and $r^{\mathrm{P}}_{ij,k}$ is the elastooptic contribution to the impermeability change.

Zinc-blende-type crystals, such as InP and GaAs, possess only a single electrooptic tensor component, r_{41} ($r_{23,1}$)[36]

$$[r] = \begin{bmatrix} 0 & 0 & 0 \\ 0 & 0 & 0 \\ 0 & 0 & 0 \\ r_{41} & 0 & 0 \\ 0 & r_{41} & 0 \\ 0 & 0 & r_{41} \end{bmatrix} \tag{9.13}$$

where $[r]$ is, in general, a *complex* third-rank tensor. Since the imaginary part of the dielectric constant may be taken as zero in the region near and below the lowest direct gap, the coefficient r_{41} can be considered as a *real* physical constant with good approximation. Then, from Eq. (9.11) [see also Eq. (9.5)], one obtains[34]

$$r^{\mathrm{S}}_{41} = \frac{\Delta(1/\epsilon_1)}{E} = -\frac{1}{\epsilon_1^2}\Delta\epsilon_1 \tag{9.14}$$

The elastooptic contribution r_{41}^{P} is now written, in the zinc-blende family, as

$$r_{41}^{\mathrm{P}} = p_{44} S_{44} e_{14} \tag{9.15}$$

where p_{44} is the photoelastic constant (see Section 9.1), S_{44} is the elastic compliance constant (Section 3.2), and e_{14} is the piezoelectric constant (Section 5.2). Since p_{44} is a wavelength-dependent quantity, r_{41}^{P} has spectral dependence.

The quantity $\Delta\epsilon_1$, required to calculate r_{41}^{S}, can be obtained by taking account of the changes in the lowest-direct-gap parameters:[34]

$$\Delta\epsilon_1 = \frac{\partial\epsilon_1}{\partial E_0}\Delta E_0 + \frac{\partial\epsilon_1}{\partial A^*}\Delta A^* + \frac{\partial^2\epsilon_1}{\partial E_0^2}(\Delta E_0)^2 + \frac{\partial^2\epsilon_1}{\partial A^{*2}}(\Delta A^*)^2 + \frac{\partial\epsilon_1^2}{\partial E_0\,\partial A^*}\Delta E_0\,\Delta A^* + \cdots \tag{9.16}$$

where ΔE_0 and ΔA^* are, respectively, changes in E_0 and A^* [see Eq. (8.52)] with the electric field E. The specific physical process considered here is the linear electrooptic effect. Thus, we can neglect the higher-order derivative terms in Eq. (9.16). Expression (9.16) can, then, be reduced to

$$\Delta\epsilon_1 = \frac{\partial\epsilon_1}{\partial E_0}\Delta E_0 + \frac{\partial\epsilon_1}{\partial A^*}\Delta A^* \tag{9.17}$$

The band-parameter changes are now written in terms of the first-order Stark-like effect as[37]

$$\Delta E_0 = a_1 E \tag{9.18a}$$

$$\Delta A^* = b_1 E \tag{9.18b}$$

Introducing Eq. (9.18) into Eq. (9.17), we obtain the field-induced changes in ϵ_1 as

$$\Delta\epsilon_1^{\mathrm{E}}(\omega) = \frac{\partial\epsilon_1}{\partial E_0}\Delta E_0 = \frac{1}{2}A^* E_0^{-1} a_1 E g(\chi) \tag{9.19a}$$

$$\Delta\epsilon_1^{\mathrm{A}}(\omega) = \frac{\partial\epsilon_1}{\partial A^*}\Delta A^* = b_1 E f(\chi) \tag{9.19b}$$

where $g(\chi)$ is function defined by Eq. (9.9). The linear electrooptic constant r_{41}^{S} can then be written from Eq. (9.14) as

$$r_{41}^{S}(\omega) = -\left(\frac{1}{\epsilon_1^2}\right)\left[A^*(\tfrac{1}{2})E_0^{-1}a_1 g(\chi) + b_1 f(\chi) + F'\right] \tag{9.20}$$

where F' represents the strength of the nondispersive, higher-lying gap contribution. In Eqs. (9.16)–(9.20), we considered only the E_0 gap contribution. This consideration is based on the fact that most of the III–V compounds (except InAs) have small Δ_0 gaps compared with their E_0 gaps. The $E_0 + \Delta_0$ gap contribution can, therefore, be successfully included into the E_0 gap one. For InAs the $E_0 + \Delta_0$ gap (0.79 eV) is much larger than the E_0 gap (0.36 eV), and, therefore, the $E_0 + \Delta_0$-gap contribution can be rightly included for this material into the higher-lying gap one (F').

As mentioned in Section 9.1 (see Fig. 9.1), $g(\chi)$ shows a very sharp dispersion near the band edge E_0 compared with that of $f(\chi)$. Thus, the contribution arising from a change ΔA^* can also be included into the contribution from the higher-lying gap. The linear electrooptic constant $r_{41}^{S}(\omega)$ in zinc-blende-type semiconductors can, then, be finally written as[34]

$$r_{41}^{S}(\omega) = -\left(\frac{1}{\epsilon_1^2}\right)[E^* g(\chi) + F^*] \tag{9.21}$$

with

$$E^* = (\tfrac{1}{2})A^* E_0^{-1} a_1 \tag{9.22}$$

The parameters E^* and F^* can be treated as adjustable parameters to fit experimental data.

9.2.2 Analyses

Linear electrooptic constant data have been measured for a variety of crystals, and for InP,[38,39] GaAs,[39–44] and GaP.[45] Table 9.2 summarizes these experimental results. But, to our knowledge, no experimental r_{41} has yet been reported for InAs. In Fig. 9.8 experimental values for r_{41} for InP,[38] GaAs,[43] and GaP[45] are plotted as a function of wavelength. The solid lines are the results calculated from Eq. (9.21). Numeric values of the strength parameters determined for these materials are given in Table 9.3.

Calculated lineshapes of r_{41}^{S} and r_{41}^{P} for InP are shown in Fig. 9.9 along

TABLE 9.2 Free (Clamped) Values of Linear Electrooptic Constant r_{41}^{T} (r_{41}^{S}) (in 10^{-12} m/V) versus Wavelength for InP, GaAs, and GaP

Wavelength (μm)	InP	GaAs	GaP
0.56			$(-1.07 \pm 0.06)^h$
0.585			$(-1.07 \pm 0.06)^h$
0.6328			$(-0.97 \pm 0.02)^h$
0.66			$(-0.98 \pm 0.06)^h$
1.0		1.2^c	
1.06	1.45^a		
1.064	-1.32^b	-1.17^b	
	$(-1.34)^b$	$(-1.33)^b$	
1.09		-1.72 ± 0.09^d	
1.15		-1.68 ± 0.09^d	
		1.43 ± 0.07^e	
1.153			$(-1.10 \pm 0.15)^h$
1.208	-1.49^b		
	$(-1.54)^b$		
1.306	-1.53^b		
	$(-1.59)^b$		
1.35	1.3^a		
1.50	-1.63^b	1.3^c	
	$(-1.68)^b$		
2.0		1.4^c	
3.0		1.6^c	
3.39		1.24 ± 0.04^e	$(-0.97 \pm 0.10)^h$
		1.6 ± 0.3^f	
10.6		1.3^c	
		1.51 ± 0.05^e	
		1.6 ± 0.3^f	
		1.6 ± 0.1^g	

[a] K. Tada and N. Suzuki, *Jpn. J. Appl. Phys.* **19,** 2295 (1980).
[b] N. Suzuki and K. Tada, *Jpn. J. Appl. Phys.* **23,** 291 (1984).
[c] T. E. Walsh, *RCA Rev.* **27,** 323 (1966).
[d] Obtained from GaAs/AlGaAs double-heterostructure waveguides [J. Faist and F. K. Reinhart, *J. Appl. Phys.* **67,** 7006 (1990)].
[e] M. Sugie and K. Tada, *Jpn. J. Appl. Phys.* **15,** 421 (1976).
[f] A. Yariv et al., *IEEE J. Quantum Electron.* **QE-2,** 243 (1966).
[g] I. P. Kaminow, *IEEE J. Quantum Electron.* **QE-4,** 23 (1968).
[h] D. F. Nelson and E. H. Turner, *J. Appl. Phys.* **39,** 3337 (1968).

with corresponding experimental data (open circles) taken from Ref. 38. For r_{41}^{P} [Eq. (9.15)], p_{44} values are taken from Section 9.1, $S_{44} = 2.193 \times 10^{-12}$ cm^2/dyn is from Table 3.1, and $e_{14} = -0.040$ C/m^2 is taken from Table 5.2. (Its sign is assumed here to be "negative.") It is clear from Fig. 9.9 that r_{41}^{P} is much smaller than r_{41}^{S}, hence $r_{41}^{T} \simeq r_{41}^{S}$ (see also

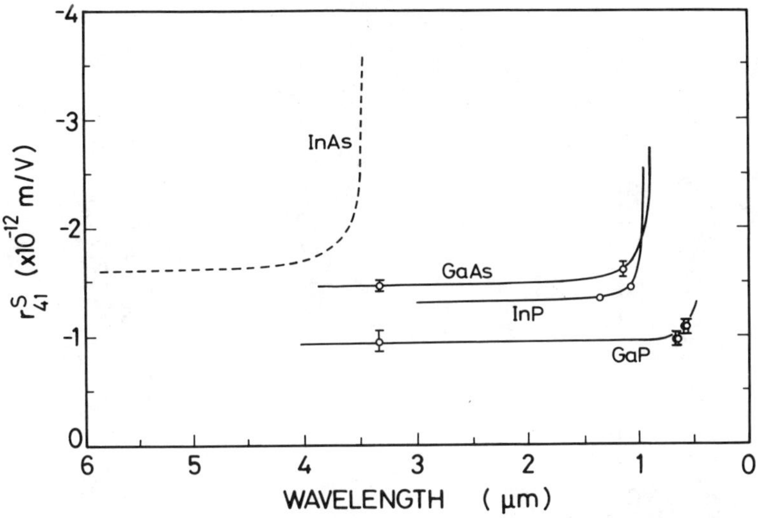

Figure 9.8 Dispersion of the linear electrooptic constants in InP, InAs, GaAs, and GaP. The experimental data are taken from Refs. 38 (InP), 43 (GaAs), and 45 (GaP). The solid lines represent the calculated results of Eq. (9.21).

Table 9.4). The calculated r_{41}^S has strong dispersion, and its value increases abruptly as the wavelength approaches the band edge [$\lambda_g \simeq 0.92$ μm].

Nonlinear optical phenomena in a zinc-blende family can be described by means of a nonlinear optical coefficient $d_{41}(\omega \pm \omega', \omega, \omega')$.[46] The optical rectification coefficient $d_{41}(\omega, \omega, 0)$ can be written by putting $\omega' = 0$ as[47]

$$d_{41}(\omega, \omega, 0) = -(\tfrac{1}{4})\epsilon_1(\omega)^2 r_{41}^S(\omega) \tag{9.23}$$

TABLE 9.3 Dispersion Parameters, E^* and F^* (in 10^{-11} m/V), Used for Calculation of Linear Electrooptic Constant $r_{41}^S(\omega)$ and Optical Rectification Coefficient $d_{41}(\omega, \omega, 0)$ of InP, InAs, GaAs, and GaP [Obtained by Fitting Experimental Data to Eq. (9.21)]

Material	E^*	F^*
InP[a]	−4.2	9.1
InAs[b]	−3.0	19.8
GaAs[c]	−7.1	12.3
GaP[d]	−8.3	1.7

[a]K. Tada and N. Suzuki, *Jpn. J. Appl. Phys.* **19,** 2295 (1980).
[b]Calculated from Eq. (9.25).
[c]M. Sugie and K. Tada, *Jpn. J. Appl. Phys.* **15,** 421 (1976).
[d]D. F. Nelson and E. H. Turner, *J. Appl. Phys.* **39,** 3337 (1968).

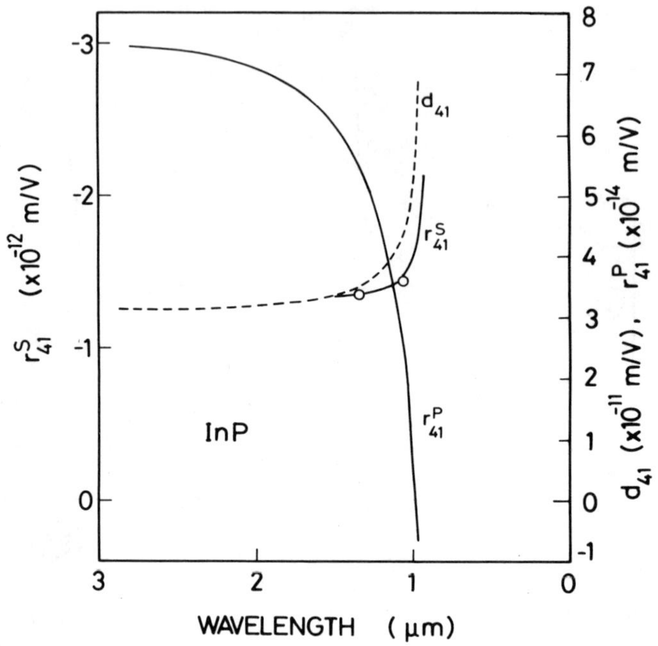

Figure 9.9 Calculated lineshapes of r_{41}^{S}, r_{41}^{P}, and d_{41} for InP. The open circles show the experimental data (r_{41}^{S}) taken from Ref. 38. (From Adachi and Oe.[34])

TABLE 9.4 Calculated Linear Electrooptic Constants, r_{41}^{S} and r_{41}^{P}, versus Photon Energy for InP

Photon Energy (eV)	r_{41}^{S} ($\times 10^{-12}$ m/V)	r_{41}^{P} ($\times 10^{-14}$ m/V)
0.5	−1.32	7.41
0.6	−1.33	7.14
0.7	−1.33	6.79
0.8	−1.34	6.34
0.9	−1.35	5.75
1.0	−1.37	4.97
1.1	−1.42	3.84
1.2	−1.53	2.02
1.3	−1.98	−2.31
1.31	−2.12	−3.28
1.32	−2.33	−4.58
1.33	−2.68	−6.54
1.34	−3.51	−10.4

The results of $d_{41}(\omega, \omega, 0)$, calculated from Eq. (9.23), are also shown in Fig. 9.9 by the dashed line.

If one of the four end-binary data is lacking (in the present case InAs), the quaternary data $Q_{\mathrm{ABCD}}(\mathrm{A}_{1-x}\mathrm{B}_x\mathrm{C}_y\mathrm{D}_{1-y})$ can be estimated from the expressions

$$Q(x, y) = (1 - x)B_{\mathrm{AC}} + (x + y - 1)B_{\mathrm{BC}} + (1 - y)B_{\mathrm{BD}} \quad (9.24a)$$

$$Q(x, y) = (1 - x)B_{\mathrm{AD}} + yB_{\mathrm{BC}} + (x - y)B_{\mathrm{BD}} \quad (9.24b)$$

$$Q(x, y) = yB_{\mathrm{AC}} + (1 - x - y)B_{\mathrm{AD}} + xB_{\mathrm{BD}} \quad (9.24c)$$

or

$$Q(x, y) = (y - x)B_{\mathrm{AC}} + (1 - y)B_{\mathrm{AD}} + xB_{\mathrm{BC}} \quad (9.24d)$$

Introducing $x = 0.0$ and $y = 1.0$ into Eq. (9.24b), we obtain

$$B_{\mathrm{InAs}} = B_{\mathrm{InP}} + B_{\mathrm{GaAs}} - B_{\mathrm{GaP}} \quad (9.25)$$

The numeric values E^* and F^* for InAs listed in Table 9.3 are estimated from this calculation. The wavelength dependence of r_{41} of InAs shown in Fig. 9.8 (dashed line) is the calculated result using these numeric parameters.

Figure 9.10 shows the dispersion parameters, E^* and F^*, as a function of y for $\mathrm{In}_{1-x}\mathrm{Ga}_x\mathrm{As}_y\mathrm{P}_{1-y}$ lattice-matched to InP. These lines are obtained from the linear interpolation scheme of Eq. (2.3). The calculated results, as a function of y, are almost linear relationships given by (in 10^{-11} m/V)

$$E^*(y) = -4.2 - 0.7y \quad (9.26a)$$

$$F^*(y) = 9.1 + 7.2y \quad (9.26b)$$

Yet in spite of the fundamental importance of the $\mathrm{In}_{1-x}\mathrm{Ga}_x\mathrm{As}_y\mathrm{P}_{1-y}$/InP system to semiconductor-based integrated optoelectronics, the only investigation that has been done to date is for the determination of the linear electrooptic constant of this alloy system.[29] Bach et al.[29] have measured the linear electrooptic constants of $\mathrm{In}_{1-x}\mathrm{Ga}_x\mathrm{As}_y\mathrm{P}_{1-y}$ lattice-matched to InP ($y \simeq 0.20$ and $y \simeq 0.34$). The quaternary layers used were unintentionally doped n type with doping levels of 3×10^{16} cm^{-3}. The data they obtained are as follows (in 10^{-12} m/V): $r_{41} = -1.34$ at $\lambda = 1.32$ μm for $y \simeq 0.20$, $r_{41} = -1.43$ at $\lambda = 1.25$ μm, and $r_{41} = -1.44$ at $\lambda = 1.32$ μm for $y \simeq 0.34$. By fitting these data with Eq. (9.21), we can obtain the dispersion

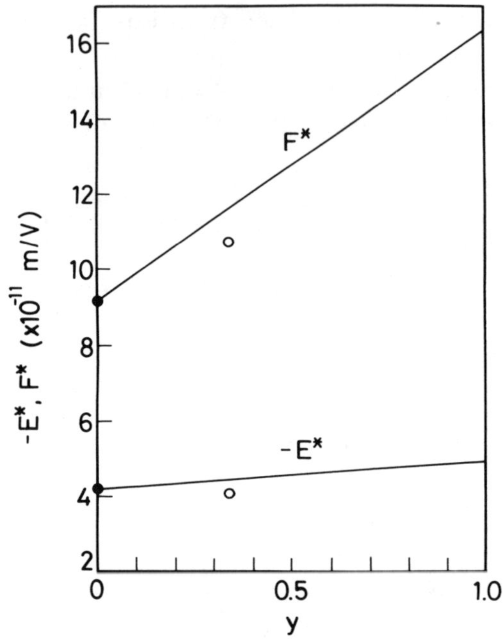

Figure 9.10 Dispersion parameters, E^* and F^*, as a function of y for $In_{1-x}Ga_xAs_yP_{1-y}$ lattice-matched to InP. The solid and open circles are, respectively, determined by fitting the experimental data of InP[38] and $In_{1-x}Ga_xAs_yP_{1-y}$($y \simeq 0.34$)[29] to Eq. (9.21).

parameters for $y \simeq 0.34$ to be $E^* = -4.04 \times 10^{-11}$ m/V and $F^* = -1.07 \times 10^{-10}$ m/V (see Fig. 9.14, below). These parameter values are also plotted in Fig. 9.9 by the open circles. It is clear from the figure that our interpolated values are in acceptable agreement with the experimental data of Bach et al.[29]

Unfortunately, however, the data of Bach et al. were limited to narrow composition and narrow wavelength ranges. We have already specified numeric parameters, which are required for calculation with Eq. (9.21) (i.e., E_0, E^*, F^*), in terms of y alone. We can, therefore, estimate the spectral dependence r_{41}^S for the optional composition of $In_{1-x}Ga_xAs_yP_{1-y}$ quaternary. It is also easy to calculate the spectral dependence of r_{41}^P of $In_{1-x}Ga_xAs_yP_{1-y}$, since we have specified p_{44}, S_{44}, and e_{14} in terms of y alone.

The calculated results for r_{41}^S as a function of wavelength with y-composition increments of 0.1 are shown in Fig. 9.11. We see from the figure that r_{41}^S has a strong dispersion only near the band-edge region. The values of r_{41}^P are much smaller than those of r_{41}^S for the complete range of y compositions.

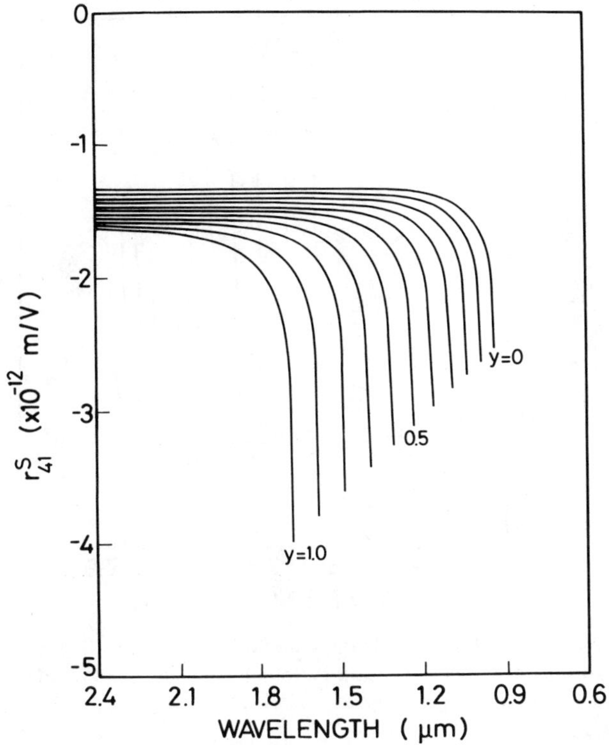

Figure 9.11 Calculated linear electrooptic constant r^S_{41} for $In_{1-x}Ga_xAs_yP_{1-y}$ lattice-matched to InP with y-composition increments of 0.1. (From Adachi and Oe.[34])

The spectral dependence of the optical rectification coefficient $d_{41}(\omega, \omega, 0)$ for $In_{1-x}Ga_xAs_yP_{1-y}/InP$ can also be obtained by using the same procedure. This result is shown in Fig. 9.12 with y-composition increments of 0.1.

9.3 QUADRATIC ELECTROOPTIC EFFECT

9.3.1 Model Representation

The electrooptic effect refers to a change in the relative optical dielectric impermeability proportional to different powers of the electric field E, whose highest frequency is well below the lattice resonance of a crystal. The effect linearly proportional to E is the linear electrooptic effect and that proportional to E^2 is so-called the quadratic electrooptic (Kerr) effect. As discussed in Section 9.2, the linear electrooptic effect was intensively investigated in several semiconductors. However, a few works have been

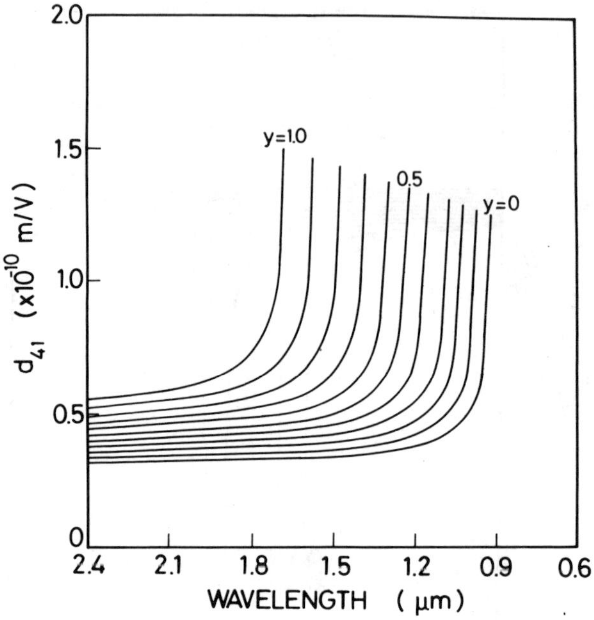

Figure 9.12 Calculated optical rectification coefficient d_{41} (ω, ω, 0) for $In_{1-x}Ga_xAs_yP_{1-y}$ lattice-matched to InP with y-composition increments of 0.1. (From Adachi and Oe.[34])

done on the quadratic electrooptic effects in semiconductors.[29] Bach et al.[29] observed the quadratic effect in the InGaAsP quaternary. Their report is the first time that the effect has been positively identified in materials of point group $\bar{4}3m$ (zinc-blende crystals).

There have also been a very few theoretical works on the quadratic electrooptic effects in semiconductors.[48] To explain the quadratic effect, we must expand the change in the dielectric impermeability of Eq. (9.11) to higher orders

$$\Delta\left(\frac{1}{\epsilon}\right)_{ij} = r_{ij,k}E_k + R_{ijkl}E_kE_l + \cdots \tag{9.27}$$

where R_{ijkl} is the quadratic electrooptic constant. In crystals with a center of symmetry, only the second-order term and higher even-order terms can exist, and so one might expect that the effects are very small in moderate fields. If a center of symmetry is lacking, as in zinc-blende-type crystals, not only even-order but also odd-order effects can exist. The effect given by the second-order term and higher even-order terms, which can occur in all substances, is known as the *Kerr effect*.

If we neglect higher-order terms than the second in the electric field, the Kerr's coefficient R_{ijkl} is given by[48]

$$R_{ijkl} = \frac{\Delta\left(\dfrac{1}{\epsilon}\right)_{ij}^{Q}}{E_k E_l} \tag{9.28}$$

where $\Delta(1/\epsilon)_{ij}^{Q}$ represents the change in the optical dielectric impermeability due to the second-order Kerr effect. The zinc-blende-type crystals, such as InP and InGaAsP, possess three nonvanishing Kerr constants, R_{11}, R_{12}, and R_{44}[49,50]

$$[R] = \begin{bmatrix} R_{11} & R_{12} & R_{12} & 0 & 0 & 0 \\ R_{12} & R_{11} & R_{12} & 0 & 0 & 0 \\ R_{12} & R_{12} & R_{11} & 0 & 0 & 0 \\ 0 & 0 & 0 & R_{44} & 0 & 0 \\ 0 & 0 & 0 & 0 & R_{44} & 0 \\ 0 & 0 & 0 & 0 & 0 & R_{44} \end{bmatrix} \tag{9.29}$$

where $[R]$ is, in general, a *complex* fourth-rank tensor. Since the imaginary part of the dielectric constant may be taken as zero in a region near and below the lowest direct gap, the coefficients R_{ijkl} are *real* physical constants to a good approximation.

Figure 9.13 shows diagrams of the electric-field and light-polarization directions used to measure the quadratic coefficients (*a*) R_{11}–R_{12} and (*b*) R_{44}. These measurements can be performed as phase-difference measurements by using a quarter-wave plate and polarizers.

The change in the inverse dielectric constant in zinc-blende crystals can now be given by

$$\Delta\left(\frac{1}{\epsilon}\right)_{ij}^{Q} = -\frac{1}{\epsilon_1^2}\,\Delta\epsilon_{1Q} \tag{9.30}$$

As in Eq. (9.18), the band-parameter changes are written in terms of the first- and second-order Stark-like effects as[37]

$$\Delta E_0 = a_1 E + a_2 E^2 \tag{9.31a}$$

$$\Delta A^* = b_1 E + b_2 E^2 \tag{9.31b}$$

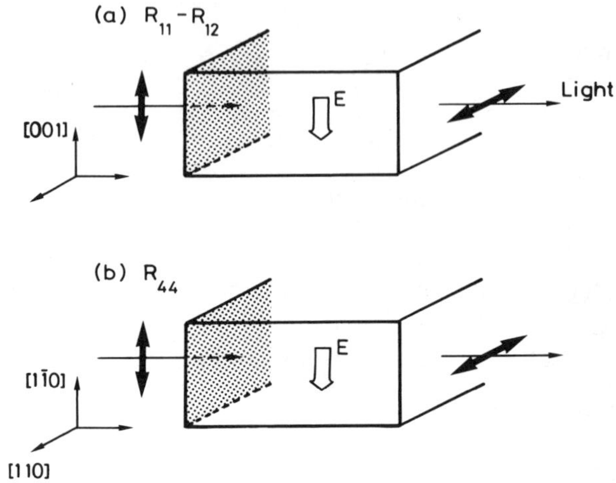

Figure 9.13 Quadratic electrooptic effects in zinc-blend-type crystals: (*a*) R_{11}–R_{12}; (*b*) R_{44}.

The specific physical process considered here is the quadratic effect. The change $\Delta\epsilon_{1Q}$ can, thus, be written as

$$\Delta\epsilon_{1Q} = \frac{\partial\epsilon_1}{\partial E_0}(\Delta E_0)_s + \frac{\partial\epsilon_1}{\partial A^*}(\Delta A^*)_s + \frac{\partial^2\epsilon_1}{\partial E_0\,\partial A^*}(\Delta E_0)_f(\Delta A^*)_f + \frac{\partial^2\epsilon_1}{\partial E_0^2}(\Delta E_0)_f^2 + \frac{\partial^2\epsilon_1}{\partial A^{*2}}(\Delta A^*)_f^2 \tag{9.32}$$

where the subscripts f and s, respectively, indicate the contributions from the first- and second-order Stark effects to the quadratic electrooptic constant.

Introducing Eqs. (8.52) and (9.31) into Eq. (9.32), one obtains the change in ϵ_1 as

$$\frac{\partial\epsilon_1}{\partial E_0}(\Delta E_0)_s = \frac{1}{2}A^*E_0^{-1}a_2E^2g(\chi) \tag{9.33a}$$

$$\frac{\partial\epsilon_1}{\partial A^*}(\Delta A^*)_s = b_2E^2f(\chi) \tag{9.33b}$$

$$\frac{\partial^2\epsilon_1}{\partial E_0\,\partial A^*}(\Delta E_0)_f(\Delta A^*)_f = \frac{1}{2}E_0^{-1}a_1b_1E^2g(\chi) \tag{9.33c}$$

$$\frac{\partial^2\epsilon_1}{\partial E_0^2}(\Delta E_0)_f^2 = -\frac{1}{4}A^*E_0^{-2}a_1^2E^2h(\chi) \tag{9.33d}$$

$$\frac{\partial^2 \epsilon_1}{\partial A^{*2}} (\Delta A^*)_{\mathrm{f}}^2 = 0 \tag{9.33e}$$

with

$$h(\chi) = \chi^{-2}[2 - (1 + \chi)^{-1.5} - (1 - \chi)^{-1.5}] \tag{9.34}$$

The expression for the quadratic coefficient, R_{11}–R_{12}, can be written from Eq. (9.28) as

$$R_{11}\text{–}R_{12} = -\frac{1}{\epsilon_1^2}\left\{\left[A^*\left(-\frac{1}{4}E_0^{-2}a_1^2\right)h(\chi) + \frac{1}{2}E_0^{-1}(A^*a_2 + a_1b_1)g(\chi) + b_2 f(\chi)\right] + H'\right\} \tag{9.35}$$

where H' represents the strength of the nondispersive, higher-lying-gap contribution [having a meaning similar to F' in Eq. (9.20)].

The calculated result of the dispersive component $h(\chi)$, which appeared in Eq. (9.35), is shown in Fig. 9.1 along with those of $g(\chi)$ and $f(\chi)$. It is worth noting from this figure that $h(\chi)$ shows very sharp dispersion near the band edge ($\chi = 1.0$) compared with those of $g(\chi)$ and $f(\chi)$. We can, thus, correctly include the contributions defined by $g(\chi)$ and $f(\chi)$ in Eq. (9.35) in the nondispersive, higher-lying-gap contribution H'. The quadratic coefficient can then, for practical purposes, be given by[48]

$$R_{11}\text{–}R_{12} = -\left(\frac{1}{\epsilon_1^2}\right)[G^*h(\chi) + H^*] \tag{9.36}$$

with

$$G^* = -\tfrac{1}{4}A^*E_0^{-2}a_1^2 \tag{9.37}$$

In Eq. (9.37) the nondispersive term H^* contains not only the higher-lying-gap contribution ($E_0 + \Delta_0$, E_1, $E_1 + \Delta_1$, E_0', etc.) but also the weaker dispersive E_0 gap contribution. The quadratic coefficient R_{44} is also given by the same form as Eq. (9.36), but one must take into consideration the corresponding Stark-shift value a_1. The parameters G^* and H^* can be treated as adjustable parameters to fit experimental data.

9.3.2 Analyses

The theoretical lineshape of R_{11}–R_{12} along with the experimental data for $In_{1-x}Ga_xAs_yP_{1-y}$ ($y \simeq 0.34$)[29] is shown in Fig. 9.14. The curve was obtained by means of a least-squares-fitting procedure. It is found from the figure that the theoretical dispersion shows excellent agreement with the experimental data. The dispersion parameters determined are as follows: $G^* = -3.43 \times 10^{-19}$ m^2/V^2 and $H^* = -2.85 \times 10^{-18}$ m^2/V^2. Bach et al.[29] have also determined the linear electrooptic constant values r_{41} in the $In_{1-x}Ga_xAs_yP_{1-y}$ quaternary (see Section 9.2). These data are also plotted in the figure along with the corresponding theoretical curve [Eq. (9.21)]. It is clear that the coefficient R_{11}–R_{12} shows stronger dispersion than the linear electrooptic constant r_{41} in a region near the lowest direct gap.

The quadratic electrooptic constant R_{ijkl} can, in principle, be written as

$$R^{T}_{ijkl} = R^{S}_{ijkl} + R^{P}_{ijkl} \tag{9.38}$$

where R^{P}_{ijkl} is the product of the photoelastic tensor p_{ijmn} and quadratic electrostrictive tensor Q_{mnkl}:

$$R^{P}_{ijkl} = \sum_{mn} p_{ijmn} Q_{mnkl} \tag{9.39}$$

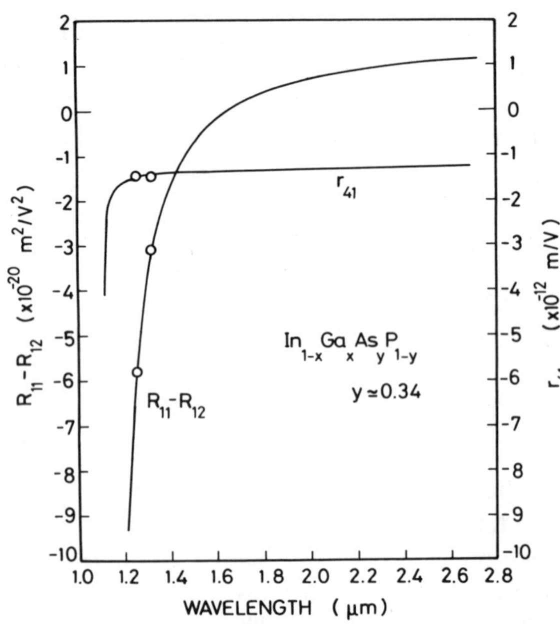

Figure 9.14 Theoretical lineshapes of R_{11}–R_{12} and r_{41} along with the experimental data[29] for $In_{1-x}Ga_xAs_yP_{1-y}$ lattice-matched to InP ($y \simeq 0.34$). (From Adachi and Oe.[48])

Expression (9.38) is, thus, essentially the same as Eq. (9.12) from a physical point of view. In the present analysis, we assumed that $R^{T}_{ijkl} \simeq R^{S}_{ijkl}$ since the detailed values of Q_{mnkl} are not well known at present.

By introducing the previous scheme (Section 9.2), one can also estimate the spectral dependence of the coefficient R_{11}–R_{12} for $In_{1-x}Ga_xAs_yP_{1-y}$ with an optional (*x*, *y*) fraction. This result is shown in Fig. 9.15 for $In_{1-x}Ga_xAs_yP_{1-y}$ layers (λ_g = 1.3 and 1.55 μm). The coefficient is nearly constant (nearly zero) for wavelengths far from λ_g, whereas it increases dramatically when the wavelength approaches the lowest direct gap. This result suggests that by properly choosing the (*x*, *y*) fraction, one may improve performance of the electrooptic light modulation in $In_{1-x}Ga_xAs_yP_{1-y}$ *p–n* junction lasers.

Zucker et al.[51] have measured the quantum-confined Stark effect in high-quality $In_{1-x}Ga_xAs_yP_{1-y}$/InP quantum well *p–i–n* heterostructures. They found that the change in refractive index Δn is predominantly a quadratic function of the applied field. A fit of the Δn as a function of the applied field yielded a value for the quadratic electrooptic constant $R = -3.01 \times 10^{-18}$ m²/V² at λ = 1.537 μm for $In_{1-x}Ga_xAs_yP_{1-y}$ wells of λ_g = 1.57-μm composition and InP barriers. The *R* they obtained is considerably

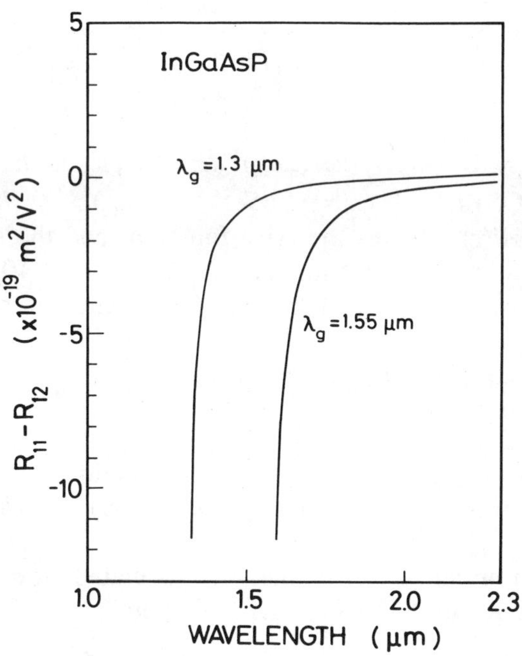

Figure 9.15 Calculated quadratic electrooptic constants R_{11}–R_{12} for $In_{1-x}Ga_xAs_yP_{1-y}$ layers (λ_g = 1.3 and 1.55 μm). (From Adachi and Oe.[48])

larger than those measured in the bulk quaternary.[29] Large quadratic electrooptic effects have also been measured in GaAs/AlGaAs multiple quantum wells.[52] Faist and Reinhart[44] have systematically investigated phase modulation in GaAs/$Al_xGa_{1-x}As$ double heterostructures with different doping profiles. An accurate estimation of the carrier effect permitted them to deduce the quadratic electrooptic coefficients for GaAs. The values they obtained are $R_{11} = -2.0 \times 10^{-20}$ m^2/V^2, $R_{12} = -1.7 \times 10^{-20}$ m^2/V^2 at $\lambda = 1.15$ μm, and $R_{11} = -2.9 \times 10^{-20}$ m^2/V^2, $R_{12} = -2.4 \times 10^{-20}$ m^2/V^2 at $\lambda = 1.09$ μm with an estimated uncertainty of $\pm 25\%$ for all values.

9.4 FRANZ–KELDYSH EFFECT

The Franz–Keldysh effect is an electric-field-induced change of the complex dielectric constant of a direct-band-gap semiconductor, occurring at photon energies close to the intrinsic band edge (E_0). This effect has two parts, electroabsorption and electrorefraction, which are respectively changes of the absorption coefficient and the refractive index due to an applied electric field E. The Franz–Keldysh electroabsorption coefficient is calculated using the expressions derived by Tharmalingam[53] and Callaway[54]

$$\alpha(\hbar\omega, E) = \sum_j A_j E^{1/3} \left[\left| \left(\frac{dAi(z)}{dz} \right)_{\beta_j} \right|^2 - \beta_j |Ai(\beta_j)|^2 \right] \tag{9.40}$$

where $\beta_j = 1.1 \times 10^5 (E_0 - \hbar\omega)(2\mu_{0j})^{1/3} E^{-2/3}$ [μ_{0j} is the combined density-of-states (DOS) mass], $A_j = 7.65 \times 10^5 (2\mu_{0j})^{4/3}/(n\hbar\omega)$ (n is the refractive index), $Ai(z)$ denotes the Airy function, and the sum is over the light- and heavy-hole valence bands. As $E \to 0$, Eq. (9.40) goes over into the familiar expression for the absorption due to direct allowed transitions

$$\alpha(\hbar\omega, 0) = \sum_j \frac{A_j}{\pi} (1.1 \times 10^5)^{1/2} (2\mu_{0j})^{1/6} (\hbar\omega - E_0)^{1/2} \tag{9.41}$$

that is, $\alpha(\hbar\omega, 0) \propto (\mu_{0j})^{3/2} (\hbar\omega - E_0)^{1/2}$ [see Section 8.3.1(a)]. The field-induced change in α is, then, given by $\Delta\alpha(\hbar\omega, E) = \alpha(\hbar\omega, E) - \alpha(\hbar\omega, 0)$.

Electrorefraction can, in principle, be calculated from this absorption change using the Kramers–Kronig (KK) relation[28]

$$\Delta n(\hbar\omega, E) = \frac{hc}{\pi} \int_0^\infty \frac{\Delta\alpha(\hbar\omega, E)}{(\hbar\omega')^2 - (\hbar\omega)^2} \, d\omega' \tag{9.42}$$

The electroabsorption effect has been measured in bulk InP,[55] InAs,[56] and GaAs,[28,55,57-60] and in $In_{1-x}Ga_xAs_yP_{1-y}$ epitaxial layers.[28,29,61,62] The Franz–Keldysh effect on indirect transitions in GaP has also been reported by Gasakov et al.[63] We show in Fig. 9.16 the results of InP and $In_{1-x}Ga_xAs_yP_{1-y}$ ($y = 0.52$) taken from Refs. 55 and 61, respectively. The dashed lines represent theoretical predictions for InP at the same electric fields. The solid lines are also the calculated predictions for GaAs at $E = 2 \times 10^5$ (#1) and 1×10^5 V/cm (#2) [since the masses and n entering into the theory are the same within better than 10% for GaAs and $In_{1-x}Ga_xAs_yP_{1-y}$ ($y = 0.52$), the theory should apply to the latter material also].[61] It can be seen from Fig. 9.16 that the experimental results are in good agreement with the theory.

Van Eck et al.[55] have also measured the electrorefraction in bulk InP and GaAs using a Mach–Zehnder interferometer. They have shown that, at least in InP, the Franz–Keldysh electrorefraction under some condition is much stronger than the linear electrooptic (Pockels) effect. Alpir and Coldren[28] have determined the electrorefraction in GaAs a

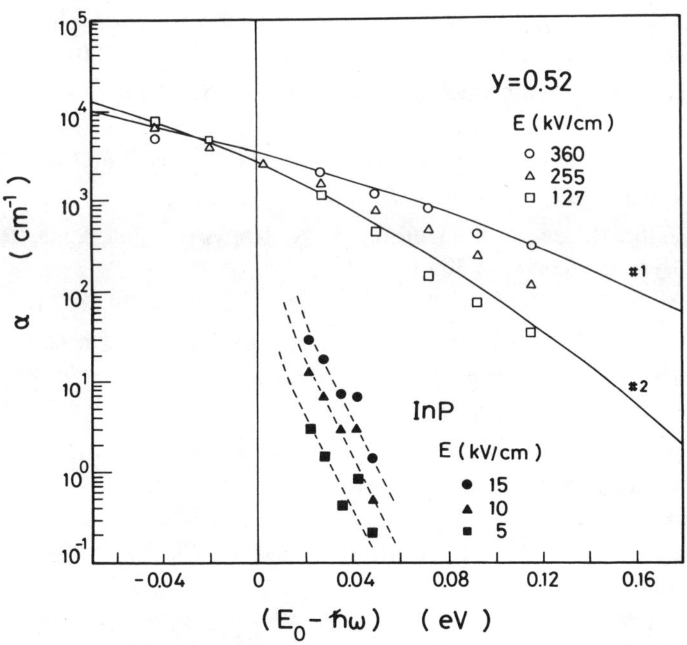

Figure 9.16 Experimental and theoretical absorption coefficients versus energy difference ($E_0 - \hbar\omega$) and electric field E for InP and $In_{1-x}Ga_xAs_yP_{1-y}$ ($y = 0.52$). The theoretical predictions and experimental data are taken for InP from Ref. 55 and for $In_{1-x}Ga_xAs_yP_{1-y}$, from Ref. 61.

$In_{1-x}Ga_xAs_yP_{1-y}$ ($y = 0.52$) from a KK analysis of Eq. (9.42). They found that the electrorefraction for $In_{1-x}Ga_xAs_yP_{1-y}$ is $\Delta n = 3.0 \times 10^{-3}$ at $\lambda = 1.3$ μm when the electric field is 400 kV/cm, but decreases to $\Delta n = 1.0 \times 10^{-3}$ at $\lambda = 1.55$ μm for the same electric field strength. They also found that for wavelengths far below E_0 the refractive index change Δn has a quadratic dependence on the applied electric field.

REFERENCES AND NOTE

1. R. W. Dixon, *J. Appl. Phys.* **38,** 5149 (1967).
2. F. K. Reinhart and R. A. Logan, *J. Appl. Phys.* **44,** 3171 (1973).
3. J. P. van der Ziel and A. C. Gossard, *J. Appl. Phys.* **48,** 3018 (1977).
4. H. Booyens and J. H. Basson, *J. Appl. Phys.* **51,** 4368 (1980).
5. C. W. Higginbotham, M. Cardona, and F. H. Pollak, *Phys. Rev.* **184,** 821 (1969).
6. S. H. Wemple and M. DiDomenico, Jr., *Phys. Rev. B* **1,** 193 (1970).
7. M. Sugie and K. Tada, *Jpn. J. Appl. Phys.* **15,** 257 (1976).
8. S. Adachi and K. Oe, *J. Appl. Phys.* **54,** 6620 (1983).
9. S. A. Geidur and A. D. Yas'kov, *Opt. Spectrosc.* (USSR) **57,** 45 (1984).
10. S. Adachi and C. Hamaguchi, *Phys. Rev. B* **21,** 1701 (1980).
11. F. Canal, M. H. Grimsditch, and M. Cardona, *Solid State Commun.* **29,** 523 (1979).
12. P. Y. Yu, M. Cardona, and F. H. Pollak, *Phys. Rev. B* **3,** 340 (1971).
13. A. Feldman and D. Horowitz, *J. Appl. Phys.* **39,** 5597 (1968).
14. L. N. Glurdzhidze, A. P. Izergin, Z. N. Kopylova, and A. D. Remenyuk, *Sov. Phys. Semicond.* **7,** 305 (1973).
15. P. Y. Yu and M. Cardona, *J. Phys. Chem. Solids* **34,** 29 (1973).
16. B. Tell, J. M. Worlock, and R. J. Martin, *Appl. Phys. Lett.* **6,** 123 (1965).
17. S. Adachi and C. Hamaguchi, *J. Phys. Soc. Jpn.* **43,** 1637 (1977); **45,** 505 (1978); *Phys. Rev. B* **19,** 938 (1979).
18. N. Suzuki and K. Tada, *Jpn. J. Appl. Phys.* **22,** 441 (1983).
19. M. Chandrasekhar, M. H. Grimsditch, and M. Cardona, *J. Opt. Soc. Am.* **68,** 523 (1978).
20. M. Chandrasekhar, M. H. Grimsditch, and M. Cardona, *Phys. Rev. B* **18,** 4301 (1978).
21. P. A. Kirkby, P. R. Selway, and L. D. Westbrook, *J. Appl. Phys.* **50,** 4567 (1979).
22. L. D. Westbrook, P. N. Robson, and A. Majerfeld, *Electron. Lett.* **15,** 99 (1979).

23. L. D. Westbrook, P. J. Fiddyment, and P. N. Robson, *Electron. Lett.* **16,** 169 (1980).

24. T. M. Benson, T. Murotani, P. N. Robson, and P. A. Houston, *IEEE Trans. Electron Dev.* **ED-19,** 1477 (1982).

25. H.-D. Liu and Z.-C. Feng, *IEEE J. Quantum Electron.* **QE-19,** 1016 (1983).

26. T. M. Benson, *J. Appl. Phys.* **54,** 6221 (1983).

27. F. J. Leonberger and J. F. Donnelly, in *Guided-Wave Optoelectronics*, T. Tamir, ed., Springer, Berlin, 1988, p. 317.

28. A. Alping and L. A. Coldren, *J. Appl. Phys.* **61,** 2430 (1987).

29. H. G. Bach, J. Krauser, H. P. Nolting, R. A. Logan, and F. K. Reinhart, *Appl. Phys. Lett.* **42,** 692 (1983).

30. A. Carenco, L. Menigaux, and N. T. Linh, *Appl. Phys. Lett.* **40,** 653 (1982).

31. M. DiDomenico, Jr. and S. H. Wemple, *J. Appl. Phys.* **40,** 720 (1969); **40,** 735 (1960).

32. C. G. B. Garrett, *IEEE J. Quantum Electron.* **QE-4,** 70 (1968).

33. M. Sugie and K. Tada, *Jpn. J. Appl. Phys.* **12,** 215 (1973).

34. S. Adachi and K. Oe, *J. Appl. Phys.* **56,** 74 (1984).

35. A. J. Hernández-Cabrera, C. Tejedor, and F. Meseguer, *J. Appl. Phys.* **58,** 4666 (1985).

36. J. F. Nye, *Physical Properties of Crystals*, Clarendon, Oxford, 1972.

37. The strength parameter A^* is a function of μ_0 and P^2 [see Eq. (8.22)]. The Stark shifts of the three band parameters E_0, μ_0, and P^2 were, then, taken into consideration in Refs. 34 and 48. The final results obtained in Refs. 34 and 48 are, however, the same as those obtained in this review [i.e., Eqs. (9.21) and (9.36)].

38. K. Tada and N. Suzuki, *Jpn. J. Appl. Phys.* **19,** 2295 (1980).

39. N. Suzuki and K. Tada, *Jpn. J. Appl. Phys.* **23,** 291 (1984).

40. A. Yariv, C. A. Mead, and J. V. Parker, *IEEE J. Quantum Electron.* **QE-2,** 243 (1966).

41. T. E. Walsh, *RCA Rev.* **27,** 323 (1966).

42. I. P. Kaminow, *IEEE J. Quantum Electron.* **QE-4,** 23 (1968).

43. M. Sugie and K. Tada, *Jpn. J. Appl. Phys.* **15,** 421 (1976).

44. J. Faist and F. K. Reinhart, *J. Appl. Phys.* **67,** 7006 (1990).

45. D. F. Nelson and E. H. Turner, *J. Appl. Phys.* **39,** 3337 (1968).

46. S. Singh, in *Handbook of Lasers*, R. J. Pressley, ed., Chemical Rubber Co., Cleveland, OH, 1971, p. 489.

47. L. Baldassarre, A. Cingolani, M. Ferrara, and M. Lugara, *Solid State Commun.* **34,** 237 (1980).

48. S. Adachi and K. Oe, *J. Appl. Phys.* **56,** 1499 (1984).

49. F. K. Reinhart, D. F. Nelson, and J. McKenna, *Phys. Rev.* **177,** 1208 (1969).

50. T. S. Narasimhamurty, *Photoelastic and Electro-Optic Properties of Crystals*, Plenum, New York, 1981.
51. J. E. Zucker, I. Bar-Joseph, B. I. Miller, U. Koren, and D. S. Chemla, *Appl. Phys. Lett.* **54,** 10 (1989).
52. See, for instance, T. H. Wood, R. W. Tkach, and A. R. Chraplyvy, *Appl. Phys. Lett.* **50,** 798 (1987).
53. K. Tharmalingam, *Phys. Rev.* **130,** 2204 (1963).
54. J. Callaway, *Phys. Rev.* **130,** 549 (1963); **134,** A998 (1964).
55. T. E. van Eck, L. M. Walpita, W. S. C. Chang, and H. H. Wieder, *Appl. Phys. Lett.* **48,** 451 (1986).
56. M. P. Mikhailova, D. N. Nasledov, and S. V. Slobodchikov, *Sov. Phys. Solid State* **7,** 1031 (1965).
57. H. D. Rees, *Solid State Commun.* **5,** 365 (1967).
58. G. E. Stillman, C. M. Wolfe, C. O. Bozler, and J. A. Rossi, *Appl. Phys. Lett.* **28,** 544 (1976).
59. R. H. Kingston, B. E. Burke, K. B. Nichols, and F. J. Leonberger, *Appl. Phys. Lett.* **41,** 413 (1982).
60. A. Partovi and E. M. Garmire, *J. Appl. Phys.* **69,** 6885 (1991).
61. R. H. Kingston, *Appl. Phys. Lett.* **34,** 744 (1979).
62. K. Okamoto, S. Matsuoka, Y. Nishiwaki, and K. Yoshida, *J. Appl. Phys.* **56,** 2595 (1984).
63. O. Gasakov, D. N. Nasledov, and S. V. Slobodchikov, *Phys. Status Solidi* **35,** 139 (1969).

10

CARRIER TRANSPORT PROPERTIES

10.1 LOW-FIELD MOBILITY

10.1.1 Electrons

A great deal has been written and published about transport phenomena in InP and other III–V compounds.[1–6] InP has become technologically important in recent years, especially for use in various optoelectronic devices. This material may be used to provide Gunn diodes for frequencies higher than that obtained by using GaAs. InAs has been used in galvanomagnetic devices, but otherwise the material is not of much commercial importance. GaAs is the most important III–V semiconductor, being a strong commercial competitor to Si for high-speed electronics applications (transistors and integrated circuits). GaP is known to be commercially important for applications to green and yellow electroluminescent devices. Because of the wider band bap, GaP is also used to make stable transistors that operate at high temperatures. $In_{0.53}Ga_{0.47}As$ is an important material for optical communication devices. Moreover, its high mobility and large drift velocity make this material very promising for high-frequency field-effect transistors and high-speed logic applications.

Hall mobility is a popular parameter used to characterize the microscopic quality of semiconductors. Accurate comparisons between experimental electron mobility and theoretical calculations are of great importance for determining a variety of fundamental material parameters and electron-scattering mechanisms. There are various electron-scattering mechanisms, such as polar optical scattering, deformation potential (acoustic phonon) scattering, piezoelectric scattering, intervalley scattering, impurity scattering, electron–electron scattering, and alloy scattering. Our knowledge of the material parameters discussed in the earlier sections enables us to calculate the strengths of various scattering mechanisms (except alloy scattering). In a case of alloy semiconductors, there is an increasing need to better understand electron scattering due to the random alloy potentials of the crystalline lattice.[7–20] This type of scattering mechanism is called the *alloy scattering*.

The semianalytic calculation has been used largely because of its simplicity. The effect of individual scattering mechanisms on the total calculated mobility can be visualized using Matthiessen's rule:

$$(\mu_{\mathrm{TOT}})^{-1} = \sum_i \mu_{\mathrm{i}}^{-1} \tag{10.1}$$

Thus, the total mobility μ_{TOT} can be calculated from the scattering-limited mobilities μ_{i} of each scattering mechanism. In InP and other direct-band-gap semiconductors at electric fields much less than the threshold field, almost all the electrons are in the lowest-conduction-band valley Γ, and so their mobility is limited by the scattering mechanisms appropriate to this valley. However, in GaP the lowest conduction band is at the *X* point. The constant-energy surfaces are, therefore, ellipsoidal, and we are required to consider intervalley phonon scattering in addition to the scattering mechanisms that are taken into account for the direct-band-gap materials.

Figure 10.1 shows the temperature dependence of the electron Hall mobility for relatively pure, *n*-type InP, InAs, GaAs, and GaP. The data are taken from Ref. 2 for InP (*a*), from Ref. 21 for InAs (*b*), from Refs. 22 and 23 for GaAs (*c*), and from Ref. 24 for GaP (*d*).

Low-field transport properties of *n*-InP have been studied in some detail.[3,5,25–39] However, there are still discrepancies in various values of the acoustic deformation potential E_1 (see Section 7.2 and Table 7.3), and the agreement between theory and experiment is not always good. The existence of ionized-impurity scattering and carrier compensation problems usually makes analyses very difficult. Anderson and Apsley[2] have studied temperature dependence of Hall mobility in *n*-InP [Fig. 10.1*a*] by taking into account different mechanisms of scattering. The mobility data increase sharply as the temperature decreases, showing a maximum at about 60 K [$\mu \simeq 1.4 \times 10^5$ cm^2/(V · s)], and then decreases with further decrease in temperature. In the range between 10 and 50 K it approximately follows $\mu \propto T^{3/2}$ (i.e., ionized-impurity scattering). Galavanov and Siukaev[34] have also tried to interpret the temperature dependence of mobility data for InP

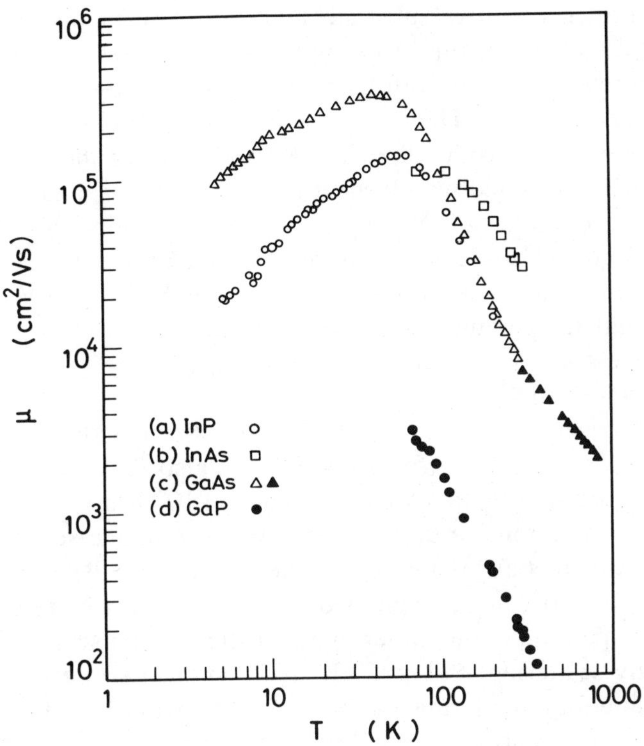

Figure 10.1 Temperature dependence of the electron Hall mobility in relatively pure, *n*-type InP, InAs, GaAs, and GaP. The data are taken for InP from Ref. 2, for InAs from Ref. 21, for GaAs from Refs. 22 (open triangles) and 23 (solid triangles), and for GaP from Ref. 24.

between 60 and 1000 K. Their data sharply dropped with increasing temperature from $T = 100$ to 300 K ($\mu \propto T^{-2}$). With the further increase in T the drop of μ slowed down, and in the range between 400 and 1000 K it followed $\mu \propto T^{-1}$. The μ in their data showed a peak at about 60 K [$\mu \simeq 1.4 \times 10^5$ cm^2/(V · s)]. They found that the combination of two lattice-scattering mechanisms (polar optical and acoustic phonon scattering) can successfully explain the experimental μ in the temperature range between 100 and 1000 K. The E_1 value (=21 eV) obtained by them was, however, considerably larger than that reported by other authors. Recent measurements of the electron Hall mobility as a function of pressure in pure InP[28] support a value of $E_1 \simeq 11.5$ eV. Lancefield et al.[28] also measured the temperature dependence of μ in the range 10–250 K and obtained good agreement between theory and experiment.

The dependence of μ on T for *n*-type InAs has been reported by several authors.[21,27,31,40–44] Cronin and Borrello[21] have studied electrical properties of n-InAs grown on InAs substrate by halogen-transport vapor growth. Their result is shown in Fig. 10.1 by the open squares. The experimental data were fitted reasonably well by a combination of polar optical and ionized impurity scattering. The computed curve was in close agreement with the experimental data from 300 to 150 K, below which the deviations were less than 30%. An effective electron mass of $0.025m_0$ and a donor-plus-acceptor concentration of 9×10^{15} cm^{-3} were determined from this fitting procedure. Rode[27] also made use of the result of Cronin and Borrello[21] and determined an acoustic deformation potential of $E_1 = 5.8$ eV. Because of the substantial temperature dependence of the small energy gap, he included temperature variations of m_e^Γ and E_0 (energy gap) measured by Stradling and Wood.[45]

Electron-transport properties of GaAs have been studied extensively in the literature[22,23,27,28,46–50] (for a review, see also Ref. 51). The mobility data obtained from different experiments above 100 K are found to be identical and may therefore be taken to represent the lattice-scattering-limited mobility. Stillman ct al.[22] showed that the temperature dependence of μ for GaAs (Fig. 10.1b) can be explained by contributions of three scattering mechanisms: polar phonon scattering, acoustic phonon scattering, and ionized-impurity scattering. Below 50 K, the dominant scattering mechanism is ionized-impurity scattering ($\mu \propto T^{3/2}$). Above 70 K, the mobility is found to be almost entirely limited to polar optical scattering. Lesser scattering processes, such as piezoelectric and carrier–carrier scattering, were omitted in their analyses. The n-GaAs layers with peak electron mobilities increasing from 1.06×10^5 to 4.02×10^5 cm^2/(V · s) at 28–40 K, the highest value every achieved in bulk n-GaAs, have been grown by Stanley et al.[50] by solid source molecular beam epitaxy using As_2.

As clearly seen in Fig. 10.1 (see also Refs. 27, 52–56) the indirect-band-gap material GaP has relatively a lower mobility value than those of InP, InAs, and GaAs. In general, a larger effective-mass material has lower electron mobility. A number of different experiments have given values for the conductivity effective mass m_c^X [Eq. (6.8)] in GaP. From plasma reflection studies different authors obtained different values for the mass $m_c^X = (0.32 \pm 0.02)m_0$,[57] $0.35m_0$,[58] and $0.35m_0$ or $0.29m_0$.[59,60] The wavelength variation of the refractive index also gave a value of $m_c^X = (0.29 \pm 0.02)m_0$.[60] More recently,[61] the cyclotron resonance in n-GaP provided $m_c^X = 0.37m_0$. These values are considerably larger than that for InP ($0.079m_0$; see Section 6.2.1), GaAs ($0.067m_0$)[62] or InAs ($0.023m_0$).[62] Intervalley phonon scattering also lowers the electron mobility of this material. Toyama et al.[53] determined the intervalley deformation potential constant D_i of GaP to be 7×10^8 eV/cm from Hall measurements. Rode[27] also obtained a value of $D_i = 1.2 \times 10^9$ eV/cm by fitting Casey's Hall data.[54] Fletcher and Butcher[56] analyzed the experimental mobility data of Taylor et al.[52] and obtained $D_i = 1.0 \times 10^9$ eV/cm. Wiley and DiDomenico[63] also successfully explained their free-carrier absorption data of *n*-GaP using numeric values of $D_i = 8 \times 10^8$ eV/cm and $E_1 \sim 10$ eV.

We show in Figs. 10.2 and 10.3 the temperature dependence of electron Hall mobility in $In_{1-x}Ga_xAs_yP_{1-y}$ lattice-matched to InP. The data are taken from Ref. 64 for $y = 0.46$ (Fig. 10.2) and Ref. 65 for $y = 1.0$ ($In_{0.53}Ga_{0.47}As$; Fig. 10.3). The samples were grown by liquid-phase epitaxy. The $In_{1-x}Ga_xAs_yP_{1-y}$/InP system has been proposed as a promising material for microwave devices because of its high mobility and large drift velocity. Especially $In_{0.53}Ga_{0.47}As$/InP has a potential advantage over either InP or GaAs in devices where the high mobility and large drift velocity are important. In fact, the Hall mobility measured at 300 K was $\mu \sim 1.4 \times 10^4$ cm^2/(V · s),[65] significantly higher than the best InP or GaAs

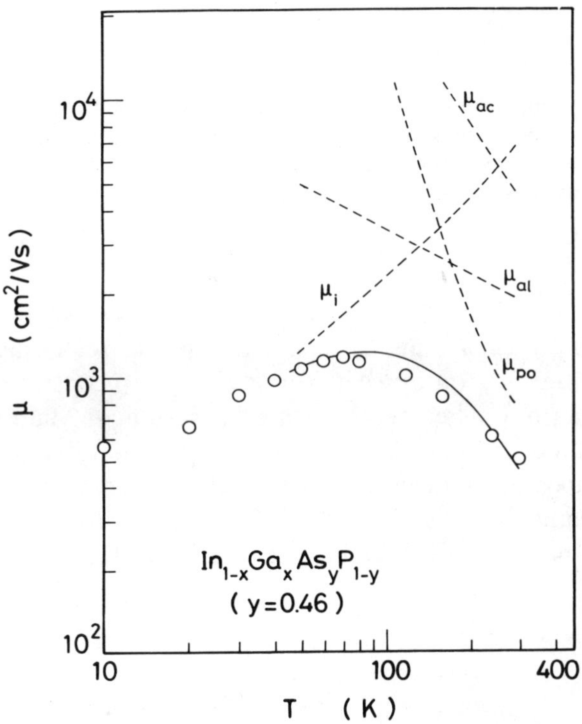

Figure 10.2 Temperature dependence of the electron Hall mobility in $In_{1-x}Ga_xAs_yP_{1-y}$ lattice-matched to InP ($y = 0.46$). The dashed lines show contributions of individual scattering mechanisms: μ_i (ionized-impurity scattering); μ_{po} (polar optical scattering); μ_{al} (alloy scattering); and μ_{ac} (acoustic phonon scattering). The solid line indicates the result of Matthiessen's rule [Eq. (10.1)]. (From Marsh et al[64])

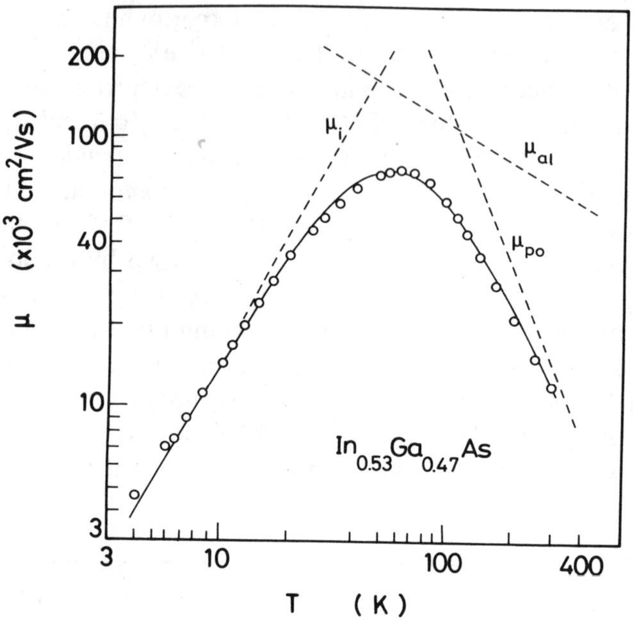

Figure 10.3 Temperature dependence of the electron Hall mobility in $In_{0.53}Ga_{0.47}As$. The dashed lines show contributions of individual scattering mechanisms: μ_i (ionized-impurity scattering); μ_{po} (polar optical scattering); and μ_{al} (alloy scattering). The solid line shows the result of Matthiessen's rule [Eq. (10.1)]. (From Oliver et al.[65])

Hall mobility. The dashed lines in the figures show contributions of individual scattering mechanisms: μ_i (ionized-impurity scattering), μ_{po} (polar optical scattering), μ_{al} (alloy scattering); and μ_{ac} [acoustic (deformation-potential) scattering]. The solid lines are the results of Matthiessen's rule.

In alloy semiconductors, the electrons see potential fluctuations as a result of the compositional disorder. This effect produces a peculiar scattering mechanism, namely, alloy scattering.[7,8,9–20] The alloy-scattering-limited mobility, as proposed by Littlejohn et al. (cited in Hayes et al.[35]), can be written as

$$\mu_{al} = \frac{8 \times 10^4 \sqrt{2}\, h^4 (m_e^\Gamma)^{-5/2}}{3\pi\Omega e(kT)^{1/2} S(\alpha)\, \Delta U_e^2} \tag{10.2}$$

where Ω is the volume of the primitive cell, ΔU_e is the alloy-scattering potential depending on the alloy composition, and $S(\alpha)$ refers to the degree of randomness of the atoms on each sublattice. [Note that $S(\alpha) = 1$ means a completely random distribution of atoms on each sublattice.]

It is clear from Eq. (10.2) that the mobility μ_{al} is strongly dependent of the scattering potential ΔU_e. Figure 10.4 shows the experimentally determined ΔU_e as a function of y for $In_{1-x}Ga_xAs_yP_{1-y}$ lattice-matched to InP. The data were gathered from Hayes et al.[35] and Marsh et al.[64] It is obvious that the ΔU_e values show large scatter across the alloy range. The solid line in the figure is drawn as a fit to the experimental data by the following quadratic equation (in electronvolts):

$$\Delta U_e(y) = 1.9y - 1.4y^2 \tag{10.3}$$

We also list in Table 10.1, the alloy-scattering potential ΔU_e for $In_{0.53}Ga_{0.47}As$ reported in the literature.[11,14,17,20,35,64–68] Three approaches can be used to determine the alloy-scattering potential:[11] from (1) the electronegativity theory, (2) the electron-affinity difference, and (3) the bandgap difference of the end-binary compounds. Littlejohn et al.[11] have discussed this problem for ternary and quaternary III–V semiconductors. The ΔU_e values obtained by these methods differ significantly (see Table 10.1). A good theoretical estimate of μ_{al} is, therefore, difficult to make at the present time. Fedders and Myles[14] have calculated the alloy-scattering po-

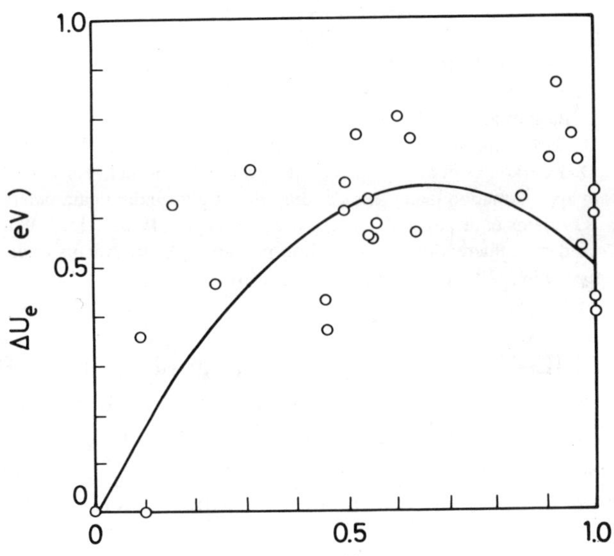

Figure 10.4 Alloy scattering potential ΔU_e for electrons as a function of y for $In_{1-x}Ga_xAs_yP_{1-y}$ lattice-matched to InP. The experimental data (open circles) are taken from Refs. 35 and 64. The solid line is drawn as a fit to the experimental data by a quadratic equation [Eq. (10.3)].

TABLE 10.1 Alloy-Scattering Potentials, ΔU_e (Electrons) and ΔU_h (Holes), for $In_{0.53}Ga_{0.47}As$ Alloy

ΔU_e (eV)	Method
1.08[a]	Band-gap difference
0.830[a]	Electron affinity difference
0.529[a]	Electronegativity difference
0.42[b]	Mobility analysis
0.65 ($m_e^\Gamma = 0.041 m_0$)[c]	Mobility analysis
0.82 ($m_e^\Gamma = 0.034 m_0$)[c]	Mobility analysis
0.60[d]	Mobility analysis
0.6–0.64[e]	Mobility analysis
0.42[f]	Mobility analysis
1.12[g]	Calculated[k]
0.7–1.0[h]	Mobility analysis
~2.5[i]	Cyclotron resonance measurements
1.07[j]	Mobility analysis
ΔU_h (eV)	**Method**
~0.2[l]	Calculated[m]

[a]M. A. Littlejohn et al., *Solid-State Electron.* **21,** 107 (1978).
[b]J. H. Marsh et al., in *Gallium Arsenide and Related Compounds*, Institute of Physics, Bristol, UK, 1981, p. 621.
[c]J. D. Oliver, Jr., et al., *J. Cryst. Growth* **54,** 64 (1981).
[d]Y. Takeda and T. P. Pearsall, *Electron. Lett.* **17,** 573 (1981).
[e]J. R. Hayes et al., *J. Electron. Mater.* **11,** 155 (1982).
[f]P. K. Basu and B. R. Nag, *Appl. Phys. Lett.* **43,** 689 (1983).
[g]P. A. Fedders and C. W. Myles, *Phys. Rev. B* **29,** 802 (1984).
[h]P. P. Basu et al., *J. Phys. C* **19,** L173 (1986).
[i]C. K. Sarkar and P. Banerji, *Phys. Status Solidi B* **156,** K145 (1989).
[j]C. D. Kourkoutas et al., *Solid State Commun.* **78,** 543 (1991).
[k]Born approximation using the semiempirical tight-binding parameters.
[l]A. R. Adams et al., *Jpn. J. Appl. Phys.* **19** (suppl. **19-3**), 315 (1980).
[m]Considering fluctuations in ionization potential [A. R. Adams et al., *Jpn. J. Appl. Phys.* **19** (suppl. **19-3**), 315 (1980)].

tentials for 18 III–V ternary semiconducting alloys in the Born approximation using semiempirical tight-binding parameters. Sarkar and Banerji[20] have recently estimated the alloy-scattering potential of $In_{0.53}Ga_{0.47}As$ using cyclotron resonance measurements. Their obtained value is ~2.5 eV, which is much larger than the accepted value of $\Delta U_e \sim 0.5$ eV for this material.

The space-charge scattering mobility μ_{sc} has the same temperature dependence as the alloy-scattering mobility μ_{al}, specifically, $\mu_{sc} \propto (m_e^\Gamma)^{-1/2} T^{-1/2}$. The piezoelectric scattering mobility μ_{pe} also has the same temperature dependence, namely, $\mu_{sc} \propto (m_e^\Gamma)^{-3/2} T^{-1/2}$. These possible scattering processes may, thus, disturb precise determination of the alloy-scattering-limited mobility in alloy materials. These three scattering mechanisms can best be distinguished through their effective mass dependence by high-pressure measurements:

$$\frac{d \ln \mu}{dp} \propto n \frac{d(m_e^\Gamma)}{dp} \tag{10.4}$$

where $n = -\frac{5}{2}$ for μ_{al}, $n = -\frac{1}{2}$ for μ_{sc}, and $n = -\frac{3}{2}$ for μ_{pe}. This approach has been made on $In_{1-x}Ga_xAs_yP_{1-y}$ alloy by Hayes et al.[35] and Patel and Spain[69] for the electron mobility, and by Adams and Shantharama[70] and Hayes et al.[35,70] for the hole mobility. Patel and Spain[69] reported that the electron mobility for $In_{0.53}Ga_{0.47}As$ is dominated by polar optical scattering with only a small fraction of alloy scattering.

In Fig. 10.5 we show the variation of room-temperature electron Hall mobility with composition y for $In_{1-x}Ga_xAs_yP_{1-y}$ lattice-matched to InP. The experimental data are gathered from various sources (see, e.g., Refs. 35, 37, and 72). The solid line is the Hall mobility with carrier compensation 1.0 in $In_{1-x}Ga_xAs_yP_{1-y}$ for $0 \leq y \leq 1.0$ calculated by Takeda[3] using the one-LO phonon model and the Phillips's electronegativity differ-

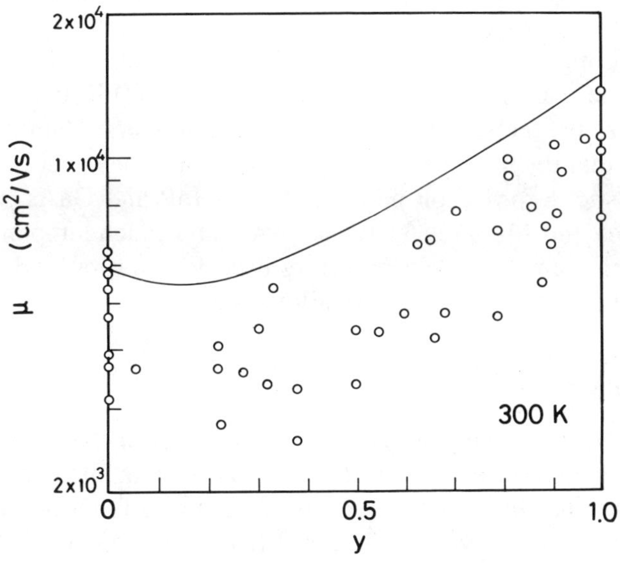

Figure 10.5 Variation of the room-temperature electron Hall mobility with composition y for $In_{1-x}Ga_xAs_yP_{1-y}$ lattice-matched to InP. The data (open circles) were gathered from various sources (see, e.g., Refs. 35, 37, and 72). The solid line represents the Hall mobility calculated using the one-LO phonon model and the Phillips' electronegativity differences as the alloy-scattering potential (see Ref. 3).

ences as the alloy-scattering potential. As we can easily understand from the figure, the carrier mobility is strongly dependent on the concentrations of impurities in specimens. The alloy scattering is seen to cause a mobility minimum around $y = 0.2$. If we accept the four-mode phonon behavior in $In_{1-x}Ga_xAs_yP_{1-y}$ alloy (see Section 3.3), we must strictly take into account the four-mode nature in the polar optical scattering mechanism. However, to our knowledge, no mobility calculation has been carried out on InGaAsP by taking the four-mode nature of the optical phonons into account. In some recent calculations,[73,74] only the two-mode nature of the optical phonons (InP-like and InGaAs-like phonons) is taken into consideration jointly with the acoustic, piezoelectric, ionized-impurity, and electron–electron-scattering data. The result of Takeda et al.[73] shows an increased downward bowing of the mobilities as compared with values calculated with a one-mode approximation of the polar optical phonons.

Many workers thus far have used Matthiessen's rule to explain the electron mobility in semiconductors. This rule is expected to be valid if (1) there is only onc dominant scattering mechanism, (2) all scattering mechanims have the same energy dependence, or (3) the electron distribution is a delta function in energy as would be the case in a metal. Several authors have discussed the validity of Matthiessen's rule for calculating electron mobility in alloy semiconductors. Takeda and Pearsall[66] have shown that Matthiessen's rule is an unacceptable approximation for analyzing scattering mechanims in $In_{0.53}Ga_{0.47}As$, and the mobility calculated using the rule was quoted to be 60% higher at 100 K than that obtained from the iterative technique with the same parameters. Hamilton[32] has also pointed out that the use of Matthiessen's rule provides unacceptable values of the acoustic deformation potential E_1 for InP and GaAs. On the other hand, Saxena and Mudares[75] have shown that a much simpler formulation using Matthiessen's rule can be safely used to interpret and analyze Hall mobility data for the $Al_xGa_{1-x}As$ alloy.

10.1.2 Holes

In direct-band-gap semiconductors, the *s*-like symmetry of the low-effective-mass conduction-band minimum (Γ) is responsible for high electron mobilities because the electron mobility is a strong function of the electron effective mass [i.e., $\mu \propto (m_e^\Gamma)^n$; $n < 0$]. As mentioned in Chapter 6, the valence bands of III–V compounds consist of three atomic *p*-like bands (light-hole, heavy-hole, and spin–orbit splitoff bands). Transport at the top of the valence band is, then, complicated by the following essential facts: (1) the degeneracy of the valence band, the influence of the light holes and interband scattering between the light and heavy hole bands; (2) the p-type symmetry of hole wavefunctions; and (3) the warping of the heavy-hole band. A dominant role of the polar optical scattering can also hardly be expected on physical grounds because large values of the heavy-hole mass favors acoustic and nonpolar optical deformation potential scattering. A useful description of the theoretical and practical aspects of transport properties in p-type semiconductors is given by Wiley,[76] Hayes et al.,[4] and Kranzer,[6] who review some of the work done on III–V compounds.

The correct way of handling the degeneracy of valence bands is by solving a set of coupled Boltzmann equations; the simplest model results from assuming that the heavy- and light-hole bands are decoupled. The mobilities of the two bands are, then, simply additive and it is easy to show that[76]

$$\mu_{\mathrm{eff}} = \frac{p_1 \mu_1 + p_2 \mu_2}{p_1 + p_2} \tag{10.5}$$

where μ_{eff} is the effective (or measured) mobility and p_1 and p_2 are the average densities of holes in the heavy- and light-hole bands (the total hole concentration is thus $p = p_1 + p_2$). If the bands are assumed to be approximately spherical, then $p_1/p_2 = (m_1/m_2)^{3/2} = (m_{\mathrm{hh}}/m_{\mathrm{lh}})^{3/2} = r^{3/2}$, where m_{hh} and m_{lh} are, respectively, the heavy- and light-hole masses (see Section 6.2). Expression (10.5) can, then, be written as

$$\mu_{\mathrm{eff}} = \frac{(\mu_2/\mu_1) + r^{3/2}}{1 + r^{3/2}} \mu_1 \tag{10.6}$$

When μ_2 and μ_1 are calculated using the standard expressions for the mobilities, the ratio μ_2/μ_1 reduces to a simple power of r depending on the mass dependence of the scattering mechanism under consideration. This approach yields

$$\mu_{\mathrm{i}}^{\mathrm{p}} = 1.5 \left(\frac{r^{1/2} + r^{3/2}}{1 + r^{3/2}} \right) \mu_{\mathrm{i}}^{\mathrm{e}} \tag{10.7}$$

$$\mu_{\mathrm{po}}^{\mathrm{p}} = 2K\mu_{\mathrm{po}}^{\mathrm{e}} \tag{10.8}$$

$$\mu_{\mathrm{sc}}^{\mathrm{p}} = \left[\frac{r^{5/2} + r^3}{(1 + r^{3/2})^2} \right] \mu_{\mathrm{sc}}^{\mathrm{e}} \tag{10.9}$$

$$\mu_{\mathrm{al}}^{\mathrm{p}} = \left[\frac{r^{5/2} + r^3}{(1 + r^{3/2})^2} \right] \mu_{\mathrm{al}}^{\mathrm{e}} \tag{10.10}$$

and

$$\mu^{p}_{ac,npo} = 3.17 \times 10^{-4} \frac{r^{5/2}(1 + r^{1/2})}{(1 + r^{3/2})^2} \frac{g\overline{v}^2}{(m_{hh}/m_0)^{5/2}} \frac{S(\theta, \eta, T)T^{-3/2}}{E^2_{ac}} \tag{10.11}$$

Where μ^e_i, μ^e_{po}, μ^e_{sc}, μ^e_{al} are obtained from the equations for electrons but substituting the heavy-hole effective mass m_{hh}. The factor 1.5 in Eq. (10.7) and the factor 2 in Eq. (10.8) take into account the *p*-like symmetry of the hole wavefunctions and *K* is a correction factor to take into account the contribution from light holes and has been shown graphically by Wiley;[76] $\mu^{p}_{ac,npo}$ is the hole mobility combined with acoustic- and nonpolar-optical-scattering-limited mobilities. The term *g* is the crystal density (see Section 2.3), and $\overline{v}$ is an average sound velocity defined, using Eq. (7.8), by

$$\overline{v}^2 = \frac{1}{3}\frac{C_l}{g} + \frac{2}{3}\frac{C_t}{g} \tag{10.12}$$

and E_{ac} is the hole deformation potential as discussed in Section 7.1. The term $S(\theta, \eta, T)$ is a function also given graphically by Wiley.[76] In $\eta = (E_{npo}/E_{ac})^2$, where E_{npo} is the phenomenologic optical deformation potential which can be related to the optical-phonon deformation potential d_o (see Section 7.1 and Table 7.2) by the following expression:

$$E_{npo} = \frac{M_1 + M_2}{2(M_1 M_2)^{1/2}} \left(\frac{C_1(\beta + 2)}{2g\omega_0^2 a_0^2 \beta} \right)^{1/2} d_o \tag{10.13}$$

where M_1 and M_2 are the masses of the atoms in a unit cell, ω_0 is the angular frequency of zone-center LO phonons, a_0 is the lattice constant, and β is defined in Eq. (7.10). The expression given in Eq. (10.11) differs from one given earlier by Wiley and DiDomenico[77] by a factor $(1 + r^{-3/2})^{-1}$, which was shown by Costato et al.[78] to take into account explicitly interband scattering. Wiley[76] has also shown that values of E_{ac} do not vary from material to material: 3.6 eV for InP, 3.2 eV for InAs, and 3.5 eV for GaAs and GaP.

Figure 10.6 shows the temperature dependence of the hole mobility for *p*-type InP, InAs, GaAs, and GaP. The experimental data are taken for InP (*a*) from Ref. 79, for InAs (*b*) from Ref. 40, for GaAs (*c*) from Refs. 80 and 81, and for GaP (*d*) from Ref. 82. The samples used in the references cited had relatively high purity, except InAs. Because of the small band gap (E_0 = 0.36 eV at 300 K),[62] InAs becomes intrinsic at high temperatures. One must, therefore, resort to rather heavy doping with acceptor

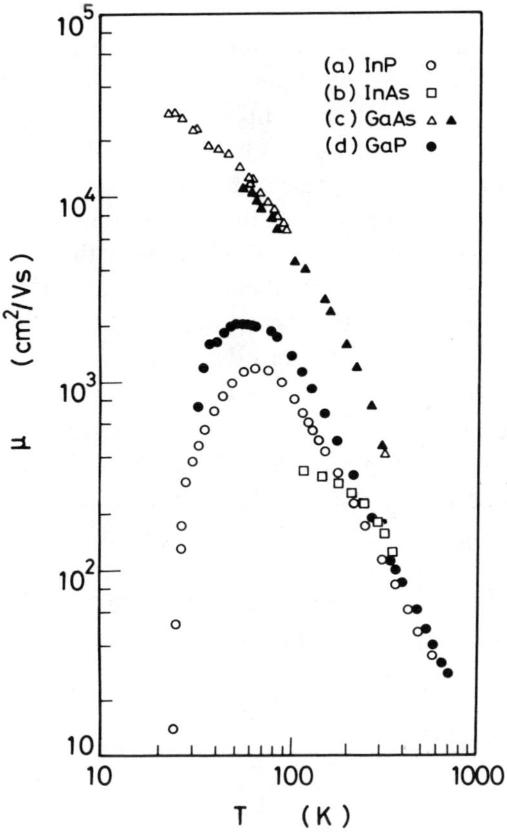

Figure 10.6 Temperature dependence of the hole mobility for *p*-type InP, InAs, GaAs, and GaP. The data are taken for InP from Ref. 79, for InAs from Ref. 40, for GaAs from Refs. 80 (open triangles) and 81 (solid triangles), and for GaP from Ref. 82.

dopants in order to obtain *p*-type InAs samples. The μ–T curve shown in Fig. 10.6 is the result from a sample with $p \simeq 2 \times 10^{17}$ cm^{-3} (at ~150 K).

For InP the empirical law $\mu \propto T^{-2}$ is found to hold at $T > 200$ K: μ decreases strongly at $T < 40$ K because of the onset of impurity conduction. The mobility in GaAs rises as the temperature is lowered, obeying the law $\mu \propto T^{-2}$ in the range 80 K $< T <$ 200 K, and tends to reach a maximum [$\mu \sim 3 \times 10^4$ cm^2/(V · s)] at 20 K. The mobility in GaP also rises as the temperature is lowered, reaches a maximum [$\mu \sim 2 \times 10^3$ cm^2/(V · s)], and decreases dramatically with further decrease of T. The temperature dependence of μ for InAs shows a $T^{-2.3}$ power law for $T >$ 250 K. Below 200 K, ionized-impurity scattering dominates other scattering mechanisms.

Because of the more complicated structure of the valence bands at $\mathbf{k} = 0$, mobility data analysis is more complicated for holes than for electrons. The scattering of electrons by polar optical phonons can be the dominant scattering mechanism in III–V compounds. This comes from the fact that *s*-like electron wavefunctions strongly couple to the polar optical mode but not to nonpolar modes. In the case of hole transport, polar optical scattering is known not to be the prominent process among various hole-scattering mechanisms. Kranzer[6] showed from his analyses that the polar optical scattering in these *p*-type materials affects, in the limit of the validity of Mattiessen's rule, the mobility by approximately 35% for InP and GaAs, 45% for InAs, and 30% for GaP at room temperature.

The hole mobility in $In_{1-x}Ga_xAs_yP_{1-y}$ lattice-matched to InP has been reported by several authors.[35,70,71,83–85] Hayes et al.[35] have measured hole mobility as a function of temperature from 77 to 300 K in a wide range of quaternary alloys grown by liquid-phase epitaxy. They also measured pressure-induced changes in the hole mobility of $In_{1-x}Ga_xAs_yP_{1-y}$ up to 15 kbar. These results suggested that alloy scattering plays an important part in the determination of low-field hole mobility of this alloy system. In Figs. 10.7 and 10.8 we reproduce their measured temperature dependence of hole

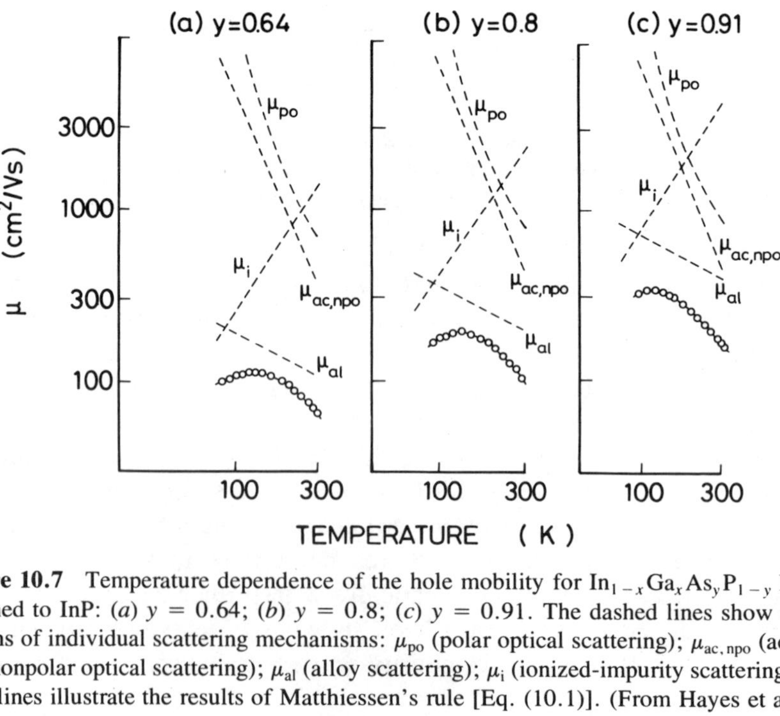

Figure 10.7 Temperature dependence of the hole mobility for $In_{1-x}Ga_xAs_yP_{1-y}$ lattice-matched to InP: (*a*) $y = 0.64$; (*b*) $y = 0.8$; (*c*) $y = 0.91$. The dashed lines show contributions of individual scattering mechanisms: μ_{po} (polar optical scattering); $\mu_{ac,npo}$ (acoustic plus nonpolar optical scattering); μ_{al} (alloy scattering); μ_i (ionized-impurity scattering). The solid lines illustrate the results of Matthiessen's rule [Eq. (10.1)]. (From Hayes et al.[35])

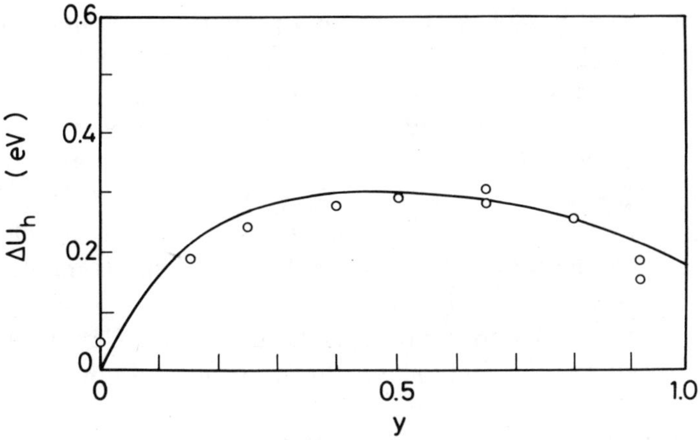

Figure 10.8 Alloy-scattering potential ΔU_h for holes as a function of y for $In_{1-x}Ga_xAs_yP_{1-y}$ lattice-matched to InP. The data (open circles) are taken from Refs. 35 and 83. The solid line represents the calculated result of Adams et al.[86]

mobility for $y = 0.64$, 0.8, and 0.91 samples (Fig. 10.7) and their determined alloy-scattering potential ΔU_h for holes as a function of y (Fig. 10.8). The dashed lines in Fig. 10.7 show the mobility limit imposed by polar optical scattering (μ_{po}), acoustic plus nonpolar optical scattering ($\mu_{ac, npo}$, μ_{po}), alloy scattering (μ_{al}), and ionized-impurity scattering (μ_i). The solid lines represent the results of Matthiessen's rule. Hayes et al.[35] found that the experimental data could not be successfully explained by combining ionized-impurity scattering with the various types of phonon scattering. However, when an additional scattering mechanism related to a mobility with a $T^{-1/2}$ temperature dependence (μ_{al}) was also included, a satisfactory fit to the experimental data was achieved (see Fig. 10.7).

The alloy-scattering potential ΔU_h can be regarded as an adjustable parameter that can be determined from the curve-fitting process. The open circles shown in Fig. 10.8 are obtained from this fitting process.[35,83] As can be seen in the figure, ΔU_h goes through a broad maximum near the center of the alloy range; the value of ΔU_h is approximately 0.3 eV at $y = 0.5$. Adams et al.[86] interpreted the values of ΔU_h in terms of the difference in ionization potential of the constituent compounds adjusted to the lattice parameter of InP. This result gives $\Delta U_h \sim 0.2$ eV for $In_{0.53}Ga_{0.47}As$ ternary (Table 10.1). The solid line in Fig. 10.8 is also the result of their calculation.

Pearsall et al.[84] have also measured the electron and hole mobilities in $In_{0.53}Ga_{0.47}As$ epitaxial films grown on InP substrates by liquid-phase epitaxy and metal–organic chemical vapor deposition (MOCVD). They found that the hole mobility μ_h is approximately proportional to the electron mobility μ_e with $(\mu_e/\mu_h)_{T=295\,K} = 25$. This high electron : hole mobility ratio may indicate that $In_{0.53}Ga_{0.47}As$ can be used to realize high-speed photoconductive devices and uncooled photoconductive arrays in the 1.5-μm atmospheric window. The highest room-temperature hole mobility they measured is $\mu_h = 300\ cm^2/(V \cdot s)$ with $N_A + N_D = 4 \times 10^{17}\ cm^{-3}$ and $N_A - N_D = 2 \times 10^{15}\ cm^{-3}$. Small concentrations (1–2%) of phosphorus were also measured in the MOCVD samples. This phosphorus may come from residual phosphine used in the pregrowth cycle. They suggested that the presence of even small amounts of phosphorus in $In_{1-x}Ga_xAs$ alloy lowers the room-temperature mobility significantly because of alloy-scattering effects.

Figure 10.9 shows the variation of room-temperature hole mobility with composition y for $In_{1-x}Ga_xAs_yP_{1-y}$ lattice-matched to InP. The experimental data are taken from Refs. 35 and 84. As in the case of electron mobility, hole mobility is strongly dependent on the concentrations of im-

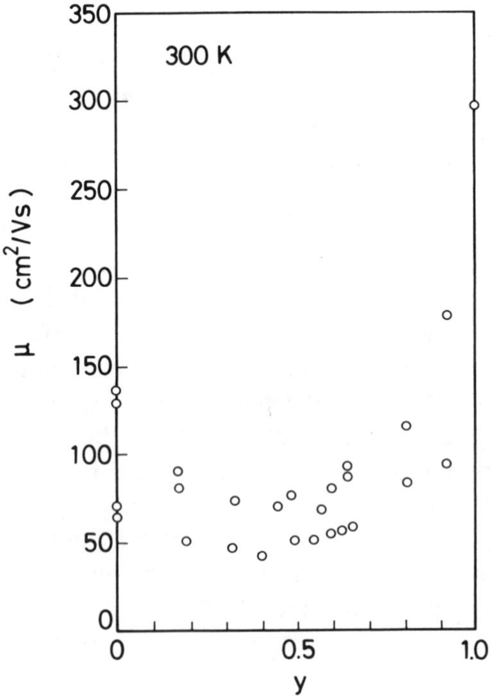

Figure 10.9 Variation of the room-temperature hole mobility with composition y for $In_{1-x}Ga_xAs_yP_{1-y}$ lattice-matched to InP. The experimental data (open circles) are taken from Refs. 35 and 84.

purities in specimens. Alloy scattering may also be seen to cause a mobility minimum around $y = 0.4$.

10.2 HIGH-FIELD TRANSPORT

10.2.1 Velocity-Field Characteristics

(a) Electrons The carrier-transport property in high electric fields is one of the most fundamental parameters in electron device design. Performance of high-speed and microwave semiconductor devices depends essentially on carrier velocity. It is, thus, very important to know the carrier drift velocity as a function of electric field in a bulk semiconductor. It is, however, very difficult to make a direct measurement of the drift velocity of electrons as a function of electric field in some III–V compounds. This is because in high-conductivity material the sample normally oscillates when the average field in the sample is above the threshold value. This type of microwave oscillation, known as the *Gunn effect*,[87] is caused by the transfer of electrons from high- to low-mobility valleys in the conduction band of III–V compounds. In lower-conductivity samples, the electric field may become highly nonuniform within the sample even if it does not oscillate.

A Monte Carlo computer technique has been used by several authors to calculate steady-state drift velocity versus electric field (v–E) relationships for n-type InP.[10,88–91] The v–E characteristics of n-InP have also been studied experimentally by many authors.[92–100] The techniques used were space-charge-wave measurements,[92] microwave measurements,[93,96,100] dipole-domain measurements,[97,98] and current–voltage measurements.[99] In Fig. 10.10 we show the results of the v–E curve for InP measured at room temperature.[93,95–97,100] These results suggest that the peak velocity occurs at the electric field $E \simeq 11$ kV/cm. However, remarkable differences in the peak velocities exist among these experimental results.

The v–E curve is strongly dependent on both the crystal quality and temperature. This is because the carrier drift mechanism is largely influenced by scattering processes in the material. When the scattering process is elastic, a relaxation time characterizing the rate at which the momentum decays can be defined. From the relaxation time, one can calculate the perturbation of an equilibrium electron distribution by a small electric field and, hence, the mobility. When the scattering is inelastic, no exact relaxation time exists, although in certain limits this approximation can be useful. The Boltzmann transport expression in which the average rates of energy loss and momentum loss through collisions are balanced by energy (ϵ) and momentum (p) gain due to the electric field provides the following

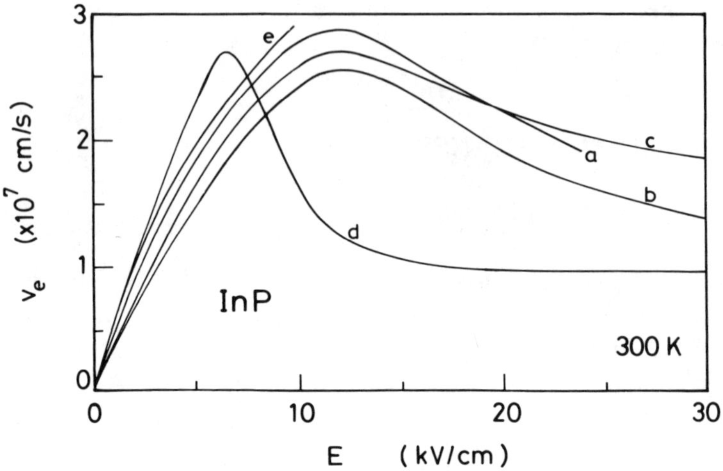

Figure 10.10 Electron velocity–field characteristics for *n*-InP measured at room temperature (*a*, Ref. 93; *b*, Ref. 95; *c*, Ref. 96; *d*, Ref. 97; and *e* Ref. 99).

energy and momentum balance equations:

$$\left\langle \frac{d\epsilon}{dt} \right\rangle_E + \sum_i \left\langle \frac{d\epsilon}{dt} \right\rangle_i = 0 \tag{10.14a}$$

$$\left\langle \frac{dp}{dt} \right\rangle_E + \sum_i \left\langle \frac{dp}{dt} \right\rangle_i = 0 \tag{10.14b}$$

where the index i refers to the ith scattering mechanism and the subscript E refers the electric field.

The drift velocity can be determined by equating the gradient of the v–E curve at the origin to the measured low-field drift mobility (μ_D), while the Hall measurement provides the Hall mobility $\mu_H = r\mu_D$ where r is the so-called Hall scattering factor. It has been pointed out by Hilsum et al.[101] that the Hall scattering factor for InP may be appreciably larger than unity and, therefore, the drift mobility is smaller than the measured Hall mobility. Takeda et al.[89] have also calculated the dependence of r on composition for $In_{1-x}Ga_xAs_yP_{1-y}$ lattice-matched to InP for a moderate electron concentration range between 1×10^{16} and 1×10^{17} cm^{-3} at 300 K by using an iterative technique in which the Boltzmann equation is solved by numeric iteration. The results show that r varies smoothly from $r = 1.18$ ($y = 0$; InP) to 1.08 ($y = 1.0$; $In_{0.53}Ga_{0.47}As$). On the other hand, Marsh et al.[64] reported that the factor r is taken to be unity for $In_{1-x}Ga_xAs_yP_{1-y}$ ($0 \lesssim y \lesssim 1.0$), which is confirmed by measurements at high magnetic fields

(9T) where $\mu B >> 1$. Majerfeld et al.[99] measured v–E curves of InP having two different mobilities: $\mu_H = 3600\ \text{cm}^2/(\text{V} \cdot \text{s})$ and $4750\ \text{cm}^2/(\text{V} \cdot \text{s})$ (at 300 K). The peak velocities they observed are 2.8×10^7 cm/s for the 3600-cm^2/(V · s) sample and 3.0×10^7 cm/s for the 4750-cm^2/(V · s) one. Tebbenham and Walsh[100] measured the v–E curves of InP at 330 and 110 K. They found that cooling the material to 110 K produces a 70% increase in the peak velocity [$(2.6 \pm 0.3) \times 10^7$ cm/s at 330 K; $(4.4 \pm 0.44) \times 10^7$ cm/s at 110 K].

There have been a fair number of publications on theoretical[88, 102–110] and experimental v–E curves[111–122] of GaAs. There are also few experimental measurements of the v–E curve of InAs (at 77 K).[123, 124] To our knowledge, however, no experimental v–E curve has been reported for GaP. In Fig. 10.11 we show a comparison of the v–E curves for InP, InAs, GaAs, and GaP. The curves are taken from Ref. 99 for InP (experimental; 300 K), from Ref. 124 for InAs (experimental; 77 K), from Ref. 122 for GaAs (experimental; 300 K), and from Ref. 125 for GaP (calculated; 300 K). One can easily recognize from Fig. 10.11 that InP may be very promising

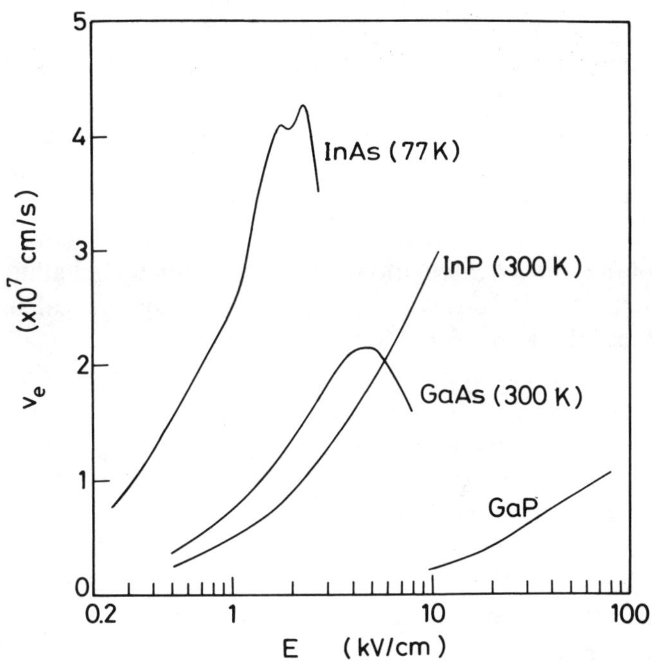

Figure 10.11 Electron velocity–field curves for n-InP, n-InAs, n-GaAs, and n-GaP. The curves are taken from Ref. 99 for InP (experimental; 300 K), Ref. 124 for InAs (experimental; 77 K), Ref. 122 for GaAs (experimental; 300 K), and Ref. 125 for GaP (calculated; 300 K).

for high-frequency transistors and high-speed logic applications (this is because the electron velocity at high electric fields of InP is considerably higher than that of GaAs). The extremely low electron velocity of GaP is due to the large electron effective mass in the lowest conduction band (X) of this material. A calculation by Ferry[125] gave a saturated drift velocity of $\sim 1 \times 10^7$ cm/s for GaP. He also suggested that the value of the saturated drift velocity for this material is sensitive only to the phonon coupling constants.

Wu and Yu[110] have studied high-electron transport in compensated GaAs by Monte Carlo calculation for various compensation ratios at lattice temperatures of 77, 160, and 300 K. They found that doping compensation has a stronger effect on electron transport at low temperatures than at high temperatures over a range of field values. They also found that the calculated v–E characteristics exhibit two maxima at both low and high temperatures for several doping compensations. This feature is rather unique as compared with other compensated semiconductors such as InP and $In_{0.53}Ga_{0.47}As$. They considered that the two-maxima feature in compensated GaAs may be attributable to the rather large electron populations in the L valleys at intermediate fields and the continuous increase of electron populations in the X valleys at high fields.

In addition to its use in optoelectronics, the $In_{1-x}Ga_xAs_yP_{1-y}$ quaternary can also be employed in fast-response electronics and microwave technology. This is because it has a high maximum drift velocity, which exceeds those in InP and GaAs, and a strong negative differential mobility. The high-field transport properties of $In_{1-x}Ga_xAs_yP_{1-y}$ quaternary have been studied both theoretically and experimentally by a number of groups. A variety of theoretical calculations has given a fundamental understanding of the subject.[11, 19, 126, 127] It was shown in Ref. 11 that the calculated values for the critical field and for the peak drift velocity in this alloy depends strongly on alloy scattering. This has been confirmed experimentally by a number of authors.[19, 64, 128] In Fig. 10.12 we show the experimentally determined v–E curves in samples of $In_{1-x}Ga_xAs_yP_{1-y}$ lattice-matched to InP ($y = 0.42, 0.44, 0.65, 0.8$) taken at 300 K. The experimental data are taken from Refs. 19 ($y = 0.42, 0.65, 0.8$) and 128 ($y = 0.44$). There is a considerable difference in the v–E curves between $y = 0.42$ and 0.44. This comes from a difference in the low-field Hall mobilities between the samples. Room-temperature mobilities of these samples, determined from Hall-effect measurements, are 3100 $cm^2/(V \cdot s)$ ($y = 0.42$) and 6000 $cm^2/(V \cdot s)$ ($y = 0.44$) [3480 $cm^2/(V \cdot s)$ ($y = 0.63$); 5800 $cm^2/(V \cdot s)$ ($y = 0.8$)]. One sees that in each case, the velocity initially rises with increasing electric field at a rate characteristic of low-field mobility [$v = (\mu_H/r)E$]. A rapid increase in the drift velocity, thus, occurs in the high-

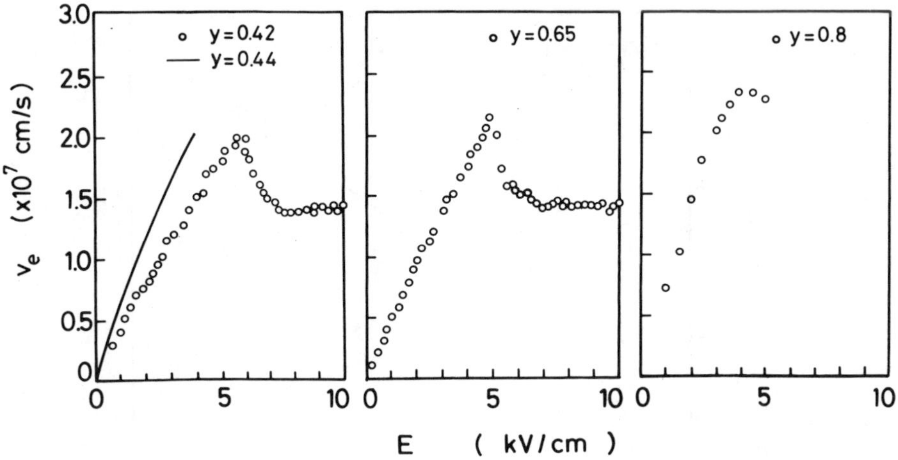

Figure 10.12 Electron velocity–field characteristics for n-$In_{1-x}Ga_xAs_yP_{1-y}$ lattice-matched to InP measured at room temperature. The data are taken from Refs. 19 ($y = 0.42$, 0.65, 0.8) and 128 ($y = 0.44$).

mobility sample. Marsh et al.[64] have also measured the peak velocity for $In_{1-x}Ga_xAs_yP_{1-y}$ over the entire range of y that lattice-match to InP. They found that for high-purity samples the peak velocity decreases with increasing y from 2.5×10^7 cm/s for InP to 1.4×10^7 cm/s at $y = 0.8$, where it starts to increase to the InP value.

Many theoretical studies[8,11,18,126,129–134] have been devoted to advancing the understanding of the high-field transport properties of the $In_{0.53}Ga_{0.47}As$ ternary since its electron mobility, small effective mass, and large intervalley separation offer the promise of providing drift velocities significantly higher than those of GaAs at room temperature. Experimental v-E curves of this material have been determined by Sasaki et al.,[135] Marsh et al.,[64] Haase et al.,[118] and Shigekawa et al.[136] Haase et al.[118] have measured the electron drift velocities in $In_{0.53}Ga_{0.47}As$ samples with carrier concentrations from 1×10^{15} to 1×10^{18} cm^{-3} for electric field up to the onset of the Gunn effect. Their results are reproduced in Fig. 10.13. The data in the figure clearly illustrate the effects of ionized-impurity scattering on the v-E characteristics. The maximum velocity observed in the most lightly doped sample is about 2.5×10^7 cm/s. Calculation[132] also predicts a peak velocity of $\sim 3.5 \times 10^7$ cm/s for this material, which is substantially higher than that for GaAs.

When a constant electric field is applied to any conducting material, or the field is changed from some previous constant value, some delay due to inertia can be expected before the average velocity of the carriers reaches

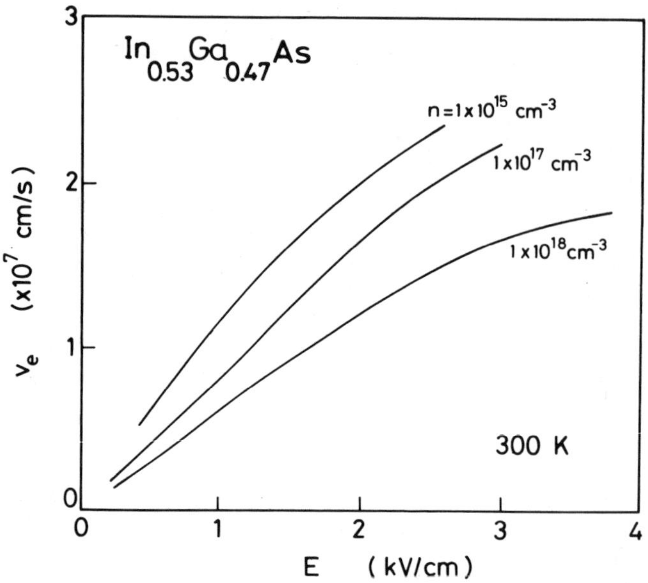

Figure 10.13 Room-temperature electron velocity–field characteristics for n-$In_{0.53}Ga_{0.47}As$. (From Haase et al.[118])

a new equilibrium value. If the carriers start from rest, then in most, if not all, semiconductors there will be, at a sufficiently high field, an acceleration of the carriers to a velocity higher than its long-term equilibrium value before the momentum and energy relaxation processes bring the velocity down to the equilibrium value. This phenomenon, called velocity "overshoot," and has been studied extensively in InP,[90,137–139] GaAs,[90,137,140–145] and $In_{0.53}Ga_{0.47}As$.[129,131,146]

It is necessary to know the v–E characteristics in the higher field regime in order to predict the frequency limitations of transit-time-limited microwave devices, such as microwave transistors, IMPATTs (impact avalanche and transit-time devices), transferred electron devices, and optical devices such as avalanche photodiodes and phototransistors. The electron velocity as a function of field in the higher-field regime has been measured for InP,[92,147–149] GaAs,[150–155] the $In_{1-x}Ga_xAS_yP_{1-y}$ quaternary,[156] and the $In_{0.53}Ga_{0.47}As$ ternary.[157,158] In Fig. 10.14 we compare the results for InP,[149] GaAs,[154] $In_{1-x}Ga_xAs_yP_{1-y}$ ($y = 0.44$[156]), and $In_{0.53}Ga_{0.47}As$.[158] We can clearly see that InP has the highest saturation velocity among these materials. We also plot, in Fig. 10.15, the saturation velocity, obtained for an average electric field of about 100 to 150 kV/cm, versus temperature for InP, $In_{1-x}Ga_xAs_yP_{1-y}$, and $In_{0.53}Ga_{0.47}As$. The reason for the saturation of the velocity can be explained in thermal equilibrium terms. At low fields

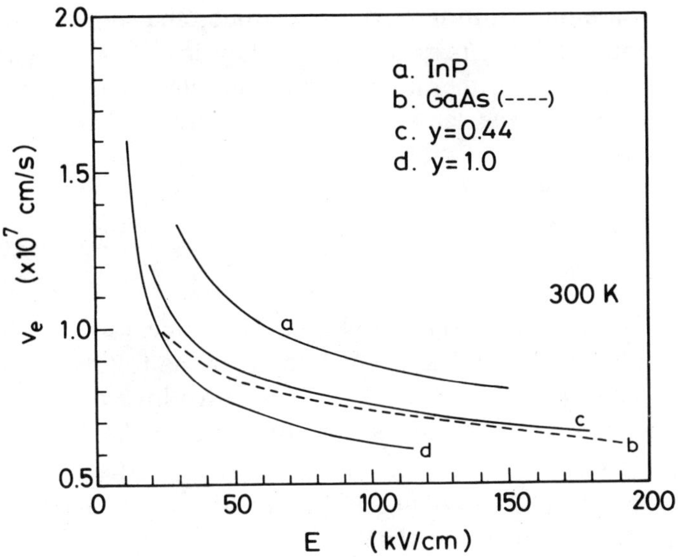

Figure 10.14 Results of the electron velocity–field characteristics for n-InP,[149] n-GaAs,[154] n-$In_{1-x}Ga_xAs_yP_{1-y}$ ($y = 0.44$[156]), and n-$In_{0.53}Ga_{0.47}As$ ($y = 1.0$[158]) measured in the higher-field regime.

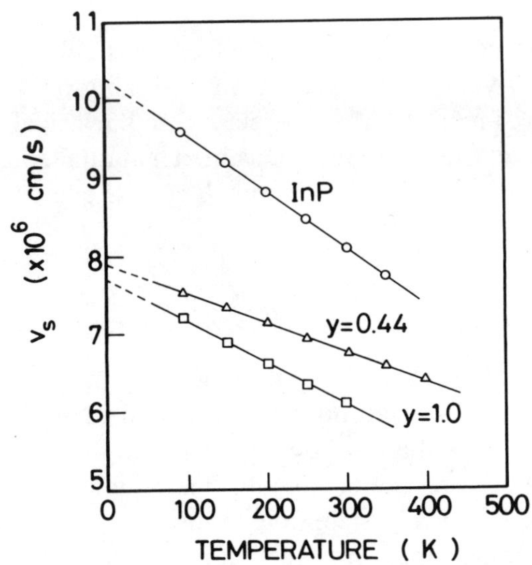

Figure 10.15 Electron saturation velocity v_s as a function of temperature for n-InP, n-$In_{1-x}Ga_xAs_yP_{1-y}$ ($y = 0.44$), and n-$In_{0.53}Ga_{0.47}As$ ($y = 1.0$).

the electrons are in equilibrium with the lattice, and, on the average, they pick up as much energy per collision as they lose. As the electron drift velocity approaches the thermal velocity, the electrons, on the average, give more energy to the lattice than they receive per collision, and the velocity saturates. We see from Fig. 10.15 that the saturation velocity v_s can be written simply as

$$v_s(T) = v_s(0) - KT \tag{10.15}$$

where $v_s(0)$ is the saturation velocity at the temperature $T = 0$ K. The parameter values $v_s(T)$ and K obtained here are as follows: $v_s(0) = 10.3 \times 10^6$ cm/s and $K = 7.3 \times 10^3$ cm/(s · K) for InP; $v_s(0) = 7.9 \times 10^6$ cm/s and $K = 3.8 \times 10^3$ cm/(s · K) for $In_{1-x}Ga_xAs_yP_{1-y}$ ($y = 0.44$); and $v_s(0) = 7.7 \times 10^6$ cm/s and $K = 5.3 \times 10^3$ cm/(s · K) for $In_{0.53}Ga_{0.47}As$. Recently, Allam and Pribetich[155] wrote their derived temperature dependence of v_s for GaAs

$$v_s(T) = v_s(T_0)\left(K_1 + \frac{K_2}{T}\right) \tag{10.16}$$

where $v_s(T_0)$ is the saturation velocity at the ambient temperature T_0 (= 300 K) and K_1 and K_2 are constants. They obtained $K_1 = 0.415$ and $K_2 = 175.4$ K at temperatures between 300 and 500 K.

(b) Holes In some semiconductor devices, particularly avalanche photodiodes, hole drift velocity is essential in analyzing their operating properties. The figures of merit for photodetectors are intrinsically limited in gain–bandwidth products and, therefore, cannot simultaneously provide large gain at high bandwidth. The bandwidth B can be expressed as

$$B = \frac{1}{2\pi m(\tau_e + \tau_h)} \tag{10.17}$$

where m is the total number of impact-ionization multiplication stages, and τ_e and τ_h are the electron and hole drift times in each stage. In most III–V compounds, τ_e is less than τ_h, owing to the higher steady-state electron drift velocity. Therefore, the bandwidth B is limited by the hole transit time or, equivalently, the steady-state hole velocity.

Despite such importance, no much attention has been paid to the high-field properties of holes in p-type III–V compounds. Only measurements of the velocity–field characteristics in GaAs have been made by Holway et al.,[159] and there are some experimental results on the characteristics of holes (minority carriers) in n-$In_{0.53}Ga_{0.47}As$ (see Section 10.3). Theoretical consideration of the high-field properties of holes has been made by Brennan and Hess[160,161] and Brennan[162] using a Monte Carlo technique.

In Fig. 10.16 we show the calculated v–E curves for InP,[160] GaAs,[160] and $In_{0.53}Ga_{0.47}As$.[162] The dashed line is the experimental result of GaAs by Holway et al.[159] As seen in the figure, velocity saturation of holes in these materials occurs at fields of about 100 kV/cm. The maximum hole velocities are $\lesssim 8 \times 10^6$ cm/s, which are considerably lower than those for the electrons.

Lower hole velocities can be understood by introducing the concept of hole effective mass. The hole effective mass at the valence-band maximum (Γ) can be written as

$$m_c^h = \frac{m_{lh}^{5/2} + m_{hh}^{5/2}}{m_{lh}^{3/2} + m_{hh}^{3/2}} \tag{10.18}$$

$$m_d^h = (m_{lh}^{3/2} + m_{hh}^{3/2})^{2/3} \tag{10.19}$$

where m_c^h and m_d^h are, respectively, the conductivity and density-of-states (DOS) effective masses. [Expression (10.18) is only a crude approximation since valence-band constant energy surfaces are not spherical.] Like m_c^α and m_e^α [see Eqs. (6.7) and (6.8)], the mass m_c^h can be used for conductivity

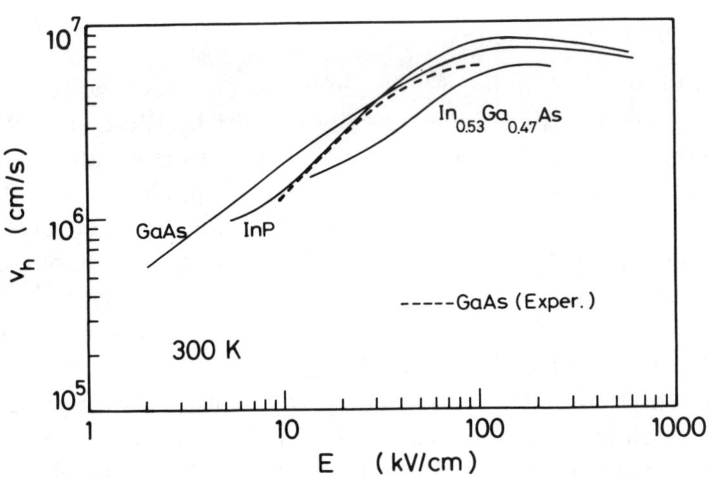

Figure 10.16 Calculated room-temperature hole velocity–field curves for p-InP,[160] p-GaAs,[160] and p-$In_{0.53}Ga_{0.47}As$.[162] The dashed line represents the measured result of GaAs by Holway et al.[159]

calculations and the m_d^h used for DOS calculations. Let us assume that the average hole velocity attained between collisions will be half their maximum velocity. Then

$$v_h = \left(\frac{1}{2}\right)\frac{eE}{m_c^h}\tau = \mu_h E \tag{10.20}$$

where τ is the average time between collisions. A similar result may be obtained for electrons. These results suggest that smaller m_c^h (m_c^α) mass carriers may have higher average drift velocities. It is a fact that m_c^α ($\alpha = \Gamma$; electrons) is considerably smaller than m_c^h (holes). The drift velocity for electrons can, therefore, be higher than that for holes.

Fig. 10.16 also indicates that the v–E characteristics do not differ very much among these materials. This is because these materials have nearly the same values of the conductivity effective mass: $m_c^h/m_0 = 0.50$ (InP; see Table 6.7), 0.47 ($In_{0.53}Ga_{0.47}As$; see Table 6.7), and 0.47 (GaAs; see Ref. 163). Brennan and Hess also calculated the impact ionization rate of holes in the valence bands of InP,[160] GaAs,[160] and $Al_xGa_{1-x}As$.[161] The ionization rate in InP is found to be much lower than in GaAs. They also carried out a Monte Carlo investigation of transient hole transport in GaAs.[164] The results reported show that a significant velocity overshoot occurs in this material at low applied electric fields over device distances greater than 1500 Å.

10.2.2 Gunn Phenomena

The possibility of obtaining a bulk semiconductor microwave negative resistance was first predicted theoretically in 1961 by Ridley and Watkins[165] and Hilsum.[166] The first experimental evidence of the negative differential-resistance effect was reported in 1963 by Gunn[87] in GaAs and InP. Besides in these binary materials, the same effect has since been observed experimentally in a variety of III–V binary, ternary, and quaternary compounds (e.g., InAs,[167] $InAs_yP_{1-y}$,[167] $Al_xGa_{1-x}As$,[168] $In_{1-x}Ga_xAs$,[64,135,169–175] and $In_{1-x}Ga_xAs_yP_{1-y}$[64,128,175–177]). The important thing to note about this effect is that the conduction bands of these compounds are composed of a low-mass, high-mobility central valley (Γ point) and of high-mass, low-mobility satellite valleys in the eight equivalent L and six equivalent X points. It is generally accepted that the negative differential mobility at high electric fields arises from the transfer of high-mobility electrons from the central valley into these satellite valleys in which their mobility is drastically reduced.

Transferred-electron effect in InP has been reported by a number of workers.[87,99,175,178–182] The high-field properties of InP as a function of pressure have been measured by several authors.[178–180] Kobayashi et al.[180] found that the threshold field (E_{th}) for transferred-electron instabilities ranges from 7.5 to 8.5 kV/cm at atmospheric pressure. The most reliable results show that E_{th} increases slightly with pressure below 40 kbar. This behavior can be successfully explained in terms of possible band-structure changes (i.e., effective masses and energy-band gaps). High efficiency transferred-electron oscillators and low-noise broad-band amplifiers have been demonstrated by Fank.[181] The devices were fabricated on n^+-InP substrates. Self-aligned notched planar InP transferred-electron oscillators have also been fabricated by Binari et al.[182] on semiinsulating InP substrates. The best pulsed result they obtained was a power output of 220 mW at a frequency of 9.1 GHz and an efficiency of 9.5%. The CW (continuous-wave) result was a power output of 80 mW with an efficiency of 5.3% at 5.5 GHz.

The transferred-electron effects in the $In_{1-x}Ga_xAs_yP_{1-y}$ alloy have been reported in Refs. 64, 128, 135, and 169–177. An important aspect of the design of Gunn-type devices is the large power dissipation. To reduce power consumption and therefore to improve the efficiency a low threshold field is desirable. In Fig. 10.17 we show the composition dependence of the threshold electric field E_{th} for the $In_{1-x}Ga_xAs_yP_{1-y}$ alloy. The experimental data are gathered from Refs. 64, 128, 175, 176, and 180. A value of E_{th} = 5.5–8.6 kV/cm was determined by Hayes and Raymond[176] for a sample of $y \simeq 0.37$. Marsh et al.[64] have reported the plots of E_{th} versus y for $In_{1-x}Ga_xAs_yP_{1-y}$ over the entire range of composition. These data indicate a decreasing E_{th} value from 11 kV/cm at $y = 0$ to 2 kV/cm at $y = 0.85$ with increasing y followed by a slight rise to 2.7 kV/cm at the ternary limit. Measurements by Kowalsky and Schlachetzki[175] also yielded E_{th} values of about 11 kV/cm for InP, 3 kV/cm for $In_{0.53}Ga_{0.47}As$, and 6 kV/cm for $In_{1-x}Ga_xAs_yP_{1-y}$ ($y = 0.64$).

Starting from a monoenergetic approximation of electron dynamics, Kowalsky and Schlachetzki[183] estimated the composition dependence of E_{th} for $In_{1-x}Ga_xAs_yP_{1-y}$ lattice-matched to InP. This result is shown in Fig. 10.17 by the solid line. They also estimated the energy relaxation time τ_{er} at threshold and, consequently, the frequency limit f_c. Their calculated results for semiconductors of particular interest are as follows: τ_{er} = 1.7 ps and f_c = 290 GHz for InP, τ_{er} = 5.1 ps and f_c = 100 GHz for $In_{1-x}Ga_xAs_yP_{1-y}$ ($y = 0.64$), and τ_{er} = 6.3 ps and f_c = 80 GHz for $In_{0.53}Ga_{0.47}As$. The values for $In_{1-x}Ga_xAs_yP_{1-y}$ and $In_{0.53}Ga_{0.47}As$ are as expected,[184] whereas the frequency limit for InP seems to be markedly higher.

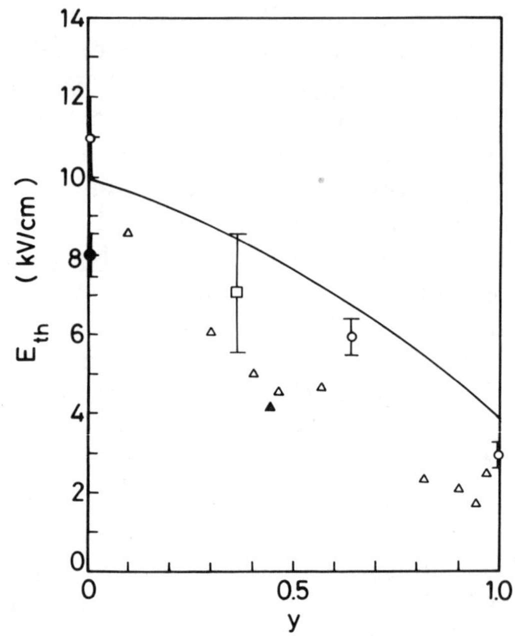

Figure 10.17 Composition dependence of the threshold electric field E_{th} for the $In_{1-x}Ga_xAs_yP_{1-y}$ lattice-matched to InP. The experimental data are taken from Refs. 64 (open triangles), 128 (solid triangle), 175 (open circles), 176 (open square), and 180 (solid circle). The solid line shows the calculated result by Kowalsky and Schlachetzki.[183]

The first experimental evidence of the transferred-electron effect in $In_{0.53}Ga_{0.47}As$ was given by Takeda et al.,[169] followed later by Marsh et al. (v–E measurements).[64] Zhao et al.[170] extended the measurements of Takeda et al., but neither group could observe coherent domains. The effect in the traveling domain mode was first demonstrated by Kowalsky and Schlachetzki.[171] From v–E characteristics they also determined the peak velocity to be $(2.2 \pm 0.3) \times 10^7$ cm/s. This value agrees with the v–E results obtained by March et al.[64] and Sasaki et al.[135] From time-resolved measurements of the device current, Kowalsky and Schlachetzki[174] also found that the current drop, usually attributed to domain nucleation, shows an unexpected behavior. These measurements resulted in a relative current drop of up to 90% at the beginning of the pulse. They discussed this finding in terms of its relation to transport simulations of Γ-valley scattering.

10.3 MINORITY-CARRIER TRANSPORT

10.3.1 Electron Mobility

Majority-carrier transport in III–V semiconductors has been the subject of intense theoretical and experimental work. Minority-carrier transport in such materials has, however, received much less attention. Short minority-carrier lifetimes and high Joule heating rates have usually hampered measurements of minority-carrier transport parameters. From a technologic point of view, however, a better understanding of minority-carrier transport is of paramount importance. Recent advances in analysis and design of heterojunction bipolar devices,[185] for example, suggest that an accurate knowledge of transit minority-carrier behavior in the base region may lead to improved design. When this information is not available, the assumption is frequently made that its parameter value (e.g., carrier mobility) is equal to the majority-carrier value. This assumption is likely to be in considerable error, particularly for III–V direct-band-gap semiconductors, because of the large differences between conduction-band and valence-band densities of state and effective masses.

Walukiewicz et al.[186] solved the Boltzmann equation without involving the relaxation-time approximation. Their results showed that the minority-electron mobility in *p*-GaAs is significantly affected by the presence of heavy holes that act as additional scattering centers. At high temperatures and at low carrier concentrations the minority-electron mobility in this material equals the electron mobility in *n*-type material. The room-temperature electron mobility at high concentration in *p*-GaAs can be significantly lower than the electron mobility in *n*-type material. The conclusion that the interaction with holes is essential for minority-electron transport in *p*-GaAs has also been obtained by Taniyama et al.[187] using a Monte Carlo method. Sadra et al.[188] have performed the Monte Carlo investigation of minority-electron mobility in *p*-GaAs. They considered the effects of multiband electron-hole scattering and hole wavefunction symmetry on transport properties and showed that a careful treatment of such scattering processes is indispensible for a quantitative understanding of minority-electron behavior. The minority-carrier (electrons and holes) mobilities in GaAs has also recently been calculated by Lowney and Bennett[189] using a quantum-mechanical phase-shift analysis for calculating minority-carrier–majority-carrier scattering.

Experimental determination of the minority-carrier mobility in III–V compounds is a difficult task since the direct Haynes–Shockley drift mobility method[190] is far from suitable (because of the short nonequilibrium carrier lifetime). The photoinduced Hall method is sufficiently sensitive only in the case of relatively pure crystals. Several experimental results on electron drift mobility in bulk, *p*-type GaAs have been reported.[191–196] Lagunova et al.[191] determined the mobilities of nonequilibrium electrons in *p*-GaAs at various hole concentrations based on a close link between the scattering processes and electron-spin relaxation. These results were in good agreement with theoretical estimates, which took into account the scattering of electrons by free holes. The time-of-flight technique was used to measure electron mobility in *p*-GaAs by Ahrenkiel et al.[192] The measured mobilities were about 2900 and 1300 $cm^2/(V \cdot s)$ at doping levels of 1×10^{17} and 2×10^{18} cm^{-3}, respectively. They concluded that the minority-carrier mobilities are lower than the expected majority-carrier mobilities at the same doping level. Nathan et al.[193] also determined the mobilities by measuring the common emitter cutoff frequency of heterojunction bipolar transistors. The room-temperature values they determined are considerably smaller than, and the 77-K value considerably larger than, the electron mobility in comparably doped *n*-type material. In GaAs quantum wells Höpfel et al.[197] observed the negative absolute mobility of minority electrons at low temperatures and under low electric fields. They concluded that such an unusual effect results from carrier drag on the electrons by the high-mobility hole plasma via electron–hole scattering.

To our knowledge, no theoretical or experimental study has been carried out on the determination of minority-carrier mobility in InP-related compounds. In the literature,[198–205] however, there is information about minority-carrier lifetimes, minority-carrier diffusion lengths, radiative and nonradiative lifetimes, and so on for $In_{1-x}Ga_xAs_yP_{1-y}/InP$ heterostructures.

10.3.2 Electron Drift Velocity

Minority-carrier drift velocities in GaAs-based semiconductors have been measured by several authors[206–209]. The samples used were epitaxial *p*-type $In_{0.53}Ga_{0.47}As$ layer,[206] graded-gap *p*-$Al_xGa_{1-x}As$ epitaxial layer,[207] *p*-type $GaAs/Al_xGa_{1-x}As$ multiple quantum well,[208] and epitaxial *n*-type $In_{0.53}Ga_{0.47}As$ layer.[209]

Levine et al.[207] measured drift velocity for electrons in a submicrometer (0.42 μm), highly doped ($p = 4 \times 10^{18}$ cm^{-3}) $Al_xGa_{1-x}As$ layer using an optical technique. A high field can be generated by compositional grading that produces a quasielectric field.[210] Their sample layer was graded from $x = 0.3$ to $x = 0$ over a distance of $L = 0.42$ μm. This band-gap difference of $\Delta E_g = 0.37$ eV corresponds to a quasifield of $E = 8.8$ kV/cm. The velocity was measured by creating carriers via absorption of pump photons at the high-gap side of the sample and timing their drift to the low-gap side. This was accomplished by measuring the time-resolved reflectivity change of the probe photon that was produced by electron accumulation at the low-gap window layer–collector interface. The electron velocity determined from this measurement is 1.8×10^7 cm/s. They concluded that this measured velocity is significantly larger than that for pure undoped GaAs where $v = 1.2 \times 10^7$ cm/s[113] for an ordinary electric field of $E = 8.8$ kV/cm.

Höpfel et al.[208] have also determined the v–E curve for minority electrons in *p*-type $GaAs/Al_xGa_{1-x}As$ multiple quantum well by a time-of-flight technique. The measured v–E curve showed that the drift velocity reaches a maximum of 1.2×10^7 cm/s at fields around 8 kV/cm. At higher fields the carrier velocity decreased and showed negative differential mobility up to the highest applied field of 14 kV/cm. This v–E curve is, thus, quite different from the ordinary one, which shows negative differential mobility at field of ~ 3 kV/cm. In order to interpret these results they considered real-space transfer and valley transfer of the hot electrons in $Al_xGa_{1-x}As$. It is known[211] that a real-space transfer of carriers is a characteristic transport mechanism of semiconductor heterostructures with a potential barrier. The mechanism of real-space transfer is based on hot-electron transfer between two adjacent layers (i.e., real-space transition). The negative differential mobility due to real-space transfer is analogous to the Gunn effect due to valley transfer of hot electrons in momentum space.

Particularly interesting is the apparent absence of negative differential mobility for minority electrons in p-$In_{0.53}Ga_{0.47}As$. Degani et al.[206] have measured the v–E characteristics for photoexcited minority electrons in bulk p-$In_{0.53}Ga_{0.47}As$. The measurement technique used consists of measuring the pulse width of current pulses induced in the photoconductor sample by laser-excited carriers. In Fig. 10.18 we plot their measured data (open cir-

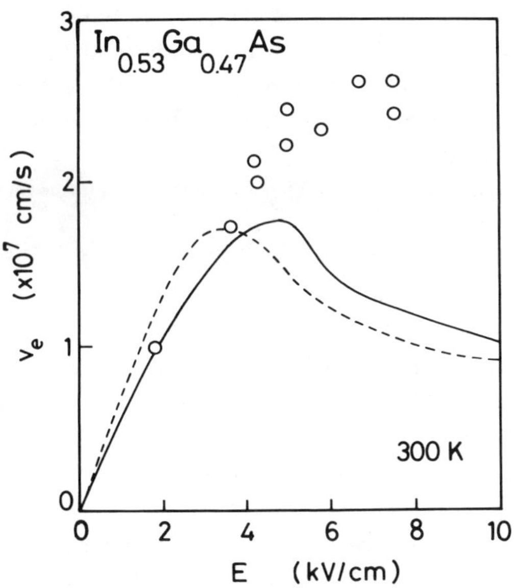

Figure 10.18 Minority-electron velocity–field characteristics for p-$In_{0.53}Ga_{0.47}As$. The open circles show the measured results by Degani et al.[206] The solid and dashed lines represent the calculated results by Osman and Grubin[213] with and without electron–hole scattering, respectively.

cles). At low fields the v–E variation corresponds to a drift mobility of 6000 $cm^2/(V \cdot s)$. At high fields the electron drift velocity continues to increase up to the maximum field they investigated (7.5 kV/cm), reaching a value of 2.6×10^7 cm/s. No evidence of the transferred-electron effect is observed in their measurements.

Otsuka[212] calculated the donor:acceptor ratio of electron scattering cross sections in p-$In_{0.53}Ga_{0.47}As$ for carrier densities between 10^{15} and 10^{20} cm^{-3}. The valley transfer of electrons requires a large momentum change that is usually accomplished by scattering. He obtained that the electron-acceptor scattering cross section σ_{eA} is by two orders of magnitude smaller than the electron-donor scattering cross section σ_{eD}. The result, thus, supports a possibility of the absence of the transferred-electron effect in this material. Monte Carlo calculations of steady-state minority-electron transport in p-$In_{0.53}Ga_{0.47}As$ have been reported by Osman and Grubin[213] and Joshi et al.[214] The solid and dashed lines in Fig. 10.18 represent the results of this calculation[213] with and without electron–hole scattering, respectively. It is shown that the principal contribution of electron–hole scattering is a more efficient channel of energy relaxation with the consequence of a significantly reduced transfer of carriers to the upper valleys for fields below 6 kV/cm. However, the calculated v–E curve (solid line) clearly exhibits a negative-differential mobility.

Hole drift velocities in n-type $In_{0.53}Ga_{0.47}As$ have also been determined experimentally by Hill et al.[209] Measurements over field strengths from 54 to 108 kV/cm showed the drift velocity to be relatively constant at $(4.8 \pm 0.2) \times 10^6$ cm/s, indicating that velocity saturation has occurred at field levels below 54 kV/cm.

REFERENCES

1. R. K. Willardson and A. C. Beer, *Semiconductors and Semimetals*, Vol. 10, Academic, New York, 1975.
2. D. A. Anderson and N. Apsley, *Semicond. Sci. Technol.* **1,** 187 (1986).
3. Y. Takeda, in *GaInAsP Alloy Semiconductors*, T. P. Pearsall, ed., Wiley, New York, 1982, p. 213.
4. J. R. Hayes, A. R. Adams, and P. D. Greene, in *GaInAsP Alloy Semiconductors*, T. P. Pearsall, ed., Wiley, New York, 1982, p. 189.
5. B. R. Nag, *Electron Transport in Compound Semiconductors*, Springer, Berlin, 1980.
6. D. Kranzer, *Phys. Status Solidi A* **26,** 11 (1974).
7. J. W. Harrison and J. R. Hauser, *J. Appl. Phys.* **47,** 292 (1976).

8. J. H. Marsh, *Appl. Phys. Lett.* **41,** 732 (1982).
9. J. W. Harrison and J. R. Hauser, *Phys. Rev. B* **13,** 5347 (1976).
10. J. R. Hauser, M. A. Littlejohn, and T. H. Glisson, *Appl. Phys. Lett.* **28,** 458 (1976).
11. M. A. Littlejohn, J. R. Hauser, T. H. Glisson, D. K. Ferry, and J. W. Harrison, *Solid-State Electron.* **21,** 107 (1978).
12. D. Chattopadhyay, S. K. Sutradhar, and B. R. Nag, *J. Phys. C* **14,** 891 (1981).
13. D. L. Rode and P. A. Fedders, *J. Appl. Phys.* **54,** 6425 (1983).
14. P. A. Fedders and C. W. Myles, *Phys. Rev. B* **29,** 802 (1984).
15. L. Makowski and M. Glicksman, *J. Phys. Chem. Solids* **34,** 487 (1973).
16. P. Blood and A. D. C. Grassie, *J. Appl. Phys.* **56,** 1866 (1984).
17. P. P. Basu, D. Chattopadhyay, and C. K. Sarker, *J. Phys. C* **19,** L173 (1986).
18. J. R. Melrose, *Semicond. Sci. Technol.* **2,** 371 (1987).
19. A. Galvanauskas, A. Gorelenok, Z. Dobrovolskis, S. Kershulis, Y. Pozhela, A. Reklaitis, and N. Shmidt, *Sov. Phys. Semicond.* **22,** 1055 (1988).
20. C. K. Sarkar and P. Banerji, *Phys. Status Solidi B* **156,** K145 (1989).
21. G. R. Cronin and S. R. Borrello, *J. Electrochem. Soc.* **114,** 1078 (1967).
22. G. E. Stillman, C. M. Wolfe, and J. O. Dimmock, *J. Phys. Chem. Solids* **31,** 1199 (1970).
23. P. Blood, *Phys. Rev. B* **6,** 2257 (1972).
24. M. G. Craford, W. O. Groves, A. H. Herzog, and D. E. Hill, *J. Appl. Phys.* **42,** 2751 (1971).
25. Y. Takeda and A. Sasaki, *Solid-State Electron.* **27,** 1127 (1984).
26. A. Taguchi and S. Yamada, *J. Appl. Phys.* **61,** 2412 (1987).
27. D. L. Rode, in *Semiconductors and Semimetals*, R. K. Willardson and A. C. Beer, eds., Vol. 10, Academic, New York, 1975, p. 1.
28. D. Lancefield, A. R. Adams, and M. A. Fisher, *J. Appl. Phys.* **62,** 2342 (1987).
29. D. A. Anderson and N. Apsley, *Semicond. Sci. Technol.* **1,** 187 (1986).
30. H. J. Lee, J. Basinski, L. Y. Juravel, and J. C. Woolley, *Can. J. Phys.* **58,** 923 (1980).
31. D. L. Rode, *Phys. Rev. B* **3,** 3287 (1971).
32. D. K. Hamilton, *Solid-State Electron.* **24,** 317 (1981).
33. B. R. Nag and G. M. Dutta, *J. Phys. C* **11,** 119 (1978).
34. V. V. Galavanov and N. V. Siukaev, *Phys. Status Solidi* **38,** 523 (1970).
35. J. R. Hayes, D. Patel, A. R. Adams, and P. D. Greene, *J. Electron. Mater.* **11,** 155 (1982).
36. J. S. Barrera and R. J. Archer, *IEEE Trans. Electron Dev.* **ED-22,** 1023 (1975).

37. R. F. Leheny, A. A. Ballman, J. C. DeWinter, R. E. Nahory, and M. A. Pollack, *J. Electron. Mater.* **9,** 561 (1980).
38. D. A. Anderson, N. Apsley, P. Davies, and P. L. Giles, *J. Appl. Phys.* **58,** 3059 (1985).
39. T. Inoue, K. Kainosho, R. Hirano, H. Shimakura, T. Kanazawa, and O. Oda, *J. Appl. Phys.* **67,** 7165 (1990).
40. O. G. Folberth, O. Madelung, and H. Weiss, *Z. Naturforsch.* **9A,** 954 (1954).
41. T. C. Harman, H. L. Goering, and A. C. Beer, *Phys. Rev.* **104,** 1562 (1956).
42. G. R. Cronin, R. W. Conrad, and S. R. Borrello, *J. Electrochem. Soc.* **113,** 1336 (1966).
43. J. P. McCarthy, *Solid State Commun.* **5,** 5 (1967); *Solid State Electron.* **10,** 649 (1967).
44. S. Kalem, J.-I. Chyi, H. Morcoç, R. Bean, and K. Zanio, *Appl. Phys. Lett.* **53,** 1647 (1988).
45. R. A. Stradling and R. A. Wood, *J. Phys. C* **3,** L94 (1970).
46. H. J. Lee, J. Basinski, L. Y. Juravel, and J. C. Woolley, *Can. J. Phys.* **57,** 233 (1979).
47. H. C. B. Hicks and D. F. Manley, *Solid State Commun.* **7,** 1463 (1969).
48. D. L. Rode and S. Knight, *Phys. Rev. B* **3,** 2534 (1971).
49. V. F. Dvoryankin, O. V. Emel'yanenko, and A. A. Telegin, *Sov. Phys. Semicond.* **5,** 1636 (1972).
50. C. R. Stanley, M. C. Holland, A. H. Kean, J. M. Chamberlain, R. T. Grimes, and M. B. Stanaway, *J. Cryst. Growth* **111,** 14 (1991).
51. J. S. Blakemore, *J. Appl. Phys.* **53,** R123 (1982).
52. R. C. Taylor, J. F. Woods, and M. R. Lorenz, *J. Appl. Phys.* **39,** 5404 (1968).
53. M. Toyama, M. Naito, and A. Kasami, *Jpn. J. Appl. Phys.* **8,** 358 (1969).
54. H. C. Casey, Jr., F. Ermanis, L. C. Luther, and L. R. Dawson, *J. Appl. Phys.* **42,** 2130 (1971).
55. S. M. Gasanli, O. V. Emel'yanenko, V. K. Ergakov, F. P. Kesamanly, T. S. Lagunova, and D. N. Nasledov, *Sov. Phys. Semicond.* **5,** 1641 (1972).
56. K. Fletcher and P. N. Butcher, *J. Phys. C* **6,** 976 (1973).
57. U. Zhumakulov, *Sov. Phys. Solid State* **8,** 2476 (1967).
58. M. Hashimoto and I. Akasaki, *Phys. Lett.* **25A,** 38 (1967).
59. S. A. Abagyan, G. A. Ivanov, R. S. Krupyshev, and Y. E. Shanurin, *Sov. Phys. Semicond.* **5,** 1444 (1972).
60. S. A. Abagyan, G. A. Ivanov, A. P. Izergin, and Y. E. Shanurin, *Sov. Phys. Semicond.* **6,** 985 (1972).
61. J. Leotin, J. C. Ousset, R. Barbaste, S. Askenazy, M. S. Skolnick, R. A. Stradling, and G. Poiblaud, *Solid State Commun.* **16,** 363 (1975).

62. S. Adachi, *J. Appl. Phys.* **53,** 8775 (1982).
63. J. D. Wiley and M. DiDomenico, Jr., *Phys. Rev. B* **1,** 1655 (1970).
64. J. H. Marsh, P. A. Houston, and P. N. Robson, in *Gallium Arsenide and Related Compounds*, Institute of Physics, Bristol, UK, 1981, p. 621.
65. J. D. Oliver, Jr., L. F. Eastman, P. D. Kirchner, and W. J. Schaff, *J. Cryst. Growth* **54,** 64 (1981).
66. Y. Takeda and T. P. Pearsall, *Electron. Lett.* **17,** 573 (1981).
67. P. K. Basu and B. R. Nag, *Appl. Phys. Lett.* **43,** 689 (1983).
68. C. D. Kourkoutas, J. Novac, M. Kuliffayova, G. J. Papaioannou, P. Kordos, and V. Ioannou-Sougleridis, *Solid State Commun.* **78,** 543 (1991).
69. D. Patel and I. L. Spain, *J. Appl. Phys.* **64,** 2790 (1988).
70. A. R. Adams and L. G. Shantharama, *Physica* **139 & 140B,** 419 (1986).
71. J. R. Hayes, H. L. Tatham, A. R. Adams, and P. D. Greene, *Electron. Lett.* **17,** 230 (1981).
72. W. T. Tsang, E. F. Schubert, T. H. Chiu, J. E. Cunningham, E. G. Burkhardt, J. A. Ditzenberger, and E. Agyekum, *Appl. Phys. Lett.* **51,** 761 (1987).
73. Y. Takeda, M. A. Littlejohn, and J. R. Hauser, *Appl. Phys. Lett.* **39,** 620 (1981).
74. S. K. Sutradhar, D. Chattopadhyay, and B. R. Nag, *Phys. Rev. B* **25,** 4069 (1982).
75. A. K. Saxena and M. A. L. Mudares, *J. Appl. Phys.* **58,** 2795 (1985).
76. J. D. Wiley, in *Semiconductors and Semimetals*, Vol. 10, R. K. Willardson and A. C. Beer, eds., Academic, New York, 1975, p. 91.
77. J. D. Wiley and M. DiDomenico, Jr., *Phys. Rev. B* **2,** 427 (1970).
78. M. Costato, G. Gagliani, C. Jacoboni, and L. Reggiani, *J. Phys. Chem. Solids* **35,** 1605 (1974).
79. D. N. Nasledov, Y. G. Popov, N. V. Siukaev, and S. P. Starosel'tseva, *Sov. Phys. Semicond.* **3,** 387 (1969).
80. K. H. Zschauer, in *Gallium Arsenide and Related Compounds*, Institute of Physics, Bristol, UK, 1973, p. 3.
81. A. L. Mears and R. A. Stradling, *J. Phys. C* **4,** L22 (1971).
82. H. C. Casey, Jr., F. Ermanis, and K. B. Wolfstirn, *J. Appl. Phys.* **40,** 2945 (1969).
83. J. R. Hayes, A. R. Adams, and P. D. Greene, *Electron. Lett.* **16,** 282 (1980).
84. T. P. Pearsall and J. P. Hirtz, *J. Cryst. Growth* **54,** 127 (1981).
85. L. G. Shantharama, A. R. Adams, E. M. Allen, and P. D. Greene, *Appl. Phys. Lett.* **47,** 843 (1985).
86. A. R. Adams, J. R. Hayes, and P. D. Greene, *Jpn. J. Appl. Phys.* **19** (suppl. **19-3**), 315 (1980).
87. J. B. Gunn, *Solid State Commun.* **1,** 88 (1963).

88. C. Hilsum and H. D. Rees, *Electron. Lett.* **6,** 277 (1970).
89. Y. Takeda, M. A. Littlejohn, J. A. Hutchby, and R. J. Trew, *Electron. Lett.* **17,** 686 (1981).
90. T. J. Maloney and J. Frey, *J. Appl. Phys.* **48,** 781 (1977).
91. J. Costa, A. Peczalski, and M. Shur, *J. Appl. Phys.* **66,** 674 (1989).
92. P. M. Boers, *Electron. Lett.* **7,** 625 (1971).
93. H. 'T Lam and G. A. Acket, *Electron. Lett.* **7,** 722 (1971).
94. L. D. Nielsen, *Solid State Commun.* **10,** 169 (1972).
95. L. D. Nielsen, *Phys. Lett.* **38A,** 221 (1972).
96. G. H. Glover, *Appl. Phys. Lett.* **20,** 224 (1972).
97. B. A. Prew, *Electron. Lett.* **8,** 592 (1972).
98. B. A. Prew, *Solid-State Electron.* **17,** 725 (1974).
99. A. Majerfeld, K. E. Potter, and P. N. Robson, *J. Appl. Phys.* **45,** 3681 (1974).
100. R. L. Tebbenham and D. Walsh, *Electron. Lett.* **11,** 96 (1975).
101. C. Hilsum, H. D. Rees, and W. Wilgoss, *Phys. Status Solidi B* **56,** K93 (1973).
102. E. M. Conwell and M. O. Vassell, *Appl. Phys. Lett.* **9,** 411 (1966).
103. E. M. Conwell and M. O. Vassell, *Phys. Rev.* **166,** 797 (1968).
104. J. G. Ruch and W. Fawcett, *J. Appl. Phys.* **41,** 3843 (1970).
105. W. Fawcett, A. D. Boardman, and S. Swain, *J. Phys. Chem. Solids* **31,** 1963 (1970).
106. M. A. Littlejohn, J. R. Hauser, and T. H. Glisson, *J. Appl. Phys.* **48,** 4587 (1977).
107. J. Pozela and A. Reklaitis, *Solid-State Electron.* **23,** 927 (1980).
108. H. Shichijo, K. Hess, and G. E. Stillman, *Appl. Phys. Lett.* **38,** 89 (1981).
109. H. Shichijo and K. Hess, *Phys. Rev. B* **23,** 4197 (1981).
110. E. Y. Wu and B. H. Yu, *Appl. Phys. Lett.* **58,** 1503 (1991).
111. J. G. Ruch and G. S. Kino, *Appl. Phys. Lett.* **10,** 40 (1967).
112. J. G. Ruch and G. S. Kino, *Phys. Rev.* **174,** 921 (1968).
113. N. Braslau and P. S. Hauge, *IEEE Trans. Electron Dev.* **ED-17,** 616 (1970).
114. M. Inoue, Y. Nakada, J. Shirafuji, and Y. Inuishi, *J. Phys. Soc. Jpn.* **32,** 1010 (1972).
115. K. Ashida, M. Inoue, J. Shirafuji, and Y. Inuishi, *J. Phys. Soc. Jpn.* **37,** 408 (1974).
116. M. H. Evanno and J. L. Vaterkowski, *Electron. Lett.* **18,** 417 (1982).
117. M. H. Evanno, B. Boittiaux, A. Safrioui, and R. Fauquembergue, in *Gallium Arsenide and Related Compounds*, Institute of Physics, Bristol, UK, 1985, p. 229.

118. M. A. Haase, V. M. Robbins, N. Tabatabaie, and G. E. Stillman, *J. Appl. Phys.* **57,** 2295 (1985).

119. J. Xu and M. Shur, *Phys. Rev. B* **36,** 1352 (1987).

120. W. T. Masselink, N. Braslau, D. LaTulipe, W. I. Wang, and S. L. Wright, *Solid-State Electron.* **31,** 337 (1988).

121. T. Furuta, N. Shigekawa, T. Mizutani, and A. Yoshii, *Appl. Phys. Lett.* **55,** 2310 (1989).

122. W. T. Masselink and T. F. Kuech, *J. Electron. Mater.* **18,** 579 (1989).

123. V. S. Komissarov and V. A. Aleksandrov, *Sov. Phys. Semicond.* **8,** 1423 (1975).

124. Z. Dobrovolskis, K. Grigoras, and A. Krotkus, *Appl. Phys. A* **48,** 245 (1989).

125. D. K. Ferry, *Phys. Rev. B* **12,** 2361 (1975).

126. M. A. Littlejohn, J. R. Hauser, and T. H. Glisson, *Appl. Phys. Lett.* **30,** 242 (1977).

127. M. D. Roy, B. R. Nag, and D. Chattopadhyay, *Appl. Phys. A* **32,** 39 (1983).

128. B. Houston, J. B. Restorff, R. S. Allgaier, J. R. Burke, D. K. Ferry, and G. A. Antypas, *Solid-State Electron.* **21,** 91 (1978).

129. A. Ghosal, D. Chattopadhyay, and N. N. Purkait, *Appl. Phys. Lett.* **44,** 773 (1984).

130. S. R. Ahmed, B. R. Nag, and M. D. Roy, *Solid-State Electron.* **28,** 1193 (1985).

131. B. R. Nag, S. R. Ahmed, and M. D. Roy, *IEEE Trans. Electron. Dev.* **ED-33,** 788 (1986).

132. S. Krishnamurthy, A. Sher, and A.-B. Chen, *J. Appl. Phys.* **61,** 1475 (1987).

133. K. S. Yoon, G. B. Stringfellow, and R. J. Huber, *J. Appl. Phys.* **62,** 1931 (1987).

134. J. Costa, A. Peczalski, and M. Shur, *J. Appl. Phys.* **65,** 5205 (1989).

135. A. Sasaki, Y. Takeda, N. Shikagawa, and T. Takagi, *Jpn. J. Appl. Phys.* **16** (suppl. **16-1**), 239 (1977).

136. N. Shigekawa, T. Furuta, and K. Arai, *Appl. Phys. Lett.* **57,** 67 (1990).

137. H. Kroemer, *Solid-State Electron.* **21,** 61 (1978).

138. K. Brennan, K. Hess, J. Y. Tang, and G. J. Iafrate, *IEEE Trans. Electron Dev.* **ED-30,** 1750 (1983).

139. B. R. Nag and S. R. Ahmed, *IEEE Trans. Electron Dev.* **ED-34,** 953 (1987).

140. A. Ghis, E. Constant, and B. Boittiaux, *J. Appl. Phys.* **54,** 214 (1983).

141. B. R. Nag and M. D. Roy, *Appl. Phys. A* **31,** 65 (1983).

142. J. Singh and K. K. Bajaj, *J. Appl. Phys.* **57,** 2048 (1985).

143. Y.-K. Feng, *Electron. Lett.* **21,** 453 (1985).

144. M. A. Osman and D. K. Ferry, *J. Appl. Phys.* **61,** 5330 (1987).

145. J. Lin and L. C. Chiu, *Appl. Phys. Lett.* **54,** 919 (1989).

146. K. S. Yoon, G. B. Stringfellow, and R. J. Huber, *J. Appl. Phys.* **63,** 1126 (1988).
147. P. N. Robson, K. E. Potter, and A. Majerfeld, *IEEE Trans. Electron Dev.* **ED-22,** 569 (1975).
148. D. C. Herbert, W. Fawcett, and C. Hilsum, *J. Phys. C* **9,** 3969 (1976).
149. T. H. Windhorn, L. W. Cook, M. A. Haase, and G. E. Stillman, *Appl. Phys. Lett.* **42,** 725 (1983).
150. B. Kramer and A. Mircea, *Appl. Phys. Lett.* **26,** 623 (1975).
151. H. Okamoto and M. Ikeda, *IEEE Trans. Electron Dev.* **ED-23,** 372 (1976).
152. P. A. Houston and A. G. R. Evans, *Solid-State Electron.* **20,** 197 (1977).
153. P. M. Smith, M. Inoue, and J. Frey, *Appl. Phys. Lett.* **37,** 797 (1980).
154. T. H. Windhorn, T. J. Roth, L. M. Zinkiewicz, O. L. Gaddy, and G. E. Stillman, *Appl. Phys. Lett.* **40,** 513 (1982).
155. R. Allam and J. Pribetich, *Electron. Lett.* **26,** 688 (1990).
156. T. H. Windhorn, L. W. Cook, and G. E. Stillman, *Appl. Phys. Lett.* **41,** 1065 (1982).
157. T. H. Windhorn, L. W. Cook, and G. E. Stillman, *IEEE Electron Dev. Lett.* **EDL-3,** 18 (1982).
158. T. H. Windhorn, L. W. Cook, and G. E. Stillman, *J. Electron. Mater.* **11,** 1065 (1982).
159. L. H. Holway, S. R. Steele, and M. G. Alderstein, in *Proc. 7th Biennial Cornell Electric. Engineer. Conf.*, Cornell University Press, Ithaca, NY, 1979, p. 199.
160. K. Brennan and K. Hess, *Phys. Rev. B* **29,** 5581 (1984).
161. K. Brennan and K. Hess, *J. Appl. Phys.* **59,** 964 (1986).
162. K. Brennan, *Appl. Phys. Lett.* **51,** 995 (1987).
163. This value is calculated from the following light- and heavy-hole masses: $m_{1h}/m_0 = 0.082$ and $m_{hh}/m_0 = 0.45$ [M. G. Wright, N. Ahmed, A. Koohian, K. Mitchell, G. R. Johnson, B. C. Cavenett, C. R. Pidgeon, C. R. Stanley, and A. H. Kean, *Semicond. Sci. Technol.* **5,** 438 (1990)].
164. K. Brennan, K. Hess, and G. J. Iafrate, *J. Appl. Phys.* **55,** 3632 (1984).
165. B. K. Ridley and T. B. Watkins, *Proc. Phys. Soc.* (London) **78,** 293 (1961).
166. C. Hilsum, *Proc. IRE* **50,** 185 (1962).
167. A. El-Sabbahy, A. R. Adams, and M. L. Young, *Solid-State Electron.* **21,** 83 (1978).
168. P. Banerjee, P. K. Bhattacharya, M. J. Ludowise, and W. T. Dietze, *IEEE Electron Dev. Lett.* **EDL-4,** 283 (1983).
169. Y. Takeda, N. Shikagawa, and A. Sasaki, *Solid-State Electron.* **23,** 1003 (1980).
170. Y. Y. Zhao, C. J. Wei, and H. Beneking, *Electron. Lett.* **18,** 835 (1982).
171. W. Kowalsky and A. Schlachetzki, *Electron. Lett.* **19,** 189 (1983).

172. W. Kowalsky, A. Schlachetzki, and H.-H. Wehmann, *Solid-State Electron.* **27,** 187 (1984).
173. W. Kowalsky and A. Schlachetzki, *Electron. Lett.* **20,** 502 (1984).
174. W. Kowalsky and A. Schlachetzki, *Electron. Lett.* **21,** 1192 (1985).
175. W. Kowalsky and A. Schlachetzki, *Solid-State Electron.* **28,** 299 (1985).
176. R. E. Hayes and R. M. Raymond, *Appl. Phys. Lett.* **31,** 300 (1977).
177. K. C. Heasman, J. R. Hayes, and A. R. Adams, *Electron. Lett.* **17,** 756 (1981).
178. G. D. Pitt and M. K. R. Vyas, *J. Phys. C* **8,** 138 (1975).
179. W. Czubatyj, M. S. Shur, and M. P. Shaw, *Solid-State Electron.* **21,** 75 (1978).
180. T. Kobayashi, K. Takahara, T. Kimura, K. Yamamoto, and K. Abe, *Solid-State Electron.* **21,** 79 (1978).
181. B. Fank, *Microwave J.* **27,** 95 (1984).
182. S. C. Binari, P. E. Thompson, and H. L. Grubin, *IEEE Electron Dev. Lett.* **EDL-6,** 22 (1985).
183. W. Kowalsky and A. Schlachetzki, *Solid-State Electron.* **30,** 161 (1987).
184. M. R. Friscourt, P. A. Rolland, and R. Fauquembergue, *IEEE Electron Dev. Lett.* **EDL-5,** 434 (1984).
185. C. M. Maziar, M. E. Klausmeier-Brown, S. Bandyopadhyay, M. S. Lundstrom, and S. Datta, *IEEE Trans. Electron Dev.* **ED-33,** 881 (1986); R. Katoh and M. Kurata, *IEEE Trans. Electron Dev.* **ED-37**, 1245 (1990).
186. W. Walukiewicz, J. Lagowski, L. Jastrzebski, and H. C. Gatos, *J. Appl. Phys.* **50,** 5040 (1979).
187. H. Taniyama, M. Tomizawa, T. Furuta, and A. Yoshii, *J. Appl. Phys.* **68,** 621 (1990).
188. K. Sadra, C. M. Maziar, B. G. Streetman, and D. S. Tang, *J. Appl. Phys.* **66,** 4791 (1989).
189. J. R. Lowney and H. S. Bennett, *J. Appl. Phys.* **69,** 7102 (1991).
190. J. R. Haynes and W. Shockley, *Phys. Rev.* **81,** 835 (1951).
191. T. S. Lagunova, V. A. Marushchak, M. N. Stepanova, and A. N. Titkov, *Sov. Phys. Semicond.* **19,** 71 (1985).
192. R. K. Ahrenkiel, D. J. Dunlavy, D. Greenberg, J. Schlupmann, H. C. Hamaker, and H. F. FacMillan, *Appl. Phys. Lett.* **51,** 776 (1987).
193. M. I. Nathan, W. P. Dumke, K. Wrenner, S. Tiwari, S. L. Wright, and K. A. Jenkins, *Appl. Phys. Lett.* **52,** 654 (1988).
194. H. Ito and T. Ishibashi, *J. Appl. Phys.* **65,** 5197 (1989).
195. S. Tiwari and S. L. Wright, *Appl. Phys. Lett.* **56,** 563 (1990).
196. T. Furuta and M. Tomizawa, *Appl. Phys. Lett.* **56,** 824 (1990).
197. R. A. Höpfel, J. Shah, P. A. Wolff, and A. C. Gossard, *Phys. Rev. Lett.* **56,** 2736 (1986).

198. S. Sakai, M. Umeno, and Y. Amemiya, *Jpn. J. Appl. Phys.* **19,** 109 (1980).
199. Y. Takeda, M. Kuzuhara, and A. Sasaki, *Jpn. J. Appl. Phys.* **19,** 899 (1980).
200. G. H. Olsen and T. J. Zamerowski, *IEEE J. Quantum Electron.* **QE-17,** 128 (1981).
201. D. Z. Garbuzov, A. T. Gorelenok, V. N. Mdivani, M. K. Trukan, V. P. Chalyi, and V. V. Agaev, *Sov. Phys. Semicond.* **15,** 218 (1981).
202. M. M. Tashima, L. W. Cook, and G. E. Stillman, *J. Electron. Mater.* **11,** 831 (1982).
203. S. Hirose, J. Pietzsch, N. Kamata, and T. Kamiya, in *Gallium Arsenide and Related Compounds*, Institute of Physics, Bristol, UK, 1983, p. 573.
204. C. H. Henry, B. F. Levine, R. A. Logan, and C. G. Bethea, *IEEE J. Quantum Electron.* **QE-19,** 905 (1983).
205. C. H. Henry, R. A. Logan, F. R. Merritt, and C. G. Bethea, *Electron. Lett.* **20,** 358 (1984).
206. J. Degani, R. F. Leheny, R. E. Nahory, and J. P. Heritage, *Appl. Phys. Lett.* **39,** 569 (1981).
207. B. F. Levine, C. G. Bethea, W. T. Tsang, F. Capasso, K. K. Thornber, R. C. Fulton, and D. A. Kleinman, *Appl. Phys. Lett.* **42,** 769 (1983).
208. R. A. Höpfel, J. Shah, A. C. Gossard, and W. Wiegmann, *Physica* **134B,** 509 (1985).
209. P. Hill, J. Schlafer, W. Powazinik, M. Urban, E. Eichen, and R. Olshansky, *Appl. Phys. Lett.* **50,** 1260 (1987).
210. H. Kroemer, *RCA Rev.* **18,** 332 (1957).
211. See, for instance, *Solid-State Electron.* **31** (special issue: ''Hot Carriers in Semiconductors''), Chapter 1 (''Real Space Transfer/Heterojunctions''), pp. 319–394.
212. E. Otsuka, *Appl. Phys. Lett.* **42,** 372 (1983).
213. M. A. Osman and H. L. Grubin, *Appl. Phys. Lett.* **51,** 1812 (1987).
214. R. P. Joshi, A. M. Kriman, and D. K. Ferry, *J. Appl. Phys.* **68,** 4322 (1990).

11

STRAIN PROBLEMS IN InGaAs(P)-BASED HETEROSTRUCTURES

11.1 STRAIN EFFECTS ON CRYSTALLOGRAPHIC PROPERTIES

11.1.1 Elastic Strain in Heterostructures

Semiconductor lasers and integrated-optics devices usually have multilayer structures. The distribution of strains in the multilayer structures is a subject of perennial, great interest since internal strains arise normally in thin epitaxial films during the preparation of the films by heteroepitaxial growth. Internal strains or stresses are then a consequence of the mismatch of lattice constants and/or the difference in the thermal expansion coefficients of the individual layers. Here, the strain and stress are related through the fourth-rank elastic compliance tensor $[S]$ [see Eq. (3.26)]. Many authors have studied this problem and have obtained approximate formulas for the curvature radius and layer stress.[1–8]

Let us now consider[8] the case of bonding a thin plate of cubic, zinc-blende crystal of size $N_e a_e \times N_e a_e \times t_e$ onto a substrate of a different size $N_s a_s \times N_s a_s \times t_s$, where N_i is the number of lattice atoms along the edge of the crystal plate, a_i is the lattice constant, and t_i is the thickness of the plate; see Fig. 11.1*a*. To achieve a coherent interface, N_e is set to equal N_s. We assume that a_e is less than a_s and thus $l_e = N_s a_e$ is less than $l_s = N_s a_s$. The bonding operation is carried out in the following way: equal and opposite forces F are applied to the plates to stretch plate "e" and compress plate "s" uniformly in the lateral direction to the same final dimension

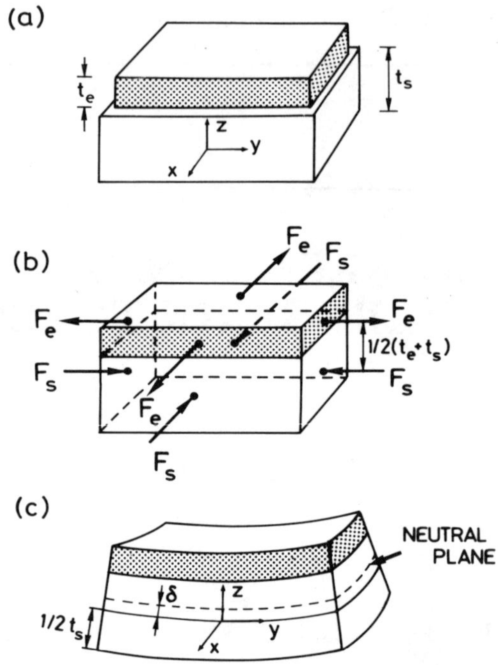

Figure 11.1 Formation of two-layer composite: (*a*) two single-crystal plates with the lattice constants a_e and a_s ($a_e < a_s$) and thicknesses t_e and t_s, respectively; (*b*) layer "e" is stretched and layer "s" is compressed to match the macroscopic dimension, and two layers are then bonded together to form a composite; (*c*) the composite bends toward the side with shorter lattice constant after removal of the external stresses F_e and F_s.

$l_f \times l_f$. The two plates are then bonded together with a perfect alignment of the atomic planes; see Fig. 11.1*b*. At the moment when the two plates are bonded together, the composite experiences an applied bending moment given by $F(t_e + t_s)/2$, which is counterbalanced by the moment resulting from the internal elastic stress. Finally, the applied forces are relieved, and the moments from the elastic stresses bend the composite in the direction shown in Fig. 11.1*c*. The bending of the composite relaxes some of the stress, and the radius of curvature is determined by the final state of the internal stress.

From this bilayer model Noyan and Segmüller[8] have obtained the coherency condition in terms of the curvature radius R:

$$\frac{a_e - a_s}{a_s} = \frac{2}{R}\left[\frac{t_s}{4}\left(1 + \frac{a_e t_e}{a_s t_s}\right) + \frac{1}{E_e t_s t_e(1 + t_e/t_s)}\frac{a_e}{a_s} \times \left(1 + \frac{a_s E_e t_e}{a_e E_s t_s}\right)\left(\frac{E_s t_s^3}{12} + \frac{E_e t_s^2 t_e}{4}\right)\right] \qquad (11.1)$$

where $E_i = [Y/(1 - P)]_i$ (Y = Young's modulus, P = poisson's ratio). We can obtain numeric values of E_i for different crystallographic orientations as given in Section 3.2.2 (also see Table 3.1). The term δ, shown in Fig. 11.1*c*, is the shift of the neutral axis of the composite from $t_s/2$. This quantity is given by

$$\delta = \frac{t_e E_e}{2E_s} \frac{1 + t_e/t_s}{1 + E_e t_e/(E_s t_s)} \tag{11.2}$$

Solving the elastic force and momentum balance equations, Noyan and Segmüller[8] gave the elastic stress component X_{xx} parallel to the film

$$X_{xx}^{e} = -\frac{E_e}{1 + (a_s/a_e)(E_e/E_s)(t_e/t_s)} \frac{a_s}{a_e} \times \left[\frac{a_e - a_s}{a_s} - \left(1 + \frac{a_e t_e}{a_s t_s}\right) \frac{t_s}{2R}\right] \tag{11.3}$$

A relationship between the film and substrate stress can be given by

$$X_{xx}^{s} = -X_{xx}^{e} \frac{t_e}{t_s} \tag{11.4}$$

The stresses' dependence on z are now given by

$$X_{xx}^{e}(z) = a_{xx}^{e} + E_e \left(\frac{z - t_e/2}{R}\right) \qquad \text{for } 0 \le z \le t_e \tag{11.5a}$$

$$X_{xx}^{s}(z) = -X_{xx}^{e} \frac{t_e}{t_s} + E_s \left(\frac{z + t_s/2}{R}\right) \qquad \text{for } 0 \ge z \ge -t_s \tag{11.5b}$$

In the case where $a_e < a_s$, these expressions give $X_{xx}^{s} < 0$, $X_{xx}^{e} > 0$, and $R < 0$.

Chu et al.[8] studied the misfit stress in $In_{1-x}Ga_xAs/InP$ heteroepitaxial structures grown by hydride-transport vapor-phase epitaxy using an X-ray technique that makes it possible to measure lattice mismatches and wafer curvatures. The composition of $In_{1-x}Ga_xAs$ was targeted for a 1.3-μm band gap ($x = 0.47$) lattice-matched to InP. Since the purpose of their work was to emphasize the relationship between lattice mismatch and misfit stress, samples with compositions ranging from $x = 0.42$ to 0.49 were used. The corresponding R values they measured ranged from 21 to -12 m, in good agreement with their theoretical prediction.

Swaminathan et al.[9] have also presented the results of measurements of the curvature radius of 1.3- and 1.5-μm wavelength $In_{1-x}Ga_xAs_yP_{1-y}/InP$

channeled substrate buried-heterostructure lasers. The objective of this investigation was to quantify the macroscopic stress present in the device and correlate it with device reliability. They suggested that it is not the macroscopic device stress that is measured by the curvature radius but localized stresses that may exist in the vicinity of the lasing active layer which would affect device reliability.

As mentioned in Section 3.3.4, the long-wavelength phonon frequency is a strong function of the elastic strain. The phonon deformation potential K values give the strain variation of the lattice spring constants. It is, therefore, possible to determine the strains in materials by measuring phonon spectral shifts. Venkateswaran et al.[10] have presented the data of strain mapping in lattice-mismatched [111]- and [100]-oriented $In_{1-x}Ga_xAs/GaAs$ superlattice wafers ($x = 0.83, 0.9$). The measurements were carried out with Raman scattering and the results checked with independent X-ray rocking curve data that yielded the average strain in the same samples. They found that the samples exhibited small but systematic variations over macroscopic dimensions, but appear locally uniform on a 30-μm scale. They also found that the LO-phonon Raman scattering is an effective technique for determining the initial strain, but TO scattering should not be used because strain alloying compensation does not apply for TO as it does for LO.

The effects of internal strain on the morphology and structural properties of $In_{1-x}Ga_xAs/InP$ grown by molecular-beam epitaxy (MBE) has also been recently discussed in Ref. 11. Internal strain energy and its effects on III–V solid–liquid equilibria (or local equilibrium) has also been considered by a number of authors.[12–24]

Since X_{zz} is equal to zero in the bilayer model discussed above, the lattices are free to expand or contract in the z direction, that is, perpendicular to the interface. As shown in Fig. 11.2, this causes the cubic structure to be tetragonally distorted. Since both X^{i}_{xx} and X^{i}_{yy} have opposite signs for $\mathrm{i} = \mathrm{s}$ compared to $\mathrm{i} = \mathrm{e}$, the two cubic lattices are tetragonally distorted in the opposite sense, one with the lattice constant perpendicular to the interface lengthened and the other shortened. Therefore, the measured vertical mismatch is enhanced by the tetragonal distortion.

Lattice deformation between the layer and substrate can be measured by conventional X-ray-diffraction measurements. Such measurements have been performed on $In_{1-x}Ga_xAs_yP_{1-y}/InP$[25–41] and $In_{1-x}Ga_xAs/InP$ heterostructures.[42–51] It has been demonstrated by Hsieh et al.[25] that $In_{1-x}Ga_xAs_yP_{1-y}$ quaternary layers equivalent to wavelengths of $\lambda_g = 1.1$–1.22 μm can be grown by liquid-phase epitaxy (LPE) with a lattice mismatch $\Delta a/a$ of up to $+0.4\%$, albeit with a certain number of cleavage steps. Feng et al.[28] have reported that mismatch effects are far more sen-

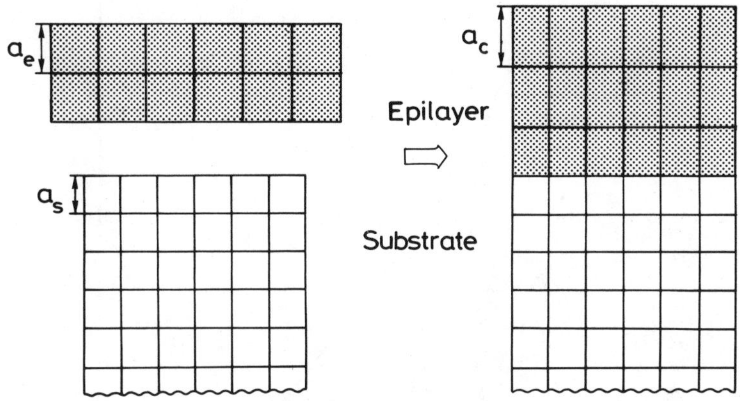

Figure 11.2 Cross section of an "epilayer–substrate" system, where a_e and a_s are, respectively, the unstrained lattice parameters of the epilayer and substrate and a_c is the strained lattice parameter of the epilayer. When $a_e > a_s$ ($a_e < a_s$), the epilayer is expected to be tetragonally distorted while the lattice parameter perpendicular to the interface is lengthened (shortened).

sitive to lattice mismatch for the 1.31-μm material than for the 1.15-μm one. Surface morphology is poorer for a longer wavelength for a given mismatch, resulting in melt-removal problems unless the mismatch is less than 0.09%. At the high-wavelength end of the $In_{1-x}Ga_xAs_yP_{1-y}$ system on InP, $In_{0.53}Ga_{0.47}As$ showed severe cracking at −0.22% mismatch and prominent cross-hatch patterns occur at +0.28%.[42]

Oe et al.[27,29] have studied lattice deformation as a function of mole fraction for LPE-grown $In_{1-x}Ga_xAs_yP_{1-y}$ layers on (100) InP with the Ga atomic fraction in liquid solution between 0.4 and 1.0%. As seen in Fig. 11.3, their measured lattice deformation $\Delta a/a$ perpendicular to the layer surface for layers with a Ga fraction lower than ~0.0065 has a positive value (i.e., compressive strain in the growth direction) while it has negative values (i.e., tensile strain) for layers with the fraction larger than 0.0065. Matsui et al.[30] have also found that for a LPE-grown $In_{1-x}Ga_xAs_yP_{1-y}$ layer ($y = 0.26$) on (100) InP the lattice deformation in the growth direction varies along the epitaxial-layer thickness while no lattice mismatch or, in other words, a coherent interface in the lateral direction is observed.

11.1.2 Misfit Dislocations and Defect Etchants

In practical cases where the thickness of the epilayer (epitaxial layer) is very small compared to that of the substrate, the lattice mismatch can be taken up totally by the tensile or compressive strain inside the epilayer, and one then has a coherent interface.[8] A discontinuity in the lattice constant

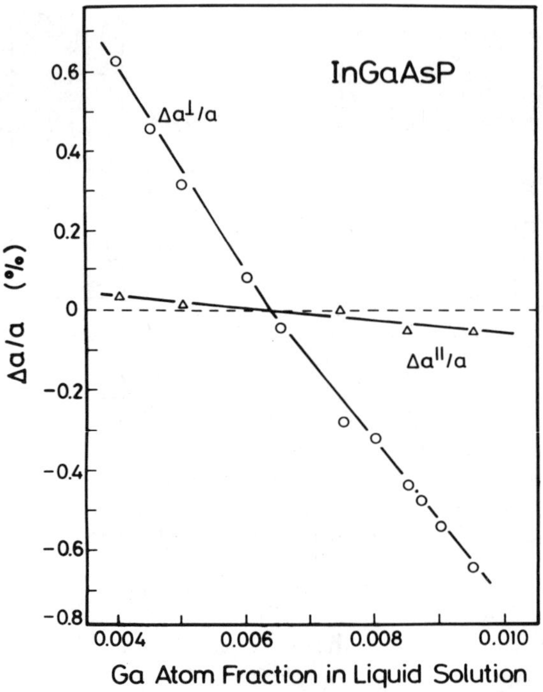

Figure 11.3 Lattice mismatch as a function of Ga atom fraction in LPE-growth liquid solution. The open circles represent the mismatches for the lattice constant perpendicular to the wafer surface and the open triangles normal to the wafer surface. (From Oe et al.[29])

parallel to the interface can exist only in the presence of misfit dislocations. Nagai[52] has studied the effect of lattice misfit on the lattice deformation for the $In_{1-x}Ga_xAs$ ternary ($0.5 < x < 1.0$) grown on GaAs by hydride-transport vapor-phase epitaxy. The thickness of the substrates was in the range 2–5 mm, and the epitaxial layer was in the range 5–10 μm. His results are shown in Fig. 11.4. Here a_s is the unstrained lattice parameter of the substrate (GaAs) and a_c is the strained lattice parameter of the epitaxial layer due to tetragonal distortion. The lattice constant of InAs (a_0 = 6.0584 Å) is larger than that of GaAs (a_0 = 5.6533 Å). The $In_{1-x}Ga_xAs$ lattice is, thus, expected to be tetragonally distorted with the lattice constant perpendicular to the interface being lengthened. His measured lattice constants perpendicular to the film are, in fact, larger than that along the surface. In the region $x > 0.8$ the degree of deformation increased with decreasing x, but in the region $x < 0.8$ it decreased with decreasing x. With the aid of a defect revealing etching technique, he concluded that in the region $x > 0.8$ the lattice mismatch is relieved only by tetragonal distortion and for $x < 0.8$ it is relieved by both misfit dislocation and tetragonal distortion.

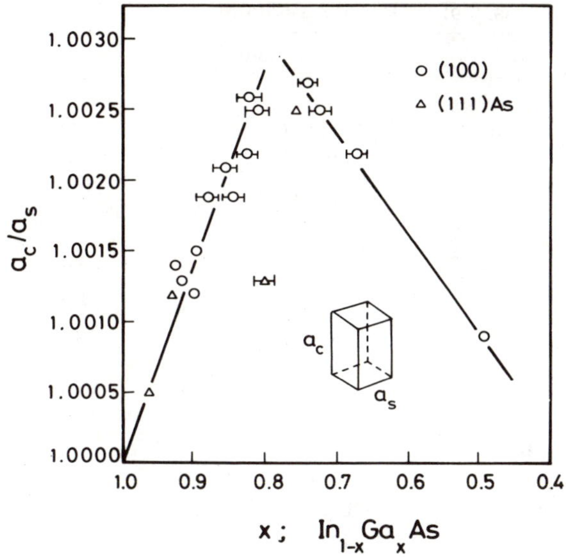

Figure 11.4 Ratio of the lattice constants a_c/a_s for hydride-transport vapor-phase-grown $In_{1-x}Ga_xAs$ epilayers on (100) and (111) AsGaAs. (From Nagai.[52])

Wagner et al.[48] showed that for a compressive strain of $\Delta a/a > -0.1\%$ (at 624°C) in the $In_{1-x}Ga_xAs$ ternary grown on (100) InP the misfit dislocations are formed as a cross-grid, where these dislocations are positioned within the InP substrate resulting from slip on inclined $\{111\}$ planes. They also studied the thickness dependence of misfit dislocations[49] and found that the formation of misfit dislocations takes place initially by half-loop nucleation at the growing surface of the ternary layer and reaches a critical height. After half-loop nucleation these loops expand, form angular half loops, and slip toward the interfacial plane, thus forming extended misfit dislocation segments. A $1H_3PO_4 : 1H_2O_2$ solution was used as a defect revealing etchant for the $In_{1-x}Ga_xAs/InP$ wafers. Using this solution it was possible to determine the exact position of the misfit dislocations within the layer system. We can also see in Ref. 53 details of misfit dislocations in the LPE-grown $In_{1-x}Ga_xAs_yP_{1-y}/InP$ system.

It is well known that wet chemical etching is by far the most convenient way to reveal structural defects in semiconductors. There are many studies on etching characteristics of such defects in InP and the $In_{1-x}Ga_xAs_yP_{1-y}$ alloy. Before closing this subsection, we will summarize various defect revealing etchants for these materials.

Table 11.1 gives a summary of dislocation etchants for InP. Since chemical polishing seldom gives a mirror-like finish on the (111) In surface, most defect etching studies were performed on the (111) P surface but not

TABLE 11.1 Summary of Dislocation Etchants for InP

Etchant	Plane	Condition	Ref.
$0.4N Fe^{3+}:HCl$	{111}	25°C, 1.5 min	*e*
$20HNO_3:10HCl:0.25Br$	{111}, (100)	<5 s	*f*
$1HCl:1HNO_3$	{111}, (100)	26°C	*g*
HCl		26°C	
$0.4N Fe^{3+}:HCl$		26°C	
1% Br in CH_3OH		26°C	
$1HCl:2H_2SO_4:3HNO_3$		26°C	
$2H_3PO_4:1HBr$	(111)P, (100), (110)	24°C, 2 min	*h*
HBr:HF	(111)P, (100)		*i*
$HBr:CH_3COOH$			
$6HCl:6H_2O:1HNO_3$	{111}, (100), (110)	60°C, 1.5 min	*j*
Modified AB[a]		75°C, 30 min	
$40HCl:80HNO_3:1Br$		25°C, 10 s	
$2H_3PO_4:1HBr$		25°C, 2 min	
$H_2SO_4:H_2O_2:H_2O$	(111)P, (100)	RT (room temp.)	*k*
$2H_3PO_4:1HBr$	(111)P, (100), (110)	RT, 1–2 min	*l*
AB etchant[b]		60°C, 30–90 min	
AB etchant[b]	(111)P, (100)	60°C, 15 min	*m*
1HBr:5HF		RT, 2 min	
$2H_3PO_4:1HBr$	(100)	Ambient, ≤2 min	*n*
$6HNO_3:6H_2O:1HCl$		60°C, 1.5 min	
$40HCl:80HNO_3:1Br$		25°C, 10 s	
Modified AB[a]		75°C, 30 min	
EE1[c]	(100)		*o*
EE2[d]	(111)P, (100)		*p*
$1HNO_3:3HBr$	(111)P, (100)	RT, 10–30 s	*q*
$HBr:CH_3COOH:K_2Cr_2O_7$	(100)	RT	*r*
$HBr:H_3PO_4:K_2Cr_2O_7$		RT	
$1H_2SO_4:1H_2O_2:1H_2O$	(100)	80–90°C	*s*
$1HNO_3:3HBr$	(100)	10 s	*t*
$20HBr:2H_2O_2:20HCl:20H_2O$	(111)P, (100)	RT, 2 min	*u*
$HBr:H_2O_2:HCl:H_2O$	(100)	RT, 2 min	*v*

[a]Modified AB (10 ml H_2O:40 mg $AgNO_3$:5 g CrO_3:8 ml HF).
[b]AB etchant (2 ml H_2O:8 mg $AgNO_3$:1 g CrO_3:1 ml HF).
[c]EE1 (electrolytic wet etching with 14*M* orthophosphoric acid).
[d]EE2 (electrolytic wet etching with 1 mol/liter NaOH).
[e]H. C. Gatos and M. C. Lavine, *J. Electrochem. Soc.* **107,** 427 (1960).
[f]R. C. Clarke et al., *J. Mater. Sci.* **8,** 1349 (1973).
[g]B. Tuck and A. J. Baker, *J. Mater. Sci.* **8,** 1559 (1973).
[h]A. Huber and N. T. Linh, *J. Cryst. Growth* **29,** 80 (1975).
[i]K. Akita et al., *J. Cryst. Growth* **46,** 783 (1979).
[j]F. A. Thiel and R. L. Barns, *J. Electrochem. Soc.* **126,** 1272 (1979).
[k]Y. Nishitani and T. Kotani, *J. Electrochem. Soc.* **126,** 2269 (1979).
[l]G. T. Brown et al., *J. Mater. Sci.* **15,** 2539 (1980).
[m]T. Kotani et al., *J. Electrochem. Soc.* **127,** 2273 (1980).
[n]S. Mahajan and A. K. Chin, *J. Cryst. Growth* **54,** 138 (1981).
[o]C. R. Elliott and J. C. Regnault, *J. Electrochem. Soc.* **128,** 113 (1981).

on (111) In. Thiel et al.[54] used Nalcoag 1060 as polishing etchant for a {111} InP wafer. This chemical resulted in an excellent surface finish on both the (111) In and the (111) P faces. By etching with a modified AB (Abrahams and Buiocchi[55]) solution, they then obtained triangular pits on the (111) In surface. Tuck and Baker[56] also obtained hexagonal etch pits on the (111) In surface with a 1% solution of Br in methanol. On the (111) P surface, the etch pits have the morphology of a pyramidal (triangular)[56–60] or a conical (circular) shape.[54,58,61,62] On the (100) surface, the etch pits have a rectangular,[56,57,59,60,63–65] oval,[60–62] or square (or circular) shape.[56,57,61–64,66,67] The rectangular pits demonstrate the nonequivalence of ⟨110⟩- and $\langle \bar{1}10 \rangle$-type directions lying in the (100) plane in III–V intermetallic compounds. This nonequivalence is known to arise from the ⟨111⟩ crystallographic polarity of a zinc-blende-type crystal. The rectangular etch pits, therefore, provide us a means for uniquely determining crystallographic directions between the ⟨110⟩ and $\langle \bar{1}10 \rangle$ directions.[60,65] Etched (110) InP surfaces revealed ridge features but some have hillocks and boat-shaped pits associated with them.[61,62]

The first degradations of semiconductor heterojunction lasers and light-emitting diodes may be caused by defects in the active layers. It is, therefore, important to investigate defects in the heterojunction layers. The difficulty in finding a suitable chemical etchant for $In_{1-x}Ga_xAs_yP_{1-y}$ layers lies in their thickness, which rarely exceeds 2 μm. A defect etchant for such layers will have to be extremely preferential and have a slow etch rate in order to reveal the defects with a minimum of material removed. Table 11.2 shows the results of defect etchant studies on the $In_{1-x}Ga_xAs_yP_{1-y}$/InP system.

11.1.3 Critical Layer Thickness

Strained-layer superlattices and quantum wells are high-quality multilayered structures grown from lattice mismatched materials.[68] The large lattice mismatch is totally accommodated by uniform elastic strains in the layers if the layer thicknesses are kept below certain critical values. As mentioned in Ref. 69, for thin films the idea emerged of a critical layer thickness

←

[p]A. Yamamoto et al., *J. Electrochem. Soc.* **128,** 1095 (1981).
[q]S. N. G. Chu et al., *J. Electrochem. Soc.* **129,** 352 (1982).
[r]S. Adachi, *J. Electrochem. Soc.* **129,** 609 (1982).
[s]E. A. Caridi and T. Y. Chang, *J. Electrochem. Soc.* **131,** 1440 (1984).
[t]J. A. Lourenco, *J. Electrochem. Soc.* **131,** 1914 (1984).
[u]D. T. C. Huo et al., *J. Electrochem. Soc.* **136,** 1804 (1989).
[v]D. T. C. Huo et al., *J. Electrochem. Soc.* **136,** 3094 (1989).

TABLE 11.2 Summary of Dislocation Etchants for $In_{1-x}Ga_xAs_yP_{1-y}$ Alloy

Etchant	Plane	Comments	Ref.
Modified AB[a] $6HCl:6H_2O:1HNO_3$	{111}	Gradient-freeze-grown bulk crystal, etch features similar to those on InP, nonuniform pit density due to Ga segregation	*d*
AB etchant[b]	(111)P	InGaAsP/InP laser wafer, reveal distinct defect structures (threading dislocations, stacking faults, dislocation rosettes and crowds)	*e*
KK etchant[c]	(100)	LPE-grown layer, etch rate of 1.5 μm/h under illumination, protrusions correlating with dislocation pits in underlying and overlying InP layers	*f*
$3Cr_3O:1HF:2H_2O$	(100)	Reveal dislocations in InGaAs, etch rate of 1.25 μm/min under He–Ne laser illumated at 0°C	*g*
$1H_3PO_4:1H_2O_2$	(100)	InGaAs/InP heterostructures, photoetching, reveal inclined and misfit dislocations	*h*

[a]Modified AB (10 ml H_2O:40 mg $AgNO_3$:5 g CrO_3:8 ml HF).
[b]AB etchant (2 ml H_2O:8 mg $AgNO_3$:1 g CrO_3:1 ml HF).
[c]KK etchant (8 g KOH:0.5 g $K_3Fe(CN)_6$:100 ml H_2O).
[d]F. A. Thiel and R. L. Barns, *J. Electrochem. Soc.* **126,** 1272 (1979).
[e]T. Kotani et al., *J. Electrochem. Soc.* **127,** 2273 (1980).
[f]J. A. Lourenco, *J. Electrochem. Soc.* **131,** 1914 (1984).
[g]C. P. Kuo et al., *J. Appl. Phys.* **57,** 5428 (1985).
[h]G. Wagner et al., *Phys. Status Solidi A* **112,** 519 (1989).

below which a dislocation-free, coherently strained interface would be stable and above which a misfit dislocation structure and semicoherent interface would be stable. The knowledge of the so-called *critical layer thickness* is therefore essential for realizing the advantages of such material systems.

Many studies have been carried out both theoretically[70–73] and experimentally[74–87] on the strained-layer $In_{1-x}Ga_xAs/GaAs$ system. This particular material system is of great current interest because of potential applications for high-speed and optoelectronic devices. In Table 11.3 we summarize the experimentally determined critical layer thickness h_c for $In_{1-x}Ga_xAs$ layers grown on (100) GaAs. Orders and Usher[78] used MBE-grown $In_{1-x}Ga_xAs/GaAs$ single heterostructures to measure critical thickness values as a function of x for $0.75 \le x \le 0.93$ by the double-crystal X-ray-diffraction technique. The experimental results were compared with critical thicknesses calculated from the energy-balance model of People and Bean[88] and the mechanical equilibrium model of Matthew and Blakeslee.[89] They found that the agreement between the energy-balance theory and ex-

TABLE 11.3 Critical Layer Thickness h_c for $In_{1-x}Ga_xAs$ Layers in $In_{1-x}Ga_xAs/GaAs(Al_{1-x}Ga_xAs)$ Strained Heterostructures (SH = Single Heterostructure; SQE = Single Quantum-Well Heterostructure; SLS = Strained-Layer Superlattices)

x	h_c (Å)	Sample
0.75	26 ± 4	SH[a]
0.86	105 ± 15	SH[a]
0.93	200 ± 30	SH[a]
0.72	100	SQW[b]
0.72	100	SQW, SLS[c]
0.8	200	SQW[d]
0.8	190	SQW[e]
0.75	120	SQW[f]
0.72	80–100	SLS[g]
0.8	250	SLS[e]

[a] P. J. Orders and B. F. Usher, *Appl. Phys. Lett.* **50,** 980 (1987).
[b] N. G. Anderson et al., *J. Appl. Phys.* **60,** 2361 (1986).
[c] T. G. Andersson et al., *Appl. Phys. Lett.* **51,** 752 (1987).
[d] I. J. Fritz et al., *Appl. Phys. Lett.* **51,** 1004 (1987).
[e] P. L. Gurley et al., *Appl. Phys. Lett.* **52,** 377 (1988).
[f] K. J. Beernink et al., *Appl. Phys. Lett.* **55,** 2167 (1989).
[g] N. G. Anderson et al., *J. Electron. Mater.* **14,** 187 (1985).

periment was very good, particularly for the experimental points at $x = 0.75$ ($h_c \sim 26$ Å) and 0.86 ($h_c \sim 105$ Å). At $x = 0.93$ the experimental value (~200 Å) was approximately a factor of 2 lower than the theoretical h_c.

Temkin et al.[90] have performed a similar study on strained $In_{1-x}Ga_xAs_yP_{1-y}/InP$ quantum wells. Well compositions covered the entire range of strain available, from −3.8% (GaAs) to +3.2% (InAs). They found that the critical layer thickness in this strained system is unambiguously described by the Matthew–Blakeslee model.[89] There have also been reports on strain problems, such as misfit dislocation nucleation and strain-relaxation mechanisms, in III–V strained-layer material systems such as $In_{1-x}Ga_xAs/GaAs$,[91-96] $In_{1-x}Ga_xAs/In_{1-x}Ga_xAs$,[97] and $In_{1-x}Ga_xAs/Al_xGa_{1-x}As$.[98]

11.2 STRAIN EFFECTS ON PHYSICAL PROPERTIES

11.2.1 Theoretical Consideration

Strains have an important influence on the physical properties of semiconductor films because they change the electronic states, which result in an electronic energy-band structure via the strain-orbital deformation potential interaction. The presence of internal strain not only shifts the band-edge energies but also modifies the effective masses associated with individual bands. The internal strain may, thus, modify transport properties of semiconductor films because carrier mobility is strongly connected with the effective mass.

As mentioned in Section 6.1, the lowest direct band gap in zinc-blende-type semiconductors occurs in the center of the Brillouin zone (BZ), where it has fourfold E_0 (counting the two spin states) and twofold $(E_0 + \Delta_0)$ gaps. The biaxial strain in the heterostructure layer can split the fourfold valence-band degeneracy into twofold light- and heavy-hole bands due to lattice deformation interactions. In unstrained zinc-blende-type materials the presence of a term linear in $\boldsymbol{k}$ also shifts band extrema slightly off $\boldsymbol{k} = 0$, but the term is small in magnitude[99] and its effects are seldom observed experimentally.

In Fig. 11.5 we show schematically the biaxial strain effects on the electronic energy-band structure near the Γ point for zinc-blende-type semiconductor heterojunctions. The directions are defined as follows: k_x lies in the growth plane (i.e., plane of biaxial strain), and k_z lies along the growth direction (i.e., perpendicular to the plane of biaxial strain). The biaxial strain in a zinc-blende-type material lowers its crystal symmetry from T_d to D_{2d}, which results in a change in the band gap [an increase (decrease) in the band gap, when the material is in compression (tension)] and a split-

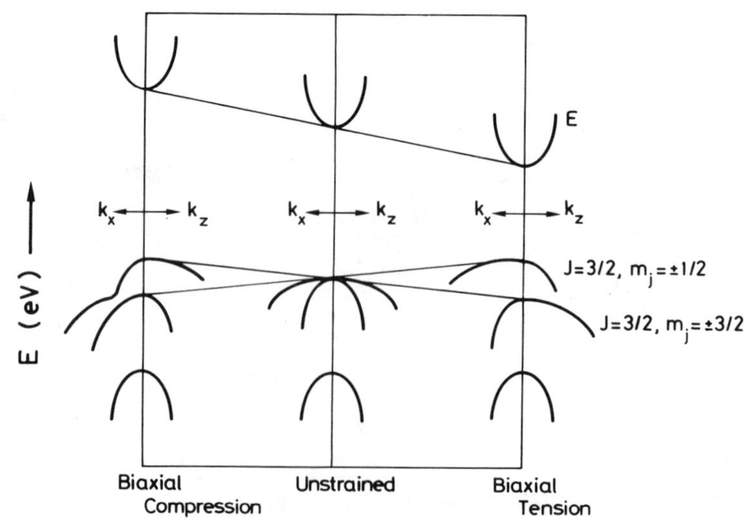

Figure 11.5 An (x, y) in-plane biaxial strain effect on the electronic energy-band structure near the Γ point in zinc-blende-type heterojunction semiconductors.

ting of the light- and heavy-hole valence bands. [The higher lying band is the heavy-hole (light-hole) band for a biaxial compression (tension).]

In the spherical model, the Luttinger–Kohn Hamiltonian calculations for strained bulk materials show that at small k_x, the band mass that is heavy along the strain direction is comparatively light perpendicular to that direction and vice versa.[100–103] For biaxial compression, the highest valence band ($|J = 3/2, m_j = \pm 3/2\rangle_z$) has heavy mass along k_z and comparatively light mass along k_x. For biaxial tension, the highest valence band ($|J = 3/2, m_j = \pm 1/2\rangle_z$) has heavy mass along k_x and comparatively light mass along k_z. The heavy-hole band then acquires a "light" in-plane mass; and the light-hole band, a "heavy" in-plane mass.

To first order in strain, the splitting energy $\Delta E_{\rm LH}$ of the $J = 3/2$ valence bands is proportional to the product of the strain and the shear deformation potential. Let us consider the effect of biaxial strain for (001) growth and define $z \parallel [001]$ (z = the rectangular coordinate axis). In this case, the splitting energy $\Delta E_{\rm LH}$ can be written as

$$\Delta E_{\rm LH} = 2b \left(\frac{C_{11} + 2C_{12}}{C_{11}} \right) e_{xx} \tag{11.6}$$

where $e_{xx} (= e_{yy})$ is the in-plane strain and b is the shear deformation potential (see Section 7.1). The in-plane strain is now given by

$$e_{xx} = e_{yy} = \frac{a_{\rm e} - a_{\rm s}}{a_{\rm s}} \tag{11.7}$$

In Eq. (11.7) $a_{\rm s}$ and $a_{\rm e}$ are the unstrained lattice parameters of substrate and epilayer, respectively. The strain along the growth direction is also given from elasticity theory by

$$e_{zz} = -2 \left(\frac{C_{12}}{C_{11}} \right) e_{xx} \tag{11.8}$$

Equations (11.7) and (11.8) show that the biaxial in-plane tensile (compressive) strain corresponds to the compressive (tensile) strain along the (001) growth direction.

11.2.2 InGaAs(P)/InP System

The effects of built-in strain on the band-gap energy of a single-heterostructure $In_{1-x}Ga_xAs$ alloy for commensurate growth on (100) InP have been studied theoretically by People.[104] The ternary experiences a biaxial in-plane tension (compression) for $x > 0.47$ ($x < 0.47$), and hence the E_0 gaps are smaller (larger) than the unstrained bulk $In_{1-x}Ga_xAs$ gaps for $x > 0.47$ ($x < 0.47$). Similar studies on $In_{1-x}Ga_xAs/InP$ strained-layer quantum wells have been carried out theoretically[105] and experimentally.[106–108] The heavy- and light-hole electronic states in quantum wells split even though well and barrier layers are lattice-matched semiconductors (e.g., $In_{0.53}Ga_{0.47}As/InP$). This effect is due to the difference in the effective quantization masses of these states. Plots of the zone-center energies for the highest-valence subbands ($n = 1$) of (20/20)-diatomic-layer superlattices versus x show that the heavy- and light-hole subbands cross at $x = 0.64$ (i.e., under large tensile strain).[105]

The effects of misfit strain on the band-gap energy in $In_{1-x}Ga_xAs_yP_{1-y}/InP$ single heterostructures have been studied experimentally by several authors.[45,50,51] Kuo et al.[45] showed that interfacial elastic strain induced by the lattice parameter mismatch between epilayer and substrate results in significant band-gap shifts for III–V compounds. The epilayers used in their study were $In_{1-x}Ga_xAs$ ($0.47 < x < 0.52$) on (100) InP and $In_{1-x}Ga_xP$ ($0.51 < x < 0.55$) on (100) GaAs prepared by metal–organic chemical vapor deposition (MOCVD). The band-gap shifts were determined from photoluminescence peak energies of the epilayers. They found that for the $In_{1-x}Ga_xAs/InP$ system the optical transition energy between the conduction and light-hole bands $[E_c^\Gamma - E_v(3/2, \pm 1/2)]$ increases with increasing biaxial tension (i.e., $x < 0.47$), in good agreement with theoretical prediction ($\Delta E = 11.38e_{xx}$ eV). Bassignana et al.[51] have also used a photoluminescence technique to study the effect of mismatch strain on the band gap in MOCVD-grown $In_{1-x}Ga_xAs$ epilayer on (100) InP. The mismatch strains studied were in the range $-0.0015 \lesssim e_{xx} \lesssim +0.001$. The strain dependences of the transition energies in the vicinity of the lattice-matched point were $5e_{xx}$ eV for $e_{xx} < 0$ $[E_c^\Gamma - E_v(3/2, \pm 3/2)]$ and $11.4e_{xx}$ eV for $e_{xx} > 0$ $[E_c^\Gamma - E_v(3/2, \pm 1/2)]$. Optical absorption measurements also clearly revealed the splitting of the light- and heavy-hole band degeneracy caused by the biaxial strain in $In_{1-x}Ga_xAs/InP$ films ($x \simeq 0.47$).[50]

Two-dimensional-electron (2DE) transport properties of MBE-grown strained $In_{1-y}Al_yAs/In_{1-x}Ga_xAs/InP$ heterostructures have been studied in Refs. 109 and 110. Despite the highly strained $In_{1-y}Al_yAs$, the electron mobility obtained by Zhang et al. ($x = 0.453$; $y = 0.421$)[109] reached 9.0×10^3, 6.5×10^4, and 8.9×10^4 $cm^2/(V \cdot s)$ at 300, 77, and 4 K, respectively, thus indicating only a minor influence of the strain of the $In_{1-y}Al_yAs$ on the 2DE mobility. Tacano et al.[110] have studied composition dependence of the 2DE mobility in $InAlAs/In_{1-x}Ga_xAs/InP$ heterostructures and obtained the highest electron mobility at $x = 0.8$ (1.5×10^4 and 1.5×10^5 $cm^2/(V \cdot s)$ at 293 and 10 K, respectively). The critical layer thickness of $In_{1-x}Ga_xAs$, which maintained a very high electron mobility of over 1×10^4 $cm^2/(V \cdot s)$, is also found to show good agreement with the energy-balance model[88] rather than the mechanical equilibrium model.[89]

Internal strain effects on Raman spectra have been studied by Emura et al.[47] and Haines et al.[111] on $In_{1-x}Ga_xAs$ epilayers grown on (100) InP. The biaxial strains investigated by Emura et al. were in the range $-0.0035 < e_{xx} < +0.004$. They found an increase (or decrease, which depends on the sign of the lattice mismatch) in the energy shift of GaAs-like LO phonons. This energy shift was also used to evaluate built-in stresses in materials in conjunction with X-ray rocking curve data. The composition range studied by Haines et al. was $0.45 < x < 0.52$. In order to interpret the Raman data successfully it was necessary to consider the effect of the strain induced by lattice mismatch.

Some authors have also studied the influence of lattice mismatch on luminescence efficiency[31,40] and the aging behavior[38] of $In_{1-x}Ga_xAs_yP_{1-y}$/InP light-emitting devices. They showed that the structures with the minimum mismatch have the highest external quantum efficiency and a long lifetime.

11.2.3 InGaAs/GaAs Lattice-Mismatched System

Prominent energy-band structures of strained $In_{1-x}Ga_xAs$/GaAs heterostructures have been studied in detail by optical techniques, such as photoluminescence,[75,77,79,82,85,112–120] absorption,[95,114,119,121–125] and reflectance measurements.[95,126,127]

The effects of built-in coherence strain on the effective mass of strained $In_{1-x}Ga_xAs$/GaAs heterostructures have been studied both theoretically[100,101,128] and experimentally.[129–133] Quantum wells of $In_{1-x}Ga_xAs$ bound by (100) GaAs layers are under biaxial compression. The highest valence band is the heavy-hole band in physical character.[134] [Note that this is the reverse situation from the usually studied case of uniaxial compression (see Refs. 135–137).] The heavy-hole electronic state (subband) is determined by the heavy-hole mass in the (001) direction (i.e., the quantitization mass m_{hh}), and its in-plane effective mass is much higher than m_{hh}.[100] Conversely, the valence band derived from the light-hole band has a heavier in-plane mass than the quantization mass m_{lh}. Recent magnetooptical measurements on $In_{0.2}Ga_{0.8}As$/GaAs samples[130] gave a value of $0.069m_0$ for the conduction-band in-plane effective mass and $0.16m_0$ for the valence-band in-plane effective mass. An electron effective mass of $0.069m_0$ is about 20% heavier than that for unstrained bulk material[138] but is consistent with the optical effective mass of $0.074m_0$ obtained in GaAs/AlGaAs superlattices by Worlock et al.[139] when it is scaled to $In_{0.2}Ga_{0.8}As$. The in-plane hole effective mass of $0.16m_0$ is considerably smaller than the unstrained GaAs-bulk masses $m_{hh}/m_0 = 0.45$,[140] 0.475,[141] and 0.73.[142]

Reed and Lee,[143] on the other hand, observed resonant tunneling in a double $Al_{0.25}Ga_{0.75}As$ barrier/single $In_{0.2}Ga_{0.8}As$ quantum-well strained-layer heterostructure and obtained an effective mass for electrons in the $In_{0.2}Ga_{0.8}As$ quantum well of approximately half the bulk effective mass. This result was considered to imply that accumulation and depletion regions in the resonant tunneling structure are significant. They were also able to fit the observed peak position for the AlGaAs/InGaAs heterojunction with a conduction band offset of nearly 100%.

Studies on electrical transport properties of the $In_{1-x}Ga_xAs$/GaAs strained-layer system have been performed by a number of authors.[76,80,129,144,145] These results are found to be comparable to reported results on similar $Al_{1-x}Ga_xAs$/GaAs heterostructure or to those of high-quality bulk alloy material.[144,145] Thobel et al.[146] presented Monte Carlo results for $In_{1-x}Ga_xAs$/GaAs and $In_{1-x}Ga_xAs$/InP strained materials. They found that the strain-induced electron velocity reduction is much more pronounced for $In_{1-x}Ga_xAs$ grown on GaAs substrate than for $In_{1-x}Ga_xAs$ grown on InP substrate.

One can expect that the low-field hole mobility in an $In_{1-x}Ga_xAs$/GaAs strained heterostructure is enhanced because it is determined by a lighter in-plane mass rather than the original heavy-hole mass m_{hh}. Transport properties of holes in $In_{1-x}Ga_xAs$/GaAs quantum-well structures have, thus, been studied by several authors.[131,144,147–153] Fritz et al.[148] achieved a 4-K Hall mobility of $\sim 3 \times 10^4$ cm^2/(V · s) in modulation-doped, *p*-type quantum wells. The value compares quite favorably with the values of up to 3.5 $\times 10^4$ cm^2/(V · s) obtained in *n*-type superlattices grown with similar techniques in the same growth chamber.[145] This value also compares with that obtained in high-purity bulk *p*-type GaAs (see Section 10.1.2). Monte Carlo calculation also suggested that for a 1.5% biaxial compressive strain the hole mobilities are increased by a factor of ≤ 2 over the unstrained values.[150] However, recent results by Reddy et al.[153] indicated that the high-field hole velocity in strained $In_{1-x}Ga_xAs$/GaAs quantum wells is seen to saturate at very low levels, particularly in highly doped samples. They concluded that this low saturation velocity will limit the performance of *p*-channel field-effect transistors (FETs) and that the high performances required for complementary logic are unlikely to be achieved in this system. Hjalmarson[151] also discussed high- and low-field transport properties of $In_{1-x}Ga_xAs$/GaAs quantum-well heterostructures. They ascertained that the small mass of light holes in strained quantum-well structures leads to high mobility at low fields, but at high fields the holes become hot and populate the heavy-hole band, thereby greatly reducing their mobilities.

The conduction- and valence-band offsets depend not only on the semiconductors involved but also on the amount of mismatch strain at the interface.[154–156] Several electrical and optical techniques have been used to determine the band offsets in the $In_{1-x}Ga_xAs$-based strained heterostructure system (also see Section 6.3.3). Arent et al.[155] concluded that the difference between band-offset ratios for the intrinsic and strained $In_{1-x}Ga_xAs/GaAs$ heterostructure is significant. Reithmaier et al.[156] determined the band-offset ratio $\Delta E_c : \Delta E_v = 0.6:0.4$ for all samples independent of composition ($In_{1-x}Ga_xAs/GaAs$; $0.75 < x < 0.82$).

The concept of strain-induced polarization fields due to piezoelectric effects in semiconductors is particularly important to device engineers. The polarization fields in piezoelectric semiconductors give rise to a charge density that causes shifts in the electrical characteristics of FETs. It has been shown[157] that for GaAs FETs shifts in threshold voltage V_{th} are sufficiently large that they are of importance in the fabrication of digital ICs (integrated circuits) requiring a high degree of V_{th} control. The internal strain in strained-layer heterostructures with a (111) growth axis is shown to be responsible for the generation of polarization fields due to the piezoelectric effects.[158,159] These fields do not occur in heterostructures with a (100) growth axis because the orientation of the lattice-mismatch-induced strains is such that they do not produce polarization fields in zinc-blende materials. Caridi et al.[160] have fabricated an $In_{1-x}Ga_xAs/GaAs$ quantum-well *p–i–n* diode on a (111) As GaAs substrate. Photoconductivity spectra for this diode under reverse-bias operation showed a quadratic blue shift of the quantum-well electroabsorption peaks, in contrast to the red shifts normally observed in the quantum-confined Stark shift. They concluded that the measured blue shifts demonstrate an electric field strength of 1.7×10^5 V/cm generated by the piezoelectric effects.

As mentioned in Section 8.3, the refractive index can be formulated by electronic energy-band parameters of the medium. These band parameters are strongly influenced by strain. Das and Bhattacharya[161] have calculated the refractive-index variation with mismatch strain and with alloying in the $In_{1-x}Ga_xAs/GaAs$ heterostructures. They found that for $0.75 < x < 1.0$ the effect of alloying is more pronounced; for example, at $x = 0.8$ the effect of alloying produces a change in the refractive index that is ~ 10 times that produced by mismatch strain.

Built-in strain effects on Raman spectra have been studied by Burns et al.[162] and Wu et al.[163] on $In_{1-x}Ga_xAs/GaAs$ strained layers. These results also demonstrated that Raman spectroscopy is a suitable method for characterizing built-in strain in strained-layer epitaxial systems. The Raman microprobe spectroscopy is also shown to be a powerful tool for probing patterned semiconductor structures on a microscopic scale using this material system [i.e., narrow $In_{1-x}Ga_xAs$ stripes on patterned GaAs (100) substrates.[164]

REFERENCES

1. F. K. Reinhart and R. A. Logan, *J. Appl. Phys.* **44,** 3171 (1973).
2. R. H. Saul, *J. Appl. Phys.* **40,** 3273 (1969).
3. R. Dingle and W. Wiegmann, *J. Appl. Phys.* **46,** 4312 (1975).
4. K. Röll, *J. Appl. Phys.* **47,** 3224 (1976).
5. G. H. Olsen and M. Ettenberg, *J. Appl. Phys.* **48,** 2543 (1977).
6. J. Vilms and D. Kerps, *J. Appl. Phys.* **53,** 1536 (1982).
7. Z.-C. Feng and H.-D. Liu, *J. Appl. Phys.* **54,** 83 (1983).
8. S. N. G. Chu, A. T. Macrander, K. E. Strege, and W. D. Johnston, Jr., *J. Appl. Phys.* **57,** 249 (1985); see also comment by I. C. Noyan and A. Segmüller, ibid. **60,** 2980 (1986).
9. V. Swaminathan, L. A. Koszi, and M. W. Focht, *J. Appl. Phys.* **66,** 1849 (1989).
10. U. D. Venkateswaran, L. J. Cui, B. A. Weinstein, K. Elcess, C. G. Fonstad, and C. Mailhiot, *Appl. Phys. Lett.* **56,** 286 (1990).
11. A. Salokatve and M. Hovinen, *J. Appl. Phys.* **67,** 3378 (1990).
12. G. B. Stringfellow, *J. Appl. Phys.* **43,** 3455 (1972).
13. J. C. Brice, *J. Cryst. Growth* **28,** 249 (1975).
14. R. E. Nahory, M. A. Pollack, E. D. Beebe, J. C. DeWinter, and M. Ilegems, *J. Electrochem. Soc.* **125,** 1053 (1978).
15. K. Nakajima, T. Tanahashi, K. Akita, and T. Yamaoka, *J. Appl. Phys.* **50,** 4975 (1979).
16. K. Takahei and H. Nagai, *Jpn. J. Appl. Phys.* **20,** L313 (1981).
17. Y. B. Bolkhovityanov, *J. Cryst. Growth* **57,** 84 (1982).
18. H. Sonomura, G. Sunatori, and T. Miyauchi, *J. Appl. Phys.* **53,** 5336 (1982).
19. P. K. Bhattacharya and S. Srinivasa, *J. Appl. Phys.* **54,** 5090 (1983).
20. H. Sonomura, *J. Appl. Phys.* **59,** 739 (1986).
21. F. C. Larché and J. W. Cahn, *J. Appl. Phys.* **62,** 1232 (1987).
22. H. Sonomura, H. Uda, A. Sugimura, A. Ashida, H. Horinaka, and T. Miyauchi, *J. Appl. Phys.* **62,** 4142 (1987).
23. F. Glas, *J. Appl. Phys.* **62,** 3201 (1987).
24. A. A. Mbaye, F. Turco, and J. Massies, *Phys. Rev. B* **37,** 10419 (1988).

25. J. J. Hsieh, M. C. Finn, and J. A. Rossi, in *Gallium Arsenide and Related Compounds*, Institute of Physics, Bristol, UK, 1977, p. 37.
26. M. Feng, T. H. Windhorn, M. M. Tashima, and G. E. Stillman, *Appl. Phys. Lett.* **32,** 758 (1978).
27. K. Oe and K. Sugiyama, *Appl. Phys. Lett.* **33,** 449 (1978).
28. M. Feng, M. M. Tashima, T. H. Windhorn, and G. E. Stillman, *Appl. Phys. Lett.* **33,** 533 (1978).
29. K. Oe, Y. Shinoda, and K. Sugiyama, *Appl. Phys. Lett.* **33,** 962 (1978).
30. J. Matsui, K. Onabe, T. Kamejima, and I. Hayashi, *J. Electrochem. Soc.* **126,** 664 (1979).
31. N. A. Bert, D. Z. Garbuzov, A. T. Gorelenok, S. G. Konnikov, V. N. Mdivani, and V. K. Tibilov, *Sov. Phys. Semicond.* **14,** 400 (1980).
32. M. M. Tashima, L. W. Cook, and G. E. Stillman, *J. Cryst. Growth* **54,** 132 (1981).
33. K. Ishida, Y. Matsumoto, and K. Taguchi, *Phys. Status Solidi A* **70,** 277 (1982).
34. K. Nakajima, S. Yamazaki, S. Komiya, and K. Akita, *J. Appl. Phys.* **52,** 4575 (1981).
35. S. Yamazaki, Y. Kishi, K. Nakajima, A. Yamaguchi, and K. Akita, *J. Appl. Phys.* **53,** 4761 (1982).
36. S. Komiya, K. Nakajima, I. Umebu, and K. Akita, *Jpn. J. Appl. Phys.* **21,** 1313 (1982).
37. S. Komiya, S. Yamazaki, Y. Kishi, I. Umebu, and T. Kotani, *J. Cryst. Growth* **61,** 362 (1983).
38. A. K. Chin, C. L. Zipfel, B. H. Chin, and M. A. DiGiuseppe, *Appl. Phys. Lett.* **42,** 1031 (1983).
39. S. Yamazaki, K. Nakajima, S. Komiya, Y. Kishi, and K. Akita, *Appl. Phys. Lett.* **43,** 82 (1983).
40. V. A. Strakhov, N. G. Yaremenko, A. A. Telegin, V. A. Ogandzhanyan, M. V. Karachevtseva, L. F. Mikhaleva, V. I. Petrov, and V. A. Prokhorov, *Sov. Phys. Semicond.* **19,** 373 (1985).
41. X. R. Wang, X. Y. Chi, H. Zheng, Z. L. Miao, J. Wang, Z. S. Zhang, and Y. S. Jin, *J. Vacuum Sci. Technol. B* **6,** 34 (1988).
42. S. B. Hyder, G. A. Antypas, J. S. Escher, and P. E. Gregory, *Appl. Phys. Lett.* **31,** 551 (1977).
43. Y. Kawamura and H. Okamoto, *J. Appl. Phys.* **50,** 4457 (1979).
44. K.-H. Goetz, D. Bimberg, H. Jürgensen, J. Selders, A. V. Solomonov, G. F. Glinskii, and M. Razeghi, *J. Appl. Phys.* **54,** 4543 (1983).
45. C. P. Kuo, S. K. Vong, R. M. Cohen, and G. B. Stringfellow, *J. Appl. Phys.* **57,** 5428 (1985).
46. P. Franzosi, G. Salviati, F. Genova, A. Stano, and F. Taiariol, *J. Cryst. Growth* **75,** 521 (1986).

47. S. Emura, S. Gonda, Y. Matsui, and H. Hayashi, *Phys. Rev. B* **38,** 3280 (1988).
48. G. Wagner, V. Gottschalch, H. Rhan, and P. Paufler, *Phys. Status Solidi A* **112,** 519 (1989).
49. G. Wagner, V. Gottschalch, H. Rhan, and P. Paufler, *Phys. Status Solidi A* **113,** 71 (1989).
50. M. A. Abdullaev, A. T. Gorelenok, S. I. Kohhanovskii, Y. M. Makushenko, D. V. Pulyaevskii, R. P. Seisyan, and K. É. Shtengel' *Sov. Phys. Semicond.* **23,** 125 (1989).
51. I. C. Bassignana, C. J. Miner, and N. Puetz, *J. Appl. Phys.* **65,** 4299 (1989).
52. H. Nagai, *J. Appl. Phys.* **45,** 3789 (1974).
53. K. Nakajima, in *Semiconductors and Semimetals*, Vol. 2, R. K. Willardson and A. C. Beer, eds. (W. T. Tsang, Vol. ed.), Academic, New York, 1985, Part A, p. 1.
54. F. A. Thiel and R. L. Barns, *J. Electrochem. Soc.* **126,** 1272 (1979).
55. M. S. Abrahams and C. J. Buiocchi, *J. Appl. Phys.* **36,** 2855 (1965).
56. B. Tuck and A. J. Baker, *J. Mater. Sci.* **8,** 1559 (1973).
57. K. Akita, T. Kusunoki, S. Komiya, and T. Kotani, *J. Cryst. Growth* **46,** 783 (1979).
58. T. Kotani, S. Komiya, S. Nakai, and Y. Yamaoka, *J. Electrochem. Soc.* **127,** 2273 (1980).
59. S. N. G. Chu, C. M. Jodlauk, and A. A. Ballman, *J. Electrochem. Soc.* **129,** 352 (1982).
60. D. T. C. Huo, J. D. Wynn, M. F. Yan, and D. P. Wilt, *J. Electrochem. Soc.* **136,** 1804 (1989).
61. A. Huber and N. T. Linh, *J. Cryst. Growth* **29,** 80 (1975).
62. G. T. Brown, B. Cockayne, and W. R. MacEvan, *J. Mater. Sci.* **15,** 2539 (1980).
63. D. T. C. Huo, M. F. Yan, J. D. Wynn, and D. P. Wilt, *J. Electrochem. Soc.* **136,** 3094 (1989).
64. S. Adachi, *J. Electrochem. Soc.* **129,** 609 (1982).
65. E. A. Caridi and T. Y. Chang, *J. Electrochem. Soc.* **131,** 1440 (1984).
66. S. Mahajan and A. K. Chin, *J. Cryst. Growth* **54,** 138 (1981).
67. J. A. Lourenco, *J. Electrochem. Soc.* **131,** 1914 (1984).
68. G. C. Osbourn, *J. Vacuum Sci. Technol. B* **4,** 1423 (1986).
69. J. W. Matthews, in *Dislocations in Solids*, Vol. 2, F. R. N. Nabarro, ed., North-Holland, Amsterdam, 1979, p. 461.
70. See Ref. 49 and references cited therein.
71. J. H. van der Merwe and W. A. Jesser, *J. Appl. Phys.* **63,** 1509 (1988).
72. G. Ji, V. M. Ristic, and U. K. Reddy, *Can. J. Phys.* **67,** 640 (1989).
73. J. P. Hirth and X. Feng, *J. Appl. Phys.* **67,** 3343 (1990).

74. S. T. Picraux, L. R. Dawson, G. C. Osbourn, and W. K. Chu, *Appl. Phys. Lett.* **43,** 930 (1983).

75. N. G. Anderson, W. D. Laidig, and Y. F. Lin, *J. Electron. Mater.* **14,** 187 (1985).

76. I. J. Fritz, S. T. Picraux, L. R. Dawson, T. J. Drummond, W. D. Laidig, and N. G. Anderson, *Appl. Phys. Lett.* **46,** 967 (1985).

77. N. G. Anderson, W. D. Laidig, R. M. Kolbas, and Y. C. Lo, *J. Appl. Phys.* **60,** 2361 (1986).

78. P. J. Orders and B. F. Usher, *Appl. Phys. Lett.* **50,** 980 (1987).

79. T. G. Andersson, Z. G. Chen, V. D. Kulakovskii, A. Uddin, and J. T. Vallin, *Appl. Phys. Lett.* **51,** 752 (1987).

80. I. J. Fritz, P. L. Gourley, and L. R. Dawson, *Appl. Phys. Lett.* **51,** 1004 (1987).

81. P. L. Gourley, I. J. Fritz, and L. R. Dawson, *Appl. Phys. Lett.* **52,** 377 (1988).

82. J.-P. Reithmaier, H. Cerva, and R. Lösch, *Appl. Phys. Lett.* **54,** 48 (1989).

83. J. Y. Tsao and B. W. Dodson, *Appl. Phys. Lett.* **53,** 848 (1988).

84. A. V. Drigo, A. Aydinli, A. Carnera, F. Genova, C. Rigo, C. Ferrari, P. Franzosi, and G. Salviati, *J. Appl. Phys.* **66,** 1975 (1989).

85. S.-L. Weng, *J. Appl. Phys.* **66,** 2217 (1989).

86. K. J. Beernink, P. K. York, J. J. Coleman, R. G. Waters, J. Kim, and C. M. Wayman, *Appl. Phys. Lett.* **55,** 2167 (1989).

87. D. C. Bertolet, J.-K. Hsu, F. Agahi, and K. M. Lau, *J. Electron. Mater.* **19,** 967 (1990).

88. R. People and J. C. Bean, *Appl. Phys. Lett.* **47,** 322 (1985); erratum, ibid. **49,** 229 (1986).

89. J. W. Matthews and A. E. Blakeslee, *J. Cryst. Growth* **27,** 118 (1974).

90. H. Temkin, D. G. Gershoni, S. N. G. Chu, J. M. Vandenberg, R. A. Hamm, and M. B. Panish, *Appl. Phys. Lett.* **55,** 1668 (1989).

91. W. D. Laidig, C. K. Peng, and Y. F. Lin, *J. Vacuum Sci. Technol. B* **2,** 181 (1984).

92. M. C. Joncour, M. N. Charasse, and J. Burgeat, *J. Appl. Phys.* **58,** 3373 (1985).

93. R. Flagmeyer, U. Pietsch, H. Rhan, M. Höricke, and B. Jenichen, *Phys. Status Solidi A* **113,** K211 (1989).

94. E. A. Fitzgerald, G. P. Watson, R. E. Proano, D. G. Ast, P. D. Kirchner, G. D. Pettit, and J. M. Woodall, *J. Appl. Phys.* **65,** 2220 (1989).

95. K. H. Chang, P. K. Bhattacharya, and R. Gibala, *J. Appl. Phys.* **65,** 3391 (1989).

96. J. Y. Yao, T. G. Andersson, and G. L. Dunlop, *Phil. Mag. A* **64,** 173 (1991).

97. M. Quillec, J. Y. Marzin, J. Primot, G. L. Roux, J. L. Benchimol, and J. Burgeat, *J. Appl. Phys.* **59,** 2447 (1986).

98. K. Kamigaki, H. Sakashita, H. Kato, M. Nakayama, N. Sano, and H. Terauchi, *Appl. Phys. Lett.* **49,** 1071 (1986).

99. C. R. Pidgeon and S. H. Groves, *Phys. Rev.* **186,** 824 (1969).

100. G. C. Osbourn, *Superlatt. Microstruct.* **1,** 223 (1985).

101. E. P. O'Reilly and G. P. Witchlow, *Phys. Rev. B* **34,** 6030 (1986).

102. E. P. O'Reilly, *Semicond. Sci. Technol.* **1,** 128 (1986).

103. E. P. O'Reilly, *Semicond. Sci. Technol.* **4,** 121 (1989).

104. R. People, *Appl. Phys. Lett.* **50,** 1604 (1987).

105. M.-P. Houng and Y.-C. Chang, *J. Appl. Phys.* **65,** 4990 (1989).

106. T. Y. Wang and G. B. Stringfellow, *J. Appl. Phys.* **67,** 344 (1990).

107. D. C. Rogers, R. J. Nicholas, J. C. Portal, and M. Razeghi, *Semicond. Sci. Technol.* **1,** 350 (1986).

108. D. Gershoni, J. M. Vandenberg, R. A. Hamm, H. Temkin, and M. B. Panish, *Phys. Rev. B* **36,** 1320 (1987).

109. Y. H. Zhang, L. Tapfer, and K. Ploog, *Semicond. Sci. Technol.* **5,** 590 (1990) and references cited therein.

110. M. Tacano, Y. Sugiyama, and Y. Takeuchi, *Appl. Phys. Lett.* **58,** 2420 (1991).

111. M. J. L. S. Haines and B. C. Cavenett, *Appl. Phys. Lett.* **55,** 849 (1989).

112. A. P. Roth, R. A. Masut, M. Sacilotti, P. J. D'Arcy, Y. L. Page, G. I. Sproule, and D. F. Mitchell, *Superlatt. Microstruct.* **2,** 507 (1986).

113. D. A. Dahl, L. J. Dries, F. A. Junga, and W. G. Opyd, *J. Appl. Phys.* **61,** 2079 (1987).

114. Y. Chen, J. Oh, J. Singh, and P. K. Bhattacharya, *J. Vacuum Sci. Technol. B* **6,** 660 (1988).

115. M. Gal, P. J. Orders, B. F. Usher, M. J. Joyce, and J. Tann, *Appl. Phys. Lett.* **53,** 113 (1988).

116. D. A. H. Mace, D. C. Rogers, K. J. Monserrat, J. N. Tothill, and S. T. Davey, *Semicond. Sci. Technol.* **3,** 597 (1988).

117. J. Woodhead, F. G. Sanz, P. A. Claxton, and J. P. R. David, *Semicond. Sci. Technol.* **3,** 601 (1988).

118. J. Y. Marzin and J. M. Gérard, *Superlatt. Microstruct.* **5,** 51 (1989).

119. D. Gershoni, J. M. Vandenberg, S. N. G. Chu, H. Temkin, T. Tanbun-Ek, and R. A. Logan, *Phys. Rev. B* **40,** 10017 (1989).

120. K. J. Moore, G. Dugaan, K. Woodbridge, and C. Roberts, *Phys. Rev. B* **41,** 1090 (1990).

121. J. Y. Marzin, M. N. Charasse, and B. Sermage, *Phys. Rev. B* **31,** 8298 (1985).

122. G. Ji, U. K. Reddy, D. Huang, T. S. Henderson, and H. Morkoç, *Superlatt. Microstruct.* **3,** 539 (1987).

123. G. Ji, D. Huang, U. K. Reddy, T. S. Henderson, R. Houdré, and H. Morkoç, *J. Appl. Phys.* **62,** 3366 (1987).

124. S.-C. Hong, G. P. Kothiyal, N. Debbar, P. K. Bhattacharya, and J. Singh, *Phys. Rev. B* **37,** 878 (1988).

125. X. Zhou, P. K. Bhattacharya, G. Hugo, S. C. Hong, and E. Gulari, *Appl. Phys. Lett.* **54,** 855 (1989).

126. W. D. Laidig, D. K. Blanks, and J. F. Schetzina, *J. Appl. Phys.* **56,** 1791 (1984).

127. U. K. Reddy, G. Ji, D. Huang, G. Munns, and H. Morkoç, *Appl. Phys. Lett.* **50,** 1748 (1987).

128. W. Pötz and D. K. Ferry, *J. Vacuum Sci. Technol. B* **4,** 1006 (1986).

129. J. E. Schirber, I. J. Fritz, L. R. Dawson, and G. C. Osbourn, *Phys. Rev. B* **28,** 2229 (1983).

130. E. D. Jones, H. Ackermann, J. E. Schirber, T. J. Drummond, L. R. Dawson, and I. J. Fritz, *Solid State Commun.* **55,** 525 (1985).

131. J. E. Schirber, I. J. Fritz, and L. R. Dawson, *Appl. Phys. Lett.* **46,** 187 (1985).

132. E. D. Jones, H. Ackermann, J. E. Schirber, T. J. Drummond, L. R. Dawson, and I. J. Fritz, *Appl. Phys. Lett.* **47,** 492 (1985).

133. J. Hwang, C. K. Shih, P. Pianetta, G. D. Kubiak, R. H. Stulen, L. R. Dawson, Y.-C. Pao, and J. S. Harris, Jr., *Appl. Phys. Lett.* **52,** 308 (1988).

134. D. A. Dahl, *Solid State Commun.* **61,** 825 (1987).

135. F. H. Pollak and M. Cardona, *Phys. Rev.* **172,** 816 (1968).

136. M. Chandrasekhar and F. H. Pollak, *Phys. Rev. B* **15,** 2127 (1977).

137. J. C. Hensel and G. Feher, *Phys. Rev.* **129,** 1041 (1963).

138. O. Berolo, J. C. Woolley, and J. A. Van Vechten, *Phys. Rev. B* **8,** 3794 (1973).

139. J. M. Worlock, A. C. Maciel, A. Petrou, C. H. Perry, R. L. Aggarwal, M. Smith, A. C. Gossard, and W. Wiegmann, *Surface Sci.* **142,** 486 (1984).

140. A. L. Mears and R. A. Stradling, *J. Phys. C* **4,** L22 (1971).

141. Q. H. F. Vrehen, *J. Phys. Chem. Solids* **29,** 129 (1968).

142. C. Neumann, A. Nöthe, and N. O. Lipari, *Phys. Rev. B* **37,** 922 (1988).

143. M. A. Reed and J. W. Lee, *Appl. Phys. Lett.* **50,** 845 (1987).

144. I. J. Fritz, L. R. Dawson, and T. E. Zipperian, *J. Vacuum Sci. Technol. B* **1,** 387 (1983).

145. I. J. Fritz, L. R. Dawson, and T. E. Zipperian, *Appl. Phys. Lett.* **43,** 846 (1983).

146. J. L. Thobel, L. Baudry, A. Cappy, P. Bourel, and R. Fauquembergue, *Appl. Phys. Lett.* **56,** 346 (1990).

147. I. J. Fritz, L. R. Dawson, T. J. Drummond, J. E. Schirber, and R. M. Biefeld, *Appl. Phys. Lett.* **48,** 139 (1986).

148. I. J. Fritz, T. J. Drummond, G. C. Osbourn, J. E. Schirber, and E. D. Jones, *Appl. Phys. Lett.* **48,** 1678 (1986).

149. I. J. Fritz, B. L. Doyle, J. E. Schirber, E. D. Jones, L. R. Dawson, and T. J. Drummond, *Appl. Phys. Lett.* **49,** 581 (1986).

150. J. M. Hinckley and J. Singh, *Appl. Phys. Lett.* **53,** 785 (1988).

151. H. P. Hjalmarson, *Appl. Phys. Lett.* **54,** 2215 (1989).

152. H. P. Hjalmarson, I. J. Fritz, and L. R. Dawson, *Solid-State Electron.* **32,** 1777 (1989).

153. M. Reddy, R. Grey, P. A. Claxton, and J. Woodhead, *Semicond. Sci. Technol.* **5,** 628 (1990).

154. D. D. Coon and H. C. Liu, *J. Appl. Phys.* **60,** 2893 (1986).

155. D. J. Arent, K. Deneffe, C. van Hoof, J. D. Boeck, and G. Borghs, *J. Appl. Phys.* **66,** 1739 (1989).

156. J.-P. Reithmaier, R. Höger, H. Riechert, A. Heberle, G. Abstreiter, and G. Weimann, *Appl. Phys. Lett.* **56,** 536 (1990).

157. P. M. Asbeck, C.-P. Lee, and M.-C. F. Chang, *IEEE Trans. Electron Dev.* **ED-31,** 1377 (1984).

158. D. L. Smith, *Solid State Commun.* **57,** 919 (1986).

159. C. Mailhiot and D. L. Smith, *J. Vacuum Sci. Technol. B* **4,** 996 (1986).

160. E. A. Caridi, T. Y. Chang, K. W. Goossen, and L. F. Eastman, *Appl. Phys. Lett.* **56,** 659 (1990).

161. U. Das and P. K. Bhattacharya, *J. Appl. Phys.* **58,** 341 (1985).

162. G. Burns, C. R. Wie, F. H. Dacol, G. D. Pettit, and J. M. Woodall, *Appl. Phys. Lett.* **51,** 1919 (1987).

163. C. Wu, P. Lao, and S. C. Shen, *Appl. Phys. Lett.* **58,** 1491 (1991).

164. W. C. Tang, H. J. Rosen, S. Guha, and A. Madhukar, *Appl. Phys. Lett.* **58,** 1644 (1991).

12

CONCLUDING REMARKS

The $In_{1-x}Ga_xAs_yP_{1-y}$/InP system has been the subject of considerable research and device development activity in the past. A number of important semiconducting properties needed for the analysis of InP, $In_{1-x}Ga_xAs_yP_{1-y}$, and $In_{1-x}Ga_xAs$ require quite a detailed and precise knowledge of material parameters, but it is at present rather difficult to obtain the most reliable values from the literature reported. The material parameters and semiconducting properties considered in this book can be classified into the following 10 groups: (1) structural properties; (2) mechanical, elastic, and lattice vibrational properties; (3) thermal properties; (4) collective effects and some response characteristics; (5) electronic energy-band-structure consequences; (6) electronic deformation potentials; (7) optical properties; (8) elastooptic and electrooptic properties; (9) carrier-transport properties; and (10) the strain problem in InGaAs(P)-based heterostructures. Various models for calculating material parameters in compound alloys were discussed, and the results for $In_{1-x}Ga_xAs_yP_{1-y}$ and $In_{1-x}Ga_xAs$ alloys have been presented. An interpolation scheme has been used for want of more detailed experimental data, and it has been assumed that the accuracy of the interpolated values are probably estimates for these alloys. The interpolation scheme necessitates known values of material parameters for the related binaries, InP, InAs, GaAs, and GaP. The emphasis is, therefore, placed on material properties of these end-binary compounds. The material parameters derived are used with wide success to make clear fundamental semiconducting properties of these alloys. A detailed discussion is given of the acceptability of the estimated values of the parameters in connection with solid-state physics. Key properties of the material parameters and constants for a variety of $In_{1-x}Ga_xAs_yP_{1-y}$/InP device applications are also discussed in detail. Much progress is still to be expected in this fascinating field of research and development.

APPENDIX

SUMMARY OF INTERPOLATION SCHEME

An interpolation scheme is known to be a useful tool for estimating some material parameters of alloy compounds. The electronic energy-band parameters of $In_{1-x}Ga_xAs_yP_{1-y}$ and their dependence on alloy composition are known to be very important device parameters, and they have received considerable attention in the past (see Chapter 6). Transport- and optoelectronic-device parameters of this alloy system have, however, been hampered by a lack of definite knowledge of many various material parameters such as effective masses, dielectric constants, and electronic deformation potentials. This necessitates the use of some sort of an interpolation scheme. Although the interpolation scheme is still open to experimental verification, it provides more useful and reliable material parameters over the entire range of alloy composition.

The quaternary material $A_{1-x}B_xC_yD_{1-y}$ is thought to be constructed of four binary compounds: AC, AD, BC, and BD. If one uses a linear interpolation scheme, the quaternary material parameter (Q) can be derived from binary parameters (B values) by [see Eq. (2.3)]

$$Q(x, y) = (1 - x)yB_{AC} + (1 - x)(1 - y)B_{AD} + xyB_{BC} + x(1 - y)B_{BD} \tag{A.1}$$

If one of the four binary parameters is lacking, the quaternary parameter can be obtained by the following expressions [see Eq. (9.24)]:

$$Q(x, y) = (1 - x)B_{AC} + (x + y - 1)B_{BC} + (1 - y)B_{BD} \tag{A.2a}$$

$$Q(x, y) = (1 - x)B_{AD} + yB_{BC} + (x - y)B_{BD} \tag{A.2b}$$

$$Q(x, y) = yB_{AC} + (1 - x - y)B_{AD} + xB_{BD} \tag{A.2c}$$

or

$$Q(x, y) = (y - x)B_{AC} + (1 - y)B_{AD} + xB_{BC} \tag{A.2d}$$

If the material parameter can be given by a specific expression owing to some physical basis, it is natural to consider that interpolation scheme may also be obeyed to this expression. The dielectric constant parameter ϵ is one such case that follows the Clausius–Mosotti relation. Then, the interpolation expression has a form given by (see Section 5.1)

$$\frac{\epsilon(x, y) - 1}{\epsilon(x, y) + 2} = (1 - x)y\frac{\epsilon(\text{AC}) - 1}{\epsilon(\text{AC}) + 2} + (1 - x)(1 - y)\frac{\epsilon(\text{AD}) - 1}{\epsilon(\text{AD}) + 2} + xy\frac{\epsilon(\text{BC}) - 1}{\epsilon(\text{BC}) + 2} + x(1 - y)\frac{\epsilon(\text{BD}) - 1}{\epsilon(\text{BD}) + 2} \tag{A.3}$$

As mentioned in Section 5.1, this interpolation scheme gives nearly the same ϵ values as does Eq. (A.1) (see Fig. 5.1).

The electron effective mass $m_e^\Gamma(x, y)$ can also be taken as

$$\frac{1}{m_e^\Gamma(x, y)} = \frac{(1 - x)y}{m_e^\Gamma(\text{AC})} + \frac{(1 - x)(1 - y)}{m_e^\Gamma(\text{AD})} + \frac{xy}{m_e^\Gamma(\text{BC})} + \frac{x(1 - y)}{m_e^\Gamma(\text{BD})} \tag{A.4}$$

Justification for this is suggested by the effective-mass transformation of the Schrödinger equation (see Ref. 18 in Chapter 5)

$$\left[-\frac{\hbar^2\nabla^2}{2m_0} + \sum_r U(r - \tau)\right] \rightarrow -\frac{\hbar^2\nabla^2}{2m_e^\Gamma} \tag{A.5}$$

The dashed-dotted line in Fig. 6.7 is the result of this interpolation. [The dashed line is also the result of Eq. (A.4) by Restorff et al. (see Ref. 70 in Chapter 6).] The binary data used here were $m_e^\Gamma/m_0 = 0.079$ for InP, 0.023 for InAs, 0.067 for GaAs, and 0.17 for GaP (see Ref. 23 in Chapter 6). It should be noted that the interpolation scheme of Eq. (A.4) gives effective mass values considerably smaller than those determined experimentally.

If relationships for the ternary parameters (T values) are available, the quaternary parameter can be given by [see Eq. (6.5)]

$$Q(x, y) = \frac{1}{x(1-x) + y(1-y)} \{x(1-x)[yT_{ABC}(x) + (1-y)T_{ABD}(x)] + y(1-y)[xT_{ACD}(y) + (1-x)T_{BCD}(y)]\}$$

Because of the polar nature of the III–V compounds ($In_{1-x}Ga_xAs_yP_{1-y}$), there may be only mixing of Ga and In on group III sites and of As and P only on group V sites. When bowing from the anion sublattice disorder is independent of the disorder in the cation sublattice, the interpolation scheme is written by incorporating these cation and anion bowing parameters ($C_{A\text{-}B}$, $C_{C\text{-}D}$) into the linear interpolation scheme as (see Section 4.3 and Ref. 44 in Chapter 4):

$$Q(x, y) = (1-x)yB_{AC} + (1-x)(1-y)B_{AD} + xyB_{BC} + x(1-y)B_{BD} + C_{A\text{-}B}x(1-x) + C_{C\text{-}D}y(1-y) \quad \text{(A.7)}$$

If we consider anion-cation disorder interactions ($C_{AB\text{-}C}$, $C_{A\text{-}CD}$, $C_{AB\text{-}D}$, $C_{A\text{-}CD}$), expression (A.7) is modified in its form by

$$Q(x, y) = (1-x)yB_{AC} + (1-x)(1-y)B_{AD} + xyB_{BC} + x(1-y)B_{BD} + C_{AB\text{-}C}x(1-x)y + C_{A\text{-}CD}xy(1-y) + C_{AB\text{-}D}x(1-x) \cdot (1-y) + C_{B\text{-}CD}(1-x)y(1-y) \quad \text{(A.8)}$$

In the case where $C_{AB\text{-}C} = C_{AB\text{-}D} (\equiv C_{A\text{-}B})$ and $C_{A\text{-}CD} = C_{B\text{-}CD} (\equiv C_{C\text{-}D})$, Eq. (A.8) agrees exactly with Eq. (A.7).

INDEX TO TABLES OF PHYSICAL CONSTANTS

AUTHOR INDEX

The names appearing in this Index are from the References at the end of each chapter and from tabular material (italics) only.

SUBJECT INDEX